An Introduction to Survival Analysis Using Stata

Using Stata

Second Edition

An Introduction to Survival Analysis Using Stata

Second Edition

MARIO A. CLEVES
Department of Pediatrics
University of Arkansas Medical Sciences

WILLIAM W. GOULD
StataCorp

ROBERTO G. GUTIERREZ
StataCorp

YULIA U. MARCHENKO
StataCorp

A Stata Press Publication
StataCorp LP
College Station, Texas

Published by Stata Press, 4905 Lakeway Drive, College Station, Texas 77845
Typeset in LaTeX 2_ε
Printed in the United States of America

10 9 8 7 6 5 4 3 2 1

ISBN-10: 1-59718-041-6
ISBN-13: 978-1-59718-041-2

Contents

Tables

Figures

Preface to the Second Edition

This second edition updates the revised edition (revised to support Stata 8) to reflect Stata 9, which was released in April 2005, and Stata 10, which was released in June 2007. The updates include the syntax and output changes that took place in both versions. For example, as of Stata 9 the `estat phtest` command replaces the old `stphtest` command for computing tests and graphs for examining the validity of the proportional-hazards assumption. As of Stata 10, all `st` commands (as well as other Stata commands) accept option `vce(vcetype)`. The old `robust` and `cluster(varname)` options are replaced with `vce(robust)` and `vce(cluster varname)`. Most output changes are cosmetic. There are slight differences in the results from `streg, distribution(gamma)`, which has been improved to increase speed and accuracy.

Chapter 8 includes a new section on nonparametric estimation of median and mean survival times. Other additions are examples of producing Kaplan–Meier curves with at-risk tables and a short discussion of the use of boundary kernels for hazard function estimation.

Stata's facility to handle complex survey designs with survival models is described in chapter 9 in application to the Cox model, and what is described there may also be used with parametric survival models.

Chapter 10 is expanded to include more model-building strategies. The use of fractional polynomials in modeling the log relative-hazard is demonstrated in chapter 10. Chapter 11 includes a description of how fractional polynomials can be used in determining functional relationships, and it also includes an example of using concordance measures to evaluate the predictive accuracy of a Cox model.

Chapter 16 is new and introduces power analysis for survival data. It describes Stata's ability to estimate sample size, power, and effect size for the following survival methods: a two-sample comparison of survivor functions and a test of the effect of a covariate from a Cox model. This chapter also demonstrates ways of obtaining tabular and graphical output of results.

College Station, Texas
March 2008

Mario A. Cleves
William W. Gould
Roberto G. Gutierrez
Yulia U. Marchenko

Preface to the Revised Edition

This revised edition updates the original text (written to support Stata 7) to reflect Stata 8, which was released in January 2003. Most of the changes are minor and include new graphics, including the appearance of the graphics and the syntax used to create them, and updated datasets.

New sections describe Stata's ability to graph nonparametric and semiparametric estimates of hazard functions. Stata now calculates estimated hazards as weighted kernel-density estimates of the times at which failures occur, where weights are the increments of the estimated cumulative hazard function. These new capabilities are described for nonparametric estimation in chapter 8 and for Cox regression in chapter 9.

Another added section in chapter 9 discusses Stata's ability to apply shared frailty to the Cox model. This section complements the discussion of parametric shared and unshared frailty models in chapter 8. Because the frailty is best understood by beginning with a parametric model, this new section is relatively brief and focuses only on practical issues of estimation and interpretation.

College Station, Texas
August 2003

Mario A. Cleves
William W. Gould
Roberto G. Gutierrez

Preface to the First Edition

We have written this book for professional researchers outside mathematics, people who do not spend their time wondering about the intricacies of generalizing a result from discrete space to $\Re_1$ but who nonetheless understand statistics. Our readers may sometimes be sloppy when they say that a probability density is a probability, but when pressed, they know there is a difference and remember that a probability density can indeed even be greater than 1. However, our readers are never sloppy when it comes to their science. Our readers use statistics as a tool, just as they use mathematics, and just as they sometimes use computer software.

This is a book about survival analysis for the professional data analyst, whether a health scientist, an economist, a political scientist, or any of a wide range of scientists who have found that survival analysis applies to their problems. This is a book for researchers who want to understand what they are doing and to understand the underpinnings and assumptions of the tools they use; in other words, this is a book for all researchers.

This book grew out of software, but nonetheless it is not a manual. That genesis, however, gives this book an applied outlook that is sometimes missing from other works. We also wrote Stata's survival analysis commands, which have had something more than modest success. Writing application software requires a discipline of authors similar to that of building of scientific machines by engineers. Problems that might be swept under the rug as mere details cannot be ignored in the construction of software, and the authors are often reminded that the devil is in the details. It is those details that cause users such grief, confusion, and sometimes pleasure.

In addition to having written the software, we have all been involved in supporting it, which is to say, interacting with users (real professionals). We have seen the software used in ways that we would never have imagined, and we have seen the problems that arise in such uses. Those problems are often not simply programming issues but involve statistical issues that have given us pause. To the statisticians in the audience, we mention that there is nothing like embedding yourself in the problems of real researchers to teach you that problems you thought unimportant are of great importance, and vice versa. There is nothing like "straightforwardly generalizing" some procedure to teach you that there are subtle issues worth much thought.

In this book, we illustrate the concepts of using Stata. Readers should expect a certain bias on our part, but the concepts go beyond our implementation of them. We will often discuss substantive issues in the midst of issues of computer use, and we do that because, in real life, that is where they arise.

This book also grew out of a course we taught several times over the web, and the many researchers who took that course will find in this book the companion text they lamented not having for that course.

We do not wish to promise more than we can deliver, but the reader of this book should come away not just with an understanding of the formulas but with an intuition of how the various survival analysis estimators work and exactly what information they exploit.

We thank all the people who over the years have contributed to our understanding of survival analysis and the improvement of Stata's survival capabilities, be it through programs, comments, or suggestions. We are particularly grateful to the following:

David Clayton of the Cambridge Institute for Medical Research
Joanne M. Garrett of the University of North Carolina
Michael Hills, retired from the London School of Hygiene and Tropical Medicine
David Hosmer, Jr., of the University of Massachusetts–Amherst
Stephen P. Jenkins of the University of Essex
Stanley Lemeshow of Ohio State University
Adrian Mander of the MRC Biostatistics Unit
William H. Rogers of The Health Institute at New England Medical Center
Patrick Royston of the MRC Clinical Trials Unit
Peter Sasieni of Cancer Research UK
Jeroen Weesie of Utrecht University

By no means is this list complete; we express our thanks as well to all those who should have been listed.

College Station, Texas   Mario A. Cleves
May 2002   William W. Gould
                                                                           Roberto G. Gutierrez

Notation and Typography

This book is an introduction to the analysis of survival data using Stata, and we assume that you are already more or less familiar with Stata.

For instance, if you had some raw data on outcomes after surgery, and we tell you to (1) enter it into Stata, (2) sort the data by patient age, (3) save the data, (4) list the age and outcomes for the 10 youngest and 10 oldest patients in the data, (5) tell us the overall fraction of observed deaths, and (6) tell us the median time to death among those who died, you could do that.

```
. infile ...
. sort age
. save mydata
. list age outcome in 1/10
. list age outcome in -10/1
. summarize died
. summarize time if died, detail
```

This text was written using Stata 10, and to ensure that you can fully replicate what we have done, you need an up-to-date Stata version 10 or later. Type

```
. update query
```

from a web-aware Stata and follow the instructions to ensure that you are up to date.

The developments in this text are largely applied, and you should read this text while sitting at a computer so that you can try to replicate our results for yourself by using the sequences of commands contained in the text. In this way, you may generalize these sequences to suit your own data analysis needs.

We use the typewriter font `command` to refer to Stata commands, syntax, and variables. When a "dot" prompt is displayed followed by a command (such as in the above sequence), it means you can type what is displayed after the dot (in context) to replicate the results in the book.

Except for some small expository datasets we use, all the data we use in this text are freely available for you to download (via a web-aware Stata) from the Stata Press web site, http://www.stata-press.com. In fact, when we introduce new datasets, we load them into Stata the same way that you would. For example,

```
. use http://www.stata-press.com/data/cggm/hip, clear   /* hip-fracture data */
```

Try this for yourself. The `cggm` part of the pathname, in case you are curious, is from the last initial of each of the four authors.

This text serves as a complement to the material in the Stata manuals, not as a substitute, and thus we often make reference to the material in the Stata manuals using the [R] , [P] , etc., notation. For example, [R] **logistic** refers to the *Stata Base Reference Manual* entry for `logistic`, and [P] **syntax** refers to the entry for `syntax` in the *Stata Programming Reference Manual*.

Survival analysis, as with most substantive fields, is filled with jargon: left truncation, right censoring, hazard rates, cumulative hazard, survivor function, etc. Jargon arises so that researchers do not have to explain the same concepts over and over again. Those of you who practice survival analysis know that researchers tend to be a little sloppy in their use of language, saying truncation when they mean censoring or hazard when they mean cumulative hazard, and if we are going to communicate by the written word, we have to agree on what these terms mean. Moreover, these words form a wall around the field that is nearly impenetrable if you are not already a member of the cognoscenti.

If you are new to survival analysis, let us reassure you: survival analysis is statistics. Master the jargon and think carefully, and you can do this.

1 The problem of survival analysis

Survival analysis concerns analyzing the time to the occurrence of an event. For instance, we have a dataset in which the times are 1, 5, 9, 20, and 22. Perhaps those measurements are made in seconds, perhaps in days, but that does not matter. Perhaps the event is the time until a generator's bearings seize, the time until a cancer patient dies, or the time until a person finds employment, but that does not matter either.

For now, we will just abstract the underlying data-generating process and say that we have some times—1, 5, 9, 20, and 22—until an event occurs. We might also have some covariates (additional variables) that we wish to use to "explain" these times. So, pretend that we have the following (completely made up) dataset:

```
time      x
  1       3
  5       2
  9       4
 20       9
 22      -4
```

Now what is to keep us from simply analyzing these data using ordinary least-squares (OLS) linear regression? Why not simply fit the model

$$\text{time}_j = \beta_0 + \beta_1 x_j + \epsilon_j, \qquad \epsilon_j \sim N(0, \sigma^2)$$

for $j = 1, \ldots, 5$, or, alternatively,

$$\ln(\text{time}_j) = \beta_0 + \beta_1 x_j + \epsilon_j, \qquad \epsilon_j \sim N(0, \sigma^2)$$

That is easy enough to do in Stata by typing

```
. regress time x
```

or

```
. generate lntime = ln(time)
. regress lntime x
```

These days, researchers would seldom analyze survival times in this manner, but why not? Before you answer too dismissively, we warn you that we can think of instances for which this would be a perfectly reasonable model to use.

1.1 Parametric modeling

The problem with using OLS to analyze survival data lies with the assumed distribution of the residuals, ϵ_j. In linear regression, the residuals are assumed to be distributed normally; i.e., time conditional on x_j is assumed to follow a normal distribution:

$$\texttt{time}_j \sim \mathrm{N}(\beta_0 + \beta_1 x_j, \sigma^2), \qquad j = 1, \ldots, 5$$

The assumed normality of time to an event is unreasonable for many events. It is unreasonable, for instance, if we are thinking about an event with an instantaneous risk of occurring that is constant over time. Then the distribution of time would follow an exponential distribution. It is also unreasonable if we are analyzing survival times following a particularly serious surgical procedure. Then the distribution might have two modes: many patients die shortly after the surgery, but if they survive, the disease might be expected to return. One other problem is that a time to failure is always positive, while theoretically, the normal distribution is supported on the entire real line. Realistically, however, this fact alone is not enough to render the normal distribution useless in this context, because σ^2 may be chosen (or estimated) to make the probability of a negative failure time virtually zero.

At its core, survival analysis concerns nothing more than making a substitution for the normality assumption characterized by OLS with something more appropriate for the problem at hand.

Perhaps, if you were already familiar with survival analysis, when we asked, "why not linear regression?" you offered the excuse of right censoring—that in real data we often do not observe subjects long enough for all of them to fail. In our data there was no censoring, but in reality, censoring is just a nuisance. We can fix linear regression easily enough to deal with right censoring. It goes under the name censored-normal regression, and Stata's `cnreg` command can fit such models; see [R] **cnreg**. The real problem with linear regression in survival applications is with the assumed normality.

Being unfamiliar with survival analysis, you might be tempted to use linear regression in the face of nonnormality. Linear regression is known, after all, to be remarkably robust to deviations from normality, so why not just use it anyway? The problem is that the distributions for time to an event might be dissimilar from the normal—they are almost certainly nonsymmetric, they might be bimodal, and linear regression is not robust to these violations.

Substituting a more reasonable distributional assumption for ϵ_j leads to parametric survival analysis.

1.2   Semiparametric modeling

That results of analyses are being determined by the assumptions and not the data is always a source of concern, and this leads to a search for methods that do not require assumptions about the distribution of failure times. That, at first blush, seems hopeless. With survival data, the key insight into removing the distributional assumption is that, because events occur at given times, these events may be ordered and the analysis may be performed exclusively using the ordering of the survival times. Consider our dataset:

```
time      x
   1      3
   5      2
   9      4
  20      9
  22     -4
```

Examine the failure that occurred at time 1. Let's ask the following what is the probability of failure after exposure to the risk of failure for 1 unit of time? At this point, observation 1 had failed and the others had not. This reduces the problem to a problem of binary-outcome analysis,

```
time      x      outcome
   1      3            1
   5      2            0
   9      4            0
  20      9            0
  22     -4            0
```

and it would be perfectly reasonable for us to analyze failure at `time` = 1 using, say, logistic regression

$$
\begin{aligned}
&= \text{Pr(failure after exposure for 1 unit of time)} \\
&= \text{Pr(outcome}_j = 1) \\
&= \frac{1}{1 + \exp(-\beta_0 - x_j \beta_x)}
\end{aligned}
$$

for $j = 1, \ldots, 5$. This is easy enough to do:

```
. logistic outcome x
```

Do not make too much of our choice of logistic regression—choose the analysis method you like. Use probit. Make a table. Whatever technique you choose, you could do all your survival analysis using this analyze-the-first-failure method. To do so would be inefficient but would have the advantage that you would be making no assumptions about the distribution of failure times. Of course, you would have to give up on being able to make predictions conditional on x, but perhaps being able to predict whether failure occurs at `time` = 1 would be sufficient.

There is nothing magical about the first death time; we could instead choose to analyze the second death time, which here is `time` = 5. We could ask about the probability of failure, given exposure of 5 units of time, in which case we would exclude the first observation (which failed too early) and fit our logistic regression model using the second and subsequent observations:

```
. drop outcome
. generate outcome = cond(time==5,1,0) if time>=5
. logistic outcome x if time>=5
```

In fact, we could use this same procedure on each of the death times, separately.

Which analysis should we use? Well, the second analysis has slightly less information than the first (because we have one less observation), and the third has less than the first two (for the same reason), and so on. So we should choose the first analysis. It is, however, unfortunate that we have to choose at all. Could we somehow combine all these analyses and constrain the appropriate regression coefficients (say, the coefficient on x) to be the same? Yes, we could, and after some math, that leads to semiparametric survival analysis and, in particular, to Cox (1972) regression if a conditional logistic model is fit for each analysis. Conditional logistic models differ from ordinary logistic models for this example in that for the former we condition on the fact that we know that outcome==1 for one and only one observation within each separate analysis.

However, for now we do not want to get lost in all the mathematical detail. We could have done each of the analyses using whatever binary analysis method seemed appropriate. By doing so, we could combine them all if we are sufficiently clever in doing the math, and because each of the separate analyses made no assumption about the distribution of failure times, the combined analysis also makes no such assumption.

That last statement is rather slippery, so it does not hurt to verify its truth. We have been considering the data

```
time       x
   1       3
   5       2
   9       4
  20       9
  22      -4
```

but now consider two variations on the data:

```
time       x
 1.1       3
 1.2       2
 1.3       4
50.0       9
50.1      -4
```

and

```
time       x
    1       3
  500       2
 1000       4
10000       9
100000     -4
```

These two alternatives have dramatically different distributions for time, yet they have the same temporal ordering and the same values of x. Think about performing the individual analyses on each of these datasets, and you will realize that the results you get will be the same. Time plays no role other than ordering the observations.

The methods described above go under the name semiparametric analysis; as far as time is concerned, they are nonparametric, but because we are still parameterizing the effect of x, there exists a parametric component to the analysis.

1.3   Nonparametric analysis

Semiparametric models are parametric in the sense that the effect of the covariates is still assumed to take a certain form. Earlier, by performing a separate analysis at each failure time and concerning ourselves only with the order in which the failures occurred, we made no assumption about the distribution of time to failure. We did, however, make an assumption about how each subject's observed x value determined the probability that subject would fail, for example, a probability determined by the logistic function.

An entirely nonparametric approach would be to go away with this assumption also and follow the philosophy of "letting the dataset speak for itself". There exists a vast literature on performing nonparametric regression using methods such as lowess or local polynomial regression; however, such methods do not adequately deal with censoring and other issues unique to survival data.

When no covariates exist, or when the covariates are qualitative in nature (gender, for instance), we can use nonparametric methods such as Kaplan and Meier (1958) or the method of Nelson (1972) and Aalen (1978) to estimate the probability of survival past a certain time or to compare the survival experiences for each gender. These methods account for censoring and other characteristics of survival data. There also exist methods such as the two-sample log-rank test, which can compare the survival experience across gender by using only the temporal ordering of the failure times. Nonparametric methods make assumptions about neither the distribution of the failure times nor how covariates serve to shift or otherwise change the survival experience.

1.4   Linking the three approaches

Going back to our original data, consider the individual analyses we performed to obtain the semiparametric (combined) results. The individual analyses were

> Pr(failure after exposure for exactly 1 unit of time)
> Pr(failure after exposure for exactly 5 units of time)
> Pr(failure after exposure for exactly 9 units of time)
> Pr(failure after exposure for exactly 20 units of time)
> Pr(failure after exposure for exactly 22 units of time)

We could omit any of the individual analyses above, and doing so would affect only the efficiency of our estimators. It is better, though, to include them all, so why not add the following to this list:

> Pr(failure after exposure for exactly 1.1 units of time)
> Pr(failure after exposure for exactly 1.2 units of time)
> ...

That is, why not add individual analyses for all other times between the observed failure times? That would be a good idea because the more analyses we can combine, the more efficient our final results will be: the standard errors of our estimated regression parameters will be smaller. We do not do this only because we do not know how to say anything about these intervening times—how to perform these analyses—unless we make an assumption about the distribution of failure time. If we made that assumption, we could perform the intervening analyses (the infinite number of them), and then we could combine them all to get superefficient estimates. We could perform the individual analyses themselves a little differently, too, by taking into account the distributional assumptions, but that would only make our final analysis even more efficient.

That is the link between semiparametric and parametric analysis. Semiparametric analysis is simply a combination of separate binary-outcome analyses, one per failure time, while parametric analysis is a combination of several analyses at *all* possible failure times. In parametric analysis, if no failures occur over a particular interval, that is informative. In semiparametric analysis, such periods are not informative. On the one hand, semiparametric analysis is advantageous in that it does not concern itself with the intervening analyses, yet parametric analysis will be more efficient if the proper distributional assumptions are made concerning those times when no failures are observed.

When no covariates are present, we hope that semiparametric methods such as Cox regression will produce estimates of relevant quantities (such as the probability of survival past a certain time) that are identical to the nonparametric estimates, and in fact, they do. When the covariates are qualitative, parametric and semiparametric methods should yield more efficient tests and comparisons of the groups determined by the covariates than nonparametric methods, and these tests should agree. Test disagreement would indicate that some of the assumptions made by the parametric or semiparametric models are incorrect.

2 Describing the distribution of failure times

The key to mastering survival analysis lies in grasping the jargon. In this chapter and the next, we describe the statistical terms unique to the analysis of survival data.

2.1 The survivor and hazard functions

These days, survival analysis is cast in a language all its own. Let T be a nonnegative random variable denoting the time to a failure event. Rather than referring to Ts probability density function, $f(t)$—or, if you prefer, its cumulative distribution function, $F(t) = \Pr(T \leq t)$—survival analysts instead talk about Ts survivor function, $S(t)$, or its hazard function, $h(t)$. There is good reason for this: it really is more convenient to think of $S(t)$ and $h(t)$ rather than $F(t)$ or $f(t)$, although all forms describe the same probability distribution for T. Translating between these four forms is simple.

The survivor function, also called the survivorship function or the survivor function, is simply the reverse cumulative distribution function of T:

$$S(t) = 1 - F(t) = \Pr(T > t)$$

The survivor function reports the probability of surviving beyond time t. Said differently, it is the probability that there is no failure event prior to t. The function is equal to one at $t = 0$ and decreases toward zero as t goes to infinity. (The survivor function is a monotone, nonincreasing function of time.)

The density function, $f(t)$, can be obtained as easily from $S(t)$ as it can from $F(t)$:

$$f(t) = \frac{dF(t)}{dt} = \frac{d}{dt}\{1 - S(t)\} = -S'(t)$$

The hazard function, $h(t)$—also known as the conditional failure rate, the intensity function, the age-specific failure rate, the inverse of the Mills' ratio, and the force of mortality—is the instantaneous rate of failure, with the emphasis on the word rate, meaning that it has units $1/t$. It is the (limiting) probability that the failure event occurs in a given interval, conditional upon the subject having survived to the beginning of that interval, divided by the width of the interval:

$$h(t) = \lim_{\Delta t \to 0} \frac{\Pr(t + \Delta t > T > t | T > t)}{\Delta t} = \frac{f(t)}{S(t)} \qquad (2.1)$$

The hazard rate (or function) can vary from zero (meaning no risk at all) to infinity (meaning the certainty of failure at that instant). Over time, the hazard rate can increase, decrease, remain constant, or even take on more serpentine shapes. There is a one-to-one relationship between the probability of survival past a certain time and the amount of risk that has been accumulated up to that time, and the hazard rate measures the rate at which risk is accumulated. The hazard function is at the heart of modern survival analysis, and it is well worth the effort to become familiar with this function.

It is, of course, the underlying process (e.g., disease, machine wear) that determines the shape of the hazard function:

- When the risk of something is zero, its hazard is zero.

- We have all heard of risks that do not vary over time. That does not mean that, as I view my future prospects, my chances of having succumbed to the risk do not increase with time. Indeed, I will succumb eventually (provided that the constant risk or hazard is nonzero), but my chances of succumbing at this instant or that are all the same.

- If the risk is rising with time, so is the hazard. Then the future is indeed bleak.

- If the risk is falling with time, so is the hazard. Here the future looks better (if only we can make it through the present).

- The human mortality pattern related to aging generates a falling hazard for a while after birth, and then a long, flat plateau, and thereafter constantly rising and eventually reaching, one supposes, values near infinity at about 100 years. Biometricians call this the "bathtub hazard".

- The risk of postoperative wound infection falls as time from surgery increases, so the hazard function decreases with time.

Given one of the four functions that describe the probability distribution of failure times, the other three are completely determined. In particular, one may derive from a hazard function the probability density function, the cumulative distribution function, and the survivor function. To show this, it is first convenient to define yet another function, the cumulative hazard function,

$$H(t) = \int_0^t h(u)du$$

and thus

$$H(t) = \int_0^t \frac{f(u)}{S(u)}du = -\int_0^t \frac{1}{S(u)}\left\{\frac{d}{du}S(u)\right\}du = -\ln\{S(t)\} \qquad (2.2)$$

The cumulative hazard function has an interpretation all its own: it measures the total amount of risk that has been accumulated up to time t, and from (2.2) we can see the (inverse) relationship between the accumulated risk and the probability of survival.

We can now conveniently write

$$
\begin{aligned}
S(t) &= \exp\{-H(t)\} \\
F(t) &= 1 - \exp\{-H(t)\} \\
f(t) &= h(t)\exp\{-H(t)\}
\end{aligned}
$$

As an example, consider the Weibull hazard function, often used by engineers,

$$
h(t) = pt^{p-1}
$$

where p is a shape parameter estimated from the data. Given this form of the hazard, we can determine the survivor, cumulative distribution, and probability density functions to be

$$
\begin{aligned}
S(t) &= \exp(-t^p) \\
F(t) &= 1 - \exp(-t^p) \\
f(t) &= pt^{p-1}\exp(-t^p)
\end{aligned}
\tag{2.3}
$$

In real datasets, we often do not observe subjects from the onset of risk. That is, rather than observing subjects from $t = 0$ until failure, we observe them from $t = t_0$ until failure, with $t_0 > 0$. When the failure event is death or some other absorbing event after which continued observation is impossible or pointless, we will instead want to deal with the conditional variants of $h()$, $H()$, $F()$, $f()$, and $S()$. Those who failed (died) during the period 0 to t_0 will never be observed in our datasets. The conditional forms of the above functions are

$$
\begin{aligned}
h(t|T > t_0) &= h(t) \\
H(t|T > t_0) &= H(t) - H(t_0) \\
F(t|T > t_0) &= \frac{F(t) - F(t_0)}{S(t_0)} \\
f(t|T > t_0) &= \frac{f(t)}{S(t_0)} \\
S(t|T > t_0) &= \frac{S(t)}{S(t_0)}
\end{aligned}
$$

Conditioning on $T > t_0$ is common; thus, in what follows, we suppress the notation so that $S(t|t_0)$ is understood to mean $S(t|T > t_0)$, for instance. The conditioning does not affect $h(t)$; it is an instantaneous rate and so is not a function of the past.

The conditional functions may also be used to describe the second and subsequent failure times for events when failing more than once is possible. For example, the survivor function describing the probability of a second heart attack would naturally have to condition on the second heart attack taking place after the first, and so one could use $S(t|t_f)$, where t_f was the time of the first heart attack.

Figure 2.1 shows some hazard functions for often-used distributions. Although determining a hazard from a density or distribution function is easy using (2.1), this is really turning the problem on its head. You want to think of hazard functions.

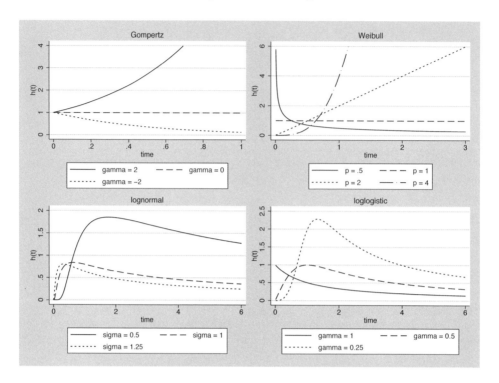

Figure 2.1. Hazard functions obtained from various parametric survival models

2.2   The quantile function

In addition to $f()$, $F()$, $S()$, $h()$, and $H()$—all different ways of summarizing the same information—a sixth way, $Q()$, is not often mentioned but is of use for those who wish to create artificial survival-time datasets. This is a book about analyzing data, not manufacturing artificial data, but sometimes the best way to understand a particular distribution is to look at the datasets it would imply.

The quantile function, $Q(u)$, is defined (for continuous distributions) to be the inverse of the cumulative distribution function; that is,

$$Q(u) = F^{-1}(u)$$

so that $Q(u) = t$ only if $F(t) = u$. Among other things, the quantile function can be used to calculate percentiles of the time-to-failure distribution. The 40th percentile, for example, is given by $Q(0.4)$.

Our interest, however, is in using $Q()$ to produce an artificial dataset. What would a survival dataset look like that was produced by a Weibull distribution with $p = 3$? Finding the quantile function associated with the Weibull lets us answer that question because the Probability Integral Transform states that if U is a uniformly distributed random variable of the unit interval, then $Q(U)$ is a random variable with cumulative distribution $F()$.

Stata has a uniform random-number generator, so with a little math by which we derive $Q()$ corresponding to $F()$, we can create artificial datasets. For the Weibull, from (2.3) we obtain $Q(u) = \{-\ln(1 - u)\}^{1/p}$. If we want 5 random deviates from a Weibull distribution with shape parameter $p = 3$, we can type

```
. set obs 5
obs was 0, now 5
. set seed 12345
. gen t = (-ln(1-uniform()))^(1/3)
. list t
```

	t
1.	.7177559
2.	1.04948
3.	.5151563
4.	.9378536
5.	.7218155

We can also generate survival times that are conditional (or given) to be greater than some value t_0, that is, those that would represent times to failure for subjects who come into observation at some time $t_0 > 0$. The conditional distribution function $F(t|t_0) = \Pr(T \le t|T > t_0) = 1 - S(t|t_0)$. The conditional quantile function, $Q(u|t_0) = F^{-1}(t|t_0)$, is obtained by setting $u = 1 - S(t)/S(t_0)$ and solving for t. For the Weibull,

$$Q(u|t_0) = \{t_0^p - \ln(1 - u)\}^{1/p}$$

so we can use the following to generate 5 observations from a Weibull distribution with $p = 3$, given to be greater than, say, $t_0 = 2$. The generated observations would thus represent observed failure times for subjects whose times to failure follow a Weibull distribution, yet these subjects are observed only past time $t_0 = 2$. If failure should occur before this time, the subjects remain unobserved.

```
. gen t2 = (2^3 - ln(1-uniform()))^(1/3)
. list t2
```

	t2
1.	2.05692
2.	2.101795
3.	2.044083
4.	2.090031
5.	2.037246

These failure times do not extend much past $t = 2$. The Weibull hazard function is an increasing function when $p > 1$, and one may gather from this small experiment that the hazard has already become large by $t = 2$. Thus those who are lucky enough to survive to the beginning of our study do not survive much longer. Subjects begin accumulating risk at time zero even though they are not initially observed until $t = 2$, and by that time they have already accumulated much risk and will continue to accumulate risk at an ever-increasing rate.

An equivalent way to generate Weibull survival times conditional to be greater than 2 would be to generate several unconditional Weibull survival times and then drop those that are less than or equal to 2. Although a direct application of the rules of conditional probability, this approach is wasteful because you would be generating many survival times only to end up dropping most of them. In addition, the number of observations that you end up with is not guaranteed to be any certain number but is instead subject to chance.

As mentioned previously, the conditional survivor function not only describes independent individuals with late entry into a study but may also describe repeated failures for an individual, where each failure time is known to be greater than the one before. We can simulate this process by conditioning each failure as being greater than the previous one. Using our Weibull with $p = 3$, we start at $t_0 = 0$ and generate the first failure time. Then we generate four subsequent failures, each conditional on being greater than the previous one:

```
. gen u = uniform()
. gen t3 = (-ln(1-u))^(1/3) in 1
(4 missing values generated)
. replace t3 = ((t3[_n-1])^3 - ln(1-u))^(1/3) in 2/L
(4 real changes made)
. list t3
```

	t3
1.	1.075581
2.	1.417904
3.	1.43258
4.	1.459457
5.	1.696586

Now we can gather that the probability of survival past $t = 2$ is really low because if this hypothetical subject were allowed to fail repeatedly, he would have already failed at least five times by $t = 2$. Of course, we would probably want to simulate various such individuals rather than consider only the one depicted above, but already we can see that survival past $t = 2$ is highly unlikely. In fact, $S(2) = \exp(-2^3) \approx 0.03\%$.

Simulation is an excellent tool not only for software developers who use it to test model estimation programs but also to illustrate and to observe in practice how theoretical functions such as the hazard and the survivor functions realize themselves into observed failure times. Simulation shows how these functions may be interpreted in more than one way, as seen above and as revisited later.

2.3   Interpreting the cumulative hazard and hazard rate

As previously mentioned, learning to think in terms of the hazard and the cumulative hazard functions, rather than the traditional density and cumulative density functions, has several advantages. Hazard functions give a more natural way to interpret the process that generates failures, and regression models for survival data are more easily grasped by observing how covariates affect the hazard.

2.3.1   Interpreting the cumulative hazard

The cumulative hazard function, $H(t)$, has much more to offer than merely an intermediate calculation to derive a survivor function from a hazard function. Hazards are rates, and in that they are not unlike the RPM—revolutions per minute—of an automobile engine.

Cumulative hazards are the integral from zero to t of the hazard rates. Because an integral is really just a sum, a cumulative hazard is like the total number of revolutions an automobile's engine makes over a given period. We could form the cumulative-revolution function by integrating RPM over time. If we let a car engine run at a constant 2,000 RPM for 2 minutes, then the cumulative revolution function at time 2 minutes would be 4,000, meaning the engine would have revolved 4,000 times over that period. Similarly, if a person faced a constant hazard rate of 2,000/minute (a big risk) for 2 minutes, he would face a total hazard of 4,000. Going back to the car engine, if we raced the engine at 3,000 RPM for 1 minute and then let it idle at 1,000 for another, the total number of revolutions would still be 4,000. Going back to our fictional risk taker, if he faced a hazard of 3,000/minute for 1 minute and then a hazard of 1,000/minute for another, his total risk would still be 4,000.

Now let's stick with our fictional friend. Whatever the profile of risk, if the cumulative hazard is the same over a 2-minute period, then the probability of the event (presumably death) occurring during that 2-minute period is the same.

Let's understand the units of this measurement of risk. In this, cumulative hazards are more easily understood than the hazard rates themselves. Remember that $S(t) = \exp\{-H(t)\}$, so our fictional friend has a probability of surviving the 2-minute interval of $\exp(-4000)$: our friend is going to die. One may similarly calculate the probability of survival given other values for the cumulative hazard.

Probabilities, however, are not the best way to think about cumulative hazards. Another interpretation of the cumulative hazard is that it records the number of times we would expect (mathematically) to observe failures over a given period, if only the failure event were repeatable. With our fictional friend, the cumulative hazard of 4,000 over the 2-minute period means that we would expect him to die 4,000 times if, as in a video game, each time he died we could instantly resurrect him and let him continue on his risky path.

This approach is called the count-data interpretation of the cumulative hazard, and learning to think this way has its advantages.

▷ Example

To see the count-data interpretation in action in Stata, let's consider an example using, as before, the Weibull distribution with shape parameter $p = 3$, which has a cumulative hazard $H(t) = t^3$. For the time interval (0,4), because $H(4) = 64$, we can interpret this to mean that if failure were repeatable, we would expect 64 failures over this period.

This fact may be verified via simulation. We proceed, as we did in section 2.2, by generating times to failure in a repeated-failure setting, where each failure time is conditional on being greater than the previous one. This time, however, we will repeat the process 1,000 times; for each replication, we observe the random quantity N, the number of failure times that are less than $t = 4$. For each replication, we count the number of failures that occur in the interval (0,4) and record this count. In Stata, this may be done via the simulate command; see [R] **simulate**.

```
. clear
. set seed 12345
. program genfail
  1.              drop _all
  2.              set obs 200
  3.              gen u = uniform()
  4.              gen t = (-ln(1-u))^(1/3) in 1
  5.              replace t = ((t[_n-1])^3 - ln(1-u))^(1/3) in 2/L
  6.              count if t<4
  7. end
. simulate nfail=r(N), reps(1000) nodots: genfail
      command:  genfail
        nfail:  r(N)
. summarize
    Variable |       Obs        Mean    Std. Dev.        Min        Max
-------------+--------------------------------------------------------
       nfail |      1000      63.788    7.792932         38         95
```

This simulation thus helps to verify that $E(N) = H(4) = 64$, and if we replicated this experiment infinitely often, the mean of our simulated N values would equal 64.

◁

❑ Technical note

In the above simulation, the line set obs 200 sets the size of the data to 200 observations for each replication of the counting failures experiment. Theoretically, any number of failures is possible in the interval (0,4), yet the probability of observing more than 200 failures is small enough to make this an acceptable upper limit.

❑

There is no contradiction between the probability-of-survival interpretation and the repeated-failure count interpretation, but be careful. Consider a cumulative hazard equal to 1. One interpretation is that we expect to observe one failure over the interval. The other interpretation is that the probability we observe no failures over the interval is $\exp(-1) = 0.368$, so the probability we observe the failure event is $1 - 0.368 = 0.632$. Are you surprised that the chances are not 100%?

More correctly, we should have said that the chance we will observe one *or more* failures per subject is 0.632, except that if this failure event is absorbing (i.e., deathlike in that it can occur only once), we will never observe the second and later failures because the first failure will prohibit these observations. Regardless of whether this is observable in the actual process being described, cumulative hazards must be interpreted in a context that allows repeated failures. The probability that we will observe zero failures is 0.368, and the probability that we will observe one or more failures is 0.632. Moreover, we can decompose that into the probability of one failure, the probability of two failures, etc., and in doing so we can compose a probability mass function for a random variable that has an expected value of 1. Because this expectation of 1 contains contributions based on two and more failures, it must be counteracted by a nonzero probability (0.368) of having no failures.

2.3.2 Interpreting the hazard rate

Hazard rates are rates; i.e., they have units $1/t$. You can interpret hazard rates just as you interpret cumulative hazards if you multiply them by t. Then you are saying, "The hazard rate is such that, were that rate to continue for 1 time unit, we would expect that ...."

For instance, if the hazard rate is 2/day, then it is such that, were that rate to continue for an entire day, you would expect two failures, or, if you prefer, the chances of observing at least one failure would be $1 - \exp(-2) = 0.8647$.

There is a subtle distinction here. If the cumulative hazard over a period is 2—if the integrated instantaneous hazard rate over the period is 2—then over that period you would expect two failures, regardless of the time profile of the hazard rates themselves. During that period, the hazard rate might be constant, increasing, decreasing, or any combination of these. If, on the other hand, the hazard rate is 2/day at some instant, then failures are happening at the rate of 2/day at that instant. However, you would expect only two failures over a period of a day if that hazard rate stayed constant over that period or varied in such a way as to integrate to 2 over that day.

Hazard rates, were they to stay constant, have a third interpretation. Hazard rates have units $1/t$; hence, the reciprocal of the hazard has units t and represents how long you would expect to have to wait for a failure if the hazard rate stayed at that level. If the hazard rate is 2/day, then were the hazard rate to remain at that level, we would expect to wait half a day for a failure. In fact, a constant hazard rate is what characterizes the classic Poisson counting process. In this process, it can be shown that if the expected time between failure is half a day with constant hazard, then the number of failures that occur in any given day (if failures were repeatable) is a Poisson random variable with expected value (and variance) equal to 2.

2.4   Means and medians

Given a random failure time T with probability density function $f(t)$, the mean time to failure, μ_T, is defined to be

$$\mu_T = \int_0^\infty t f(t) dt$$

Equivalently, one can show that $\mu_T = \int_0^\infty S(t) dt$, which often simplifies the calculation.

The median failure time, $\widetilde{\mu}_T$, is defined as the 50th percentile of the failure time distribution: that value such that half of all failure times are less than $\widetilde{\mu}_T$ and half are greater than $\widetilde{\mu}_T$. Thus for continuous distributions, $\widetilde{\mu}_T$ is the value such that $F(\widetilde{\mu}_T) = S(\widetilde{\mu}_T) = 0.5$, and $\widetilde{\mu}_T$ can be obtained directly from the quantile function

$$\widetilde{\mu}_T = Q(0.5)$$

because $Q(u)$ may be used to obtain any percentile, the 50th being just a special case. For the Weibull distribution with shape parameter p,

$$\begin{aligned}
\mu_T = \int_0^\infty t f(t) dt &= \int_0^\infty p t^p \exp(-t^p) dt \\
&= \Gamma(1 + 1/p)
\end{aligned}$$

where $\Gamma()$ is the gamma function. The median $\widetilde{\mu}_T = Q(0.5) = \{\ln(2)\}^{1/p}$.

The mean and median formulas above are often directly applied in parametric regression to obtain predictions of failure times for survivor functions that are specified given the values of certain predictors and estimated regression coefficients. In semiparametric or nonparametric schemes, predicting the mean and median is possible through some adaptation of these formulas, taking into account that for these models, the estimated survivor function is not continuous but rather a step function where the steps occur at each observed failure time. In any case, predictions of the mean and the median failure times are useful in that they give a sense of the typical time to failure for a particular distribution. Survival-time distributions can have long right tails, in which case some care is required when interpreting the mean failure time. For instance, with a constant hazard of 0.01 per day, the mean time to failure is 100 days, yet the median is only 69 days. The difference is caused by those few who survive much longer than 100 days.

Survival regression models are often fit in the ln(time) metric: one forms the model by hypothesizing a distribution for the natural logarithm of time to failure. Defining the random variable $Y = \ln(T)$ with probability distribution $f_Y(y)$, one can obtain predictions of the mean and median of Y using the above techniques and then transform the results back into the time metric by exponentiating the obtained values. In general, the exponentiated mean of Y is not equal to μ_T and thus would represent yet another flavor for a prediction of the typical survival time. For medians, however, the transformation is invariant. Taking the median of the distribution of Y and exponentiating will produce $\widetilde{\mu}_T$. When obtaining a "predicted" time to failure from statistical software, be sure you know what you are getting, or be prepared to specify what you want and see if it is available.

3 Hazard models

We began our discussion in chapter 1 by writing models of the form $t_j = \beta_0 + \beta_1 x_j + \epsilon_j$ or $\ln(t_j) = \beta_0 + \beta_1 x_j + \epsilon_j$, where ϵ_j is suitably distributed. These days, people seldom write models in this form. Instead, they write

$$h_j(t) = g(t, \beta_0 + \mathbf{x}_j \boldsymbol{\beta}_x)$$

that is, the hazard—the intensity with which the event occurs—for person j is some function $g()$ of $\beta_0 + \mathbf{x}_j \boldsymbol{\beta}_x$, where we now allow for the presence of multiple predictors via the row vector $\mathbf{x}_j$, in which case $\boldsymbol{\beta}_x$ is a column vector of regression coefficients. For instance, we might write

$$h_j(t) = (\beta_0 + \mathbf{x}_j \boldsymbol{\beta}_x)t$$

or something more complicated.

Regardless, this change from $t_j = \beta_0 + \beta_1 x_j + \epsilon_j$ to $h_j(t) = g(t, \beta_0 + \mathbf{x}_j \boldsymbol{\beta}_x)$ is just a change in notation and has no substantive implications. Remember, there is a one-to-one mapping from distributions to hazard functions. The distributional assumption we make for ϵ_j is now wrapped up in the hazard function that we choose. If we choose $g()$ appropriately, the likelihood function we obtain is the same, and therefore the resulting estimates are the same. We can even write linear regression using this notation; we just need to work out what the hazard function is for linear regression. Remember that, in general, the hazard function $h(t) = f(t)/S(t)$.

So, why should we write our models in this form? This notation can also incorporate the semiparametric models that we discussed. The idea here is to write the model as

$$h_j(t) = somefunction(h_0(t), \beta_0 + \mathbf{x}_j \boldsymbol{\beta}_x)$$

where $h_0(t)$ is called the *baseline hazard*. That is, the hazard subject j faces is *somefunction*() of the hazard everyone faces, modified by $\mathbf{x}_j$. A particularly popular way to parameterize these models is

$$h_j(t) = h_0(t) \exp(\beta_0 + \mathbf{x}_j \boldsymbol{\beta}_x)$$

These are called proportional hazards models. They are proportional in that the hazard subject j faces is multiplicatively proportional to the baseline hazard, and the function $\exp()$ was chosen simply to avoid the problem of $h_j()$ ever turning negative. Actually, even if we choose some function different from $\exp()$, it is still called the proportional hazards model.

There is nothing magical about proportional hazards models and, for some problems, the proportional-hazards assumption may be inappropriate. One could just as easily write an additive-hazards model; for instance,

$$h_j(t) = h_0(t) + \exp(\beta_0 + \mathbf{x}_j\boldsymbol{\beta}_x)$$

3.1   Parametric models

Any parametric survival model can be written in the hazard notation (although not necessarily the proportional-hazards notation), and doing that is just an exercise in translating from one notation to another. All parametric models have a corresponding hazard function.

It turns out that several popular survival parametric models, which are naturally written in the form $\ln(t_j) = \beta_0 + \mathbf{x}_j\boldsymbol{\beta}_x + \epsilon_j$, have corresponding $h()$ functions that naturally decompose into the $h_0(t)\exp(\beta_0 + \mathbf{x}_j\boldsymbol{\beta}_x)$ notation. The exponential and Weibull models are two examples. When that occurs, the model is said to have both a (log-)time metric and a proportional hazards metric formulation.

Familiarity, plus the fact that some $\ln(t)$ models have a proportional hazards interpretation, has caused many researchers to focus on this model exclusively:

$$h_j(t) = h_0(t)\exp(\beta_0 + \mathbf{x}_j\boldsymbol{\beta}_x)$$

The proportional hazards parametric model is just a matter of picking a functional form for $h_0(t)$. The exp() part is just how researchers have chosen to parameterize in all proportional hazards models the shift caused by subjects having different covariate ($\mathbf{x}$) values.

We can pick any positive function for $h_0(t)$ that we wish, although if we pick a strange one we will probably have to write our own maximum likelihood estimator. Certain functions for $h_0(t)$ are popular, and those are preprogrammed in Stata; for instance, if we choose

$$h_0(t) = c$$

for some constant c, then we fit what is called the exponential regression model. It is so called because, were we to translate the model back to the time metric, we would find that t, the time to failure, follows the exponential waiting-time distribution. If we worked out t's distribution $f(t)$—which we could easily do because it is always true that $f(t) = h(t)\exp\{-H(t)\}$—we would discover that $f(t)$ is the exponential density.

In any case, parameter estimates for these models are obtained by maximum likelihood. The likelihood of the data (with no censored observations, which we will discuss later) is

$$L(\boldsymbol{\beta}|t_1, t_2, \dots) = f(t_1|\boldsymbol{\beta}, \mathbf{x}_1)f(t_2|\boldsymbol{\beta}, \mathbf{x}_2)\dots$$

for $\boldsymbol{\beta} = (\beta_0, \boldsymbol{\beta}_x)$. Loosely, $L()$ can be taken to be the "probability" of observing a failure time for the first subject, t_1, given the value of $\mathbf{x}_1$, times the probability of observing a failure time, t_2, for the second subject given $\mathbf{x}_2$, etc. The basic idea behind maximum likelihood estimation is that, given a set of observations $(t_1, t_2, \dots, t_n)$, the best estimate of $\boldsymbol{\beta}$ is the one that maximizes the probability, or likelihood, of observing those particular data. Maximum likelihood estimates also have nice statistical properties that make inference and testing analogous to what you would see in simple OLS regression; see Casella and Berger (2002) or another graduate text on mathematical statistics for a thorough discussion of likelihood theory.

And so, despite our modern way of thinking of hazard functions, the likelihood above is still for the density function of t. Mathematically, we could just as well have thought about this for densities from the outset. We can write this more modernly as

$$L(\boldsymbol{\beta}|t_1, t_2, \dots) = S(t_1|\boldsymbol{\beta}, \mathbf{x}_1)h(t_1|\boldsymbol{\beta}, \mathbf{x}_1)S(t_2|\boldsymbol{\beta}, \mathbf{x}_2)h(t_2|\boldsymbol{\beta}, \mathbf{x}_2)\dots$$

because $f(t) = S(t)h(t)$. However we write it, we maximize this likelihood to estimate $\boldsymbol{\beta}$.

3.2   Semiparametric models

Semiparametric models, as mentioned in section 1.2, amount to combining individual binary-outcome analyses at each failure time. We want to show you exactly how that works and how it fits into the proportional-hazards way of thinking.

We will start with the same proportional hazards model that we started with in the parametric case:

$$h(t) = h_0(t)\exp(\beta_0 + \mathbf{x}_j\boldsymbol{\beta}_x)$$

This time, however, rather than specifying a function for $h_0(t)$, we will leave it unspecified, and it will cancel from our calculations when we perform the binary-outcome analyses at the individual failure times. This method is called the Cox proportional hazards model, discussed more later.

The likelihood function is calculated over the separate binary-outcome analyses that we perform:

$$L(\boldsymbol{\beta}|\text{data}) = L(\text{analysis 1})L(\text{analysis 2})\dots \tag{3.1}$$

We perform one analysis per failure time, and each analysis returns the probability of failing for those who did in fact fail at that time. For instance, in our small dataset,

```
subject    time    x
      1       1    3
      2       5    2
      3       9    4
      4      20    9
      5      22   -4
```

we would perform five analyses.

1. *Analysis at time 1:* the probability that subject 1 is the one that fails in a dataset containing all five subjects

2. *Analysis at time 5:* the probability that subject 2 is the one that fails in a dataset containing subject 2 and subjects 3, 4, and 5

3. *Analysis at time 9:* the probability that subject 3 is the one that fails in a dataset containing subject 3 and subjects 4 and 5

4. *Analysis at time 20:* the probability that subject 4 is the one that fails in a dataset containing subject 4 and subject 5

5. *Analysis at time 22:* the probability that subject 5 is the one that fails in a dataset containing only subject 5

Well, at least the last analysis is easy; the probability is 1. The next-to-last analysis is the next easiest.

At time 20 in our data, only subjects 4 and 5 survived, and subject 4 failed. Per our model, the hazard of failure at time t is

$$h_j(t) = h_0(t) \exp(\beta_0 + \mathbf{x}_j \boldsymbol{\beta}_x)$$

and so the hazard at time 20, the only time we care about for this analysis, is

$$h_j(20) = h_0(20) \exp(\beta_0 + \mathbf{x}_j \boldsymbol{\beta}_x)$$

The only subjects we care about are 4 and 5, who have scalar $\mathbf{x}$ values of 9 and -4, respectively:

$$h(\text{subject 4 at time 20}) = h_4(20) = h_0(20) \exp(\beta_0 + 9\beta_x)$$
$$h(\text{subject 5 at time 20}) = h_5(20) = h_0(20) \exp(\beta_0 + -4\beta_x)$$

Given that we observe one failure at this time, the probability that the failure is subject 4 is given by

$$\Pr(4 \text{ fails} | \text{a failure}) \quad = \quad \frac{h_4(20)}{h_4(20) + h_5(20)} \tag{3.2}$$

$$= \quad \frac{h_0(20)\exp(\beta_0 + 9\beta_x)}{h_0(20)\exp(\beta_0 + 9\beta_x) + h_0(20)\exp(\beta_0 + -4\beta_x)}$$

$$= \quad \frac{\exp(9\beta_x)}{\exp(9\beta_x) + \exp(-4\beta_x)}$$

So, tell us a value of β_x, and we will tell you a probability. If $\beta_x = -0.1$, then the probability is 0.214165, and we can tell you that, even though we make no assumptions about the shape of the baseline hazard $h_0(t)$.

Where does the value of β_x come from? We find the value of β_x that maximizes the overall likelihood, shown in (3.1). We do analyses just as the one shown for each of the failure times, and then we find the value of β_x that maximizes $L(\beta_x | \text{data})$. The intercept, β_0, drops out of the above calculations. This is a property of the semiparametric proportional hazards (Cox) model, which we cover more in chapter 9. For now, we merely take satisfaction in the fact that we have one less parameter to estimate.

The story we told you earlier about performing separate analyses was correct, although the jargon usually used to describe this is different. Rather than calling the ingredients separate *analyses*, they are called the "risk groups based on ordered survival times", and we obtain our estimates by "pooling over the risk groups based on ordered survival times".

❑ **Technical note**

Arriving at (3.2) required a bit of hand waving, so here we fill in the details for the interested reader. We are interested in the probability that subject 4 fails at $t = 20$, given that (1) subjects 4 and 5 survive to just before $t = 20$, and (2) either subject 4 or subject 5 will fail at $t = 20$. Let Δ be some arbitrarily small number, and let T_4 and T_5 denote the failure times of subjects 4 and 5, respectively. Then the probability that only subject 4 fails within Δ after time 20, given that both subjects survive to time 20 and that one of the subjects fails within Δ after time 20, is

$$\Pr\{T_4 < 20 + \Delta, T_5 > 20 + \Delta | T_4 > 20, T_5 > 20,$$
$$\text{one failure in } (20, 20 + \Delta)\}$$

which equals

$$\frac{\Pr(20 < T_4 < 20 + \Delta, T_5 > 20 + \Delta | T_4 > 20, T_5 > 20)}{\Pr\{\text{one failure in } (20, 20 + \Delta) | T_4 > 20, T_5 > 20\}} \tag{3.3}$$

Define $\mathcal{E}_i$ to be the event $20 < T_i < 20 + \Delta$, $\mathcal{F}_i$ to be the event $T_i > 20 + \Delta$, and $\mathcal{G}_i$ to be the event $T_i > 20$. Then (3.3) becomes

$$
\begin{aligned}
&\frac{\Pr(\mathcal{E}_4, \mathcal{F}_5 | \mathcal{G}_4, \mathcal{G}_5)}{\Pr(\mathcal{E}_4, \mathcal{F}_5 | \mathcal{G}_4, \mathcal{G}_5) + \Pr(\mathcal{E}_5, \mathcal{F}_4 | \mathcal{G}_4, \mathcal{G}_5)} \\
=\ &\frac{\Pr(\mathcal{E}_4 | \mathcal{G}_4)\Pr(\mathcal{F}_5 | \mathcal{G}_5)}{\Pr(\mathcal{E}_4 | \mathcal{G}_4)\Pr(\mathcal{F}_5 | \mathcal{G}_5) + \Pr(\mathcal{E}_5 | \mathcal{G}_5)\Pr(\mathcal{F}_4 | \mathcal{G}_4)} \\
=\ &\frac{\{\Pr(\mathcal{E}_4 | \mathcal{G}_4)/\Delta\}\Pr(\mathcal{F}_5 | \mathcal{G}_5)}{\{\Pr(\mathcal{E}_4 | \mathcal{G}_4)/\Delta\}\Pr(\mathcal{F}_5 | \mathcal{G}_5) + \{\Pr(\mathcal{E}_5 | \mathcal{G}_5)/\Delta\}\Pr(\mathcal{F}_4 | \mathcal{G}_4)}
\end{aligned} \qquad (3.4)
$$

Because $\lim_{\Delta \to 0}\{\Pr(\mathcal{E}_i | \mathcal{G}_i)/\Delta\} = h_i(20)$ by definition and $\lim_{\Delta \to 0} \Pr(\mathcal{F}_i | \mathcal{G}_i) = 1$, taking the limit as Δ goes to zero of (3.4) yields (3.2).

❏

3.3   Analysis time (time at risk)

We have considered hazard models of the form

$$h_j(t) = g(t, \beta_0 + \mathbf{x}_j \boldsymbol{\beta}_x)$$

and, for the proportional hazards model, we write

$$h_j(t) = h_0(t) \exp(\beta_0 + \mathbf{x}_j \boldsymbol{\beta}_x) \qquad (3.5)$$

Either way, we are assigning explanatory power to time. That is absurd. If we are analyzing a health outcome, the accumulation of toxins in someone's system might make the subject more likely to die—the accumulation is correlated with time—but time is not the cause. Or, if we are studying employment, it may be that the more information an unemployed person has about the job market—information he collects on the days he looks for employment—the more likely he is finally to accept a job offer. It then appears as if duration of unemployment is a function of time, but time is not the cause. To assume that the ticking of the clock somehow, by itself, changes the hazard is absurd unless we are physicists analyzing the true nature of time.

If we fully understand the process and have sufficient data, there will be no role for time in our model, and as a matter of fact, our current models can deal with that. In the proportional hazards case, we would simply define $h_0(t)$ as a constant (call it c) so that

$$h_j(t) = c \exp(\beta_0 + \mathbf{x}_j \boldsymbol{\beta}_x) = c^* \exp(\mathbf{x}_j \boldsymbol{\beta}_x)$$

for some other constant c^*. This model is known as the exponential model, and it is the only proportional hazards model that leaves nothing unsaid.

▷ **Example**

Consider the case where our baseline hazard function is the Weibull hazard with shape parameter $p = 3$; that is, $h_0(t) = 3t^2$. The proportional hazards model (3.5) then becomes

$$
\begin{aligned}
h_j(t) &= 3t^2 \exp(\beta_0 + \mathbf{x}_j \boldsymbol{\beta}_x) \\
&= \exp\{\beta_0 + \ln(3) + \mathbf{x}_j \boldsymbol{\beta}_x + 2\ln(t)\}
\end{aligned}
$$

If we define the time-varying covariate $z(t) = \ln(t)$, then this model may be reformulated as an exponential (constant baseline hazard) model with covariates $\mathbf{x}$ and $z(t)$, resulting in equivalent estimates of $\boldsymbol{\beta}_x$ and an intercept term that is shifted by $\ln(3)$.

Including a time-varying covariate among the explanatory variables is more easily said than done—in Stata, you must split observations into pieces, something we will discuss in section 13.1.1. In any case, it is not our intention to recommend this as a way to fit Weibull models. Weibull models assign a role to time, a fact demonstrated by reformulating the model for the exponential model—the model that leaves nothing unsaid—and in the reformulation, time itself appears as an explanatory variable.

◁

In the fully defined hazard model, we would just write $h_j(t, \mathbf{x}_j) = h_j(\mathbf{x}_j)$. The shift in the hazard would be fully explained by our $\mathbf{x}$ variables.

Except for the exponential models, every other model assigns a role to time, and that includes the semiparametric models. In semiparametric models, we get around ever having to define $h_0(t)$ because the individual analyses are performed on subjects at the same value of t, and therefore the same value of $h_0(t)$. The important part is that we compare subjects when their values of t are the same, meaning that subjects with equal t values (and $\mathbf{x}$ values) must share the same risk of the event.

When we assign a role to time, we are doing that to proxy other effects that we do not fully understand, cannot measure, are too expensive to measure, or are unknown. We need to consider how we are going to measure t so that it properly fulfills its role as a proxy. There are two properties to the definition of t:

1. Ensuring that whenever two subjects have equal t values, the risk they face would be the same if they also shared the same $\mathbf{x}$ values

2. Deciding which particular value of t should be labeled $t = 0$, denoting the onset of risk

Property 1 is important to all models. Property 2 matters only when we are fitting parametric models.

So, when do two subjects have the same risk? You have to think carefully about your problem. Here are three situations:

1. Say that you are analyzing the association between lung cancer and smoking. It would be unreasonable to define t as age. You would be stating that two persons

with equal characteristics would have the same risk of cancer when their ages are equal, even if one had been smoking for 20 years and the other for 1 month.

2. Continuing with the analysis of the association between lung cancer and smoking, it might be reasonable to assume that two smokers face the same risk when they have been smoking for the same amount of time. That would be reasonable if all smokers smoked with the same frequency. If there are large variations, and we had some smokers who smoked one cigarette per day and others who smoked 60 per day, time since onset of smoking might not be a good proxy for what we want to measure (assuming we did not account for the discrepancy with our **x** variables). If all our subjects were smokers, we might think about measuring time as total cigarettes, so that a person who smokes 1 cigarette per day for 20 days is at $t = 20$ just as a person at day 1 who smokes 20 cigarettes per day.

3. You are studying the time to death for a certain type of cancer patient. When do two identical patients face the same risk? At the same time since the onset of cancer, you decide. Fine, but when is the onset of the cancer? At detection? Are two patients really the same when, at $t = 0$, one reports to the clinic with metastatic cancer invading other organs and the other with a small, barely detectable tumor? Or is the onset of risk when the first cancer cell differentiates, and then how would you know when that was?

Defining t is worth the thought, and it is usually easiest to think in terms of the onset of risk—to find the time at which risk began and before which the event could not happen for the reason under analysis. If that onset of risk is well defined in the data, then defining $t = 0$ is easy enough. You will probably let t increment in lock step with calendar time although you may want to think about time-intensity measures such as total cigarettes.

If there is no well-defined onset of risk, think about when two subjects with the same **x** values would face the same risk and define t so that they have the same value of t at those times. That will not uniquely label one of the times $t = 0$, but that may not matter.

Pretend that, for some problem, you have defined t such that identical subjects facing the same risk do have equal t values. We could define $t' = t - 5$, and this new time would work just as well for matching subjects. So, is the onset of risk with $t = 0$ or $t' = 0$? That is a substantive question that your science must answer.

If you engage in semiparametric modeling, however, how you answer that question does not matter because semiparametric results are determined only by the matching and ordering of failure times. Time is used only to order the data, and no special significance is attached to $t = 0$.

In many parametric survival models, however, special significance is attached to $t = 0$. For these hazard functions, you can fit a parametric model, write down the results, add 5 to every t value, reestimate the parameters, and get different results. The results will not be merely apparently different—somehow undoing the $+5$ addition you just made. The results will be substantively different in that predictions made using the model will differ.

Although the value chosen for $t = 0$ matters in most parametric models, the units in which t is measured (minutes, hours, days, years) do not matter. More correctly, for no popularly used hazard function $h_0(t)$ do the units of t matter. Because you can choose any nonnegative function for $h_0(t)$, you could choose one in which the units do matter, but that would be a poor choice.

From now on, we will refer to t as *analysis time* to emphasize that t is as defined above and is not just some time or date variable you happen to have lying around in your data:

$$t = somefunction(\text{various time or date variables in your data})$$

For instance,

$$t = \frac{\texttt{date\_of\_failure} - \texttt{date\_of\_birth}}{365.25}$$

defines t as years of age and defines birth as corresponding to the onset of risk.

$$t = \texttt{date\_of\_failure} - \texttt{date\_of\_diagnosis}$$

defines t as days since diagnosis and defines diagnosis as corresponding to the onset of risk.

$$t = \texttt{hour\_of\_failure} - \texttt{hour\_of\_start}$$

defines t as hours from start and defines start as corresponding to the onset of risk.

$$t = \texttt{cigs\_per\_day} \times (\texttt{date\_of\_failure} - \texttt{date\_started\_smoking})$$

defines t as total cigarettes since the start of smoking and defines the start of smoking as corresponding to the onset of risk. Just remember, this choice of analysis time is not arbitrary.

Analysis time t does not have to be time. For instance, we have a machine that produces bolts of cloth, and we then examine those bolts for defects. It might be reasonable to measure "time" as the linear distance from the beginning of the bolt so that, if we were to find a defect at 4 meters into the bolt, the bolt "failed" at $t = 4$ meters.

4 Censoring and truncation

4.1 Censoring

In real data-analysis situations, we often do not know when failures occurred, at least not for every observation in the dataset. Rather than knowing that failures occurred at analysis times 1, 5, 9, 20, and 22, we might know that failures occurred at 1, 5, 9, and 20, and that failure had not yet occurred by analysis time 22, when we stopped our experiment:

```
time     failed      x
  1          1        3
  5          1        2
  9          1        4
 20          1        9
 22          0       -4
```

This is called *censoring*, or more precisely, right censoring. This happens so commonly that some researchers will write their datasets like this:

```
time    censored      x
  1          0        3
  5          0        2
  9          0        4
 20          0        9
 22          1       -4
```

Stata prefers the first form. `failed==1` means the observation was observed to fail at that time, and `failed==0` means the observation was not observed to fail, the implication being that the observation was right-censored.

Censoring is defined as when the failure event occurs and the subject is not under observation. Think of censoring as being caused by a censor standing between you and reality, preventing you from seeing the exact time of the event that you know occurs.

Before we get too deeply into this, let us define some popularly used jargon. We are following subjects over time, and that data collection effort is typically called a *study*. During the study period, subjects are *enrolled* and data are collected for a while, called the subject's *follow-up* period, or if you prefer, the period during which the subject was under observation. Data collection stops on a subject because the subject fails,

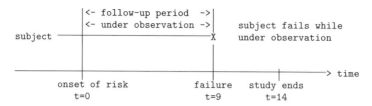

the study ends,

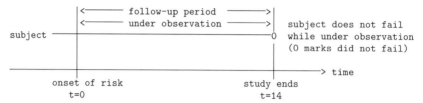

or the subject leaves the study for other reasons (or is lost to follow up):

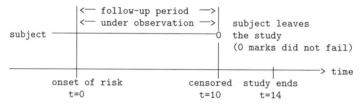

In the above diagrams, we drew the beginning of "under observation" to coincide with the onset of risk; however, the subject may come at risk for the failure event before enrollment, at the instant of enrollment, or afterward. From a statistical perspective, we are interested in the period when the subject is at risk.

In the discussions that follow, we will assume that failure is a culminating or absorbing event such as death. The event can occur only once, and once it does occur, the subject can no longer be observed. Censoring can also be extended to repeatable failures.

4.1.1   Right censoring

When most investigators say censoring, they mean right censoring. In this type of censoring, the subject participates in the study for a time and, thereafter, is no longer observed. This can occur, for example,

1. when one runs a study for a prespecified length of time, and by the end of that time, the failure event has not yet occurred for some subjects (this is common in studies with limited resources or with time constraints),

2. when a subject in the study withdraws prematurely (in trials of a new drug therapy, a patient might experience intolerable side effects and thus cease participation), or

3. when a subject is lost to follow-up, meaning that he or she disappears for unknown
    reasons (for example, in a longitudinal study a subject moves, say, because of
    marriage, and cannot be located).

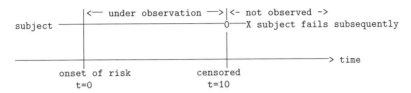

The analytic tools we use assume that any censoring occurs randomly and is unrelated to the reason for failure. It will not do at all if, just prior to failure, subjects
are highly likely to disappear from our view. In parametric models, right censoring is
easily dealt with. First, start by thinking of a dataset with no censoring in which all
failure times are measured exactly. The likelihood function is (for a sample of size n,
failure times $t_1, \ldots, t_n$)

$$L\left\{\boldsymbol{\beta}|(t_1, \mathbf{x}_1), \ldots, (t_n, \mathbf{x}_n)\right\} = \prod_{i=1}^{n} S(t_i|\mathbf{x}_i, \boldsymbol{\beta}) h(t_i|\mathbf{x}_i, \boldsymbol{\beta}) \tag{4.1}$$

because the probability density function of t_i, $f(t_i) = S(t_i)h(t_i)$.

Equation (4.1) actually becomes simpler in the presence of censoring because, for
those failure times that are censored, the density of t_i, $S(t_i|x_i, \boldsymbol{\beta})h(t_i|x_i, \boldsymbol{\beta})$, is replaced
by only the survivor function $S(t_i|x_i, \boldsymbol{\beta})$. All that is known about a censored observation
is that failure occurs *after* time t_i, yet exactly when remains unknown [by definition
$S(t) = \Pr(T > t)$]. Our software has to be up to making this substitution, but that is
all.

Right censoring is also easily dealt with in semiparametric models. Think of the
individual binary-outcome analyses that we perform. If subject i is censored at t_i, then
that subject enters all the individual failure-time studies up to and including t_i (the
subject did not fail at that time) and after that is simply ignored. The only difference
between a subject that fails and one that does not is that the latter never appears in an
individual study while being marked as having failed. Again our software has to know
not to perform certain individual analyses, but that is all.

(Continued on next page)

4.1.2   Interval censoring

With interval censoring, rather than knowing the exact time of failure or that failure occurred past a certain point, all we know is that the failure event occurred between two known time points—perhaps a short interval or a long one.

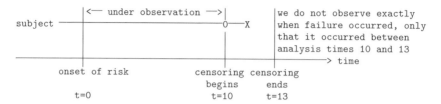

Interval censoring can result when subjects are evaluated or inspected at fixed time points throughout their follow-up period. In a clinical study, patients may be required to visit the clinic once per month for several years. Assume that a patient visits a clinic at month 6 and then at month 7. At the 6-month evaluation, the patient was negative for the event of interest; at month 7, she was positive. That is, the failure event occurred at some point between the sixth and seventh evaluations. Because we do not know exactly when the failure event occurred, this observation is interval-censored.

Stata does not directly handle this kind of censoring, but a strong argument can be made that it should. This kind of censoring is easy to handle in parametric models and difficult in semiparametric models.

Let's start with parametric models. For interval-censored observations, we know that subject i failed between times t_{0i} and t_{1i}. The probability that the failure occurred sometime before t_{1i} is $1 - S(t_{1i})$, and the probability that a failure occurred before t_{0i} is $1 - S(t_{0i})$. Thus

$$\Pr(\text{failure between } t_{0i} \text{ and } t_{1i} | \mathbf{x}_i, \boldsymbol{\beta}) = S(t_{0i} | \mathbf{x}_i, \boldsymbol{\beta}) - S(t_{1i} | \mathbf{x}_i, \boldsymbol{\beta})$$

and this would be subject i's contribution to the likelihood given by (4.1).

If we have a dataset in which the first subject failed at t_1, the second subject was right-censored at t_2, and the third subject was only known to fail between t_{03} and t_{13}, then the likelihood would be

$$
\begin{aligned}
L\{\boldsymbol{\beta} | (t_1, \mathbf{x}_1), \ldots, (t_n, \mathbf{x}_n)\} &= S(t_1 | \mathbf{x}_1, \boldsymbol{\beta}) h(t_1 | \mathbf{x}_1, \boldsymbol{\beta}) S(t_2 | \mathbf{x}_2, \boldsymbol{\beta}) \times \\
&\quad \{S(t_{03} | \mathbf{x}_3, \boldsymbol{\beta}) - S(t_{13} | \mathbf{x}_3, \beta)\} \times \cdots
\end{aligned}
$$

In semiparametric analysis, the problem is more difficult. Remember, we approach semiparametric analyses by performing separate binary-outcome analyses for each of the failure times and then combining those studies. Interval censoring is a problem only if it brings into dispute the ordering of the observations.

Pretend that we are uncertain about the failure time of subject 3, knowing only that he failed in the interval (10,14). If no subject has an observed or censored failure time in that interval, then our uncertainty about subject 3 makes no difference. Whether death occurs at time 10, 14, or any time between would still place subject 3 in the same temporal ordering, so binary analysis would be unaffected. All the other binary analyses would, of course, be unaffected, so combining the analyses must yield the same overall result.

This fact is a real lifesaver. Few of us have datasets that record the exact time of failure—the failure occurred at 12:51:02.395852 on 12Feb2000. Say that you have data that records the time of failure in days. You can still order the events—and therefore the analyses—except for the failures that happen on the same day. If subjects 3 and 4 both failed on day 20, you do not know which failed first, and so you do not know in which order to perform your individual binary-outcome analyses. That problem is easily solved, and as we discuss in chapter 9, the software can handle this.

You are equally saved if you record your data in, say, integer number of months. Here you do not have the detail of information that you would have were the times recorded in, say, days. This means only that you will get less efficient estimates should more precise time measurements prove effective in breaking the (apparent) tied failure times caused by less precise measurements of time to failure. You can still perform the analysis using standard tools.

The difficulty arises when exact times are uncertain and overlap only partially. For example, subject 3 is known to fail within the interval (10,14), and subject 4 is known to fail in (12,16):

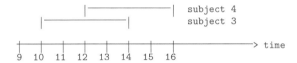

The best way to handle this is still a subject of research, but one solution would be to assume that the distribution of failure time is uniform (flat) within each interval and to break the combined interval into three sections. Within (10,12), subject 3 fails first; within (12,14), the two subjects are "tied", and we can handle that appropriately in Stata; and within (14,16), subject 3 fails first again. Assuming uniformly distributed survival times within each interval, the probability that we observe a tie [both subjects fail in (12,14)] is 1/4, and thus we can weight the analysis that treats these as tied values by 1/4, and the analysis that has subject 3 failing first by 3/4.

Of course, this system of weighting relied on the failure time having a uniform distribution within the interval—meaning the hazard must be falling in the interval. As such, these weights may prove to be invalid for certain types of data, in which case we would need to adjust our weighting scheme.

4.1.3   Left censoring

Mathematically, left censoring is no different from interval censoring. The event occurred at some time when the subject was not under observation, here prior to the subject being under observation, but that makes no mathematical difference. The event happened in an interval and the solution is the same.

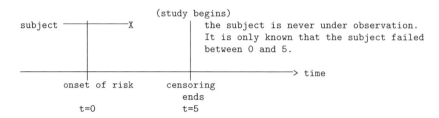

How left censoring arises in reality is probably different from interval censoring. Left censoring means that the failure event occurs prior to the subject's entering the study. For example, in a study of time to employment, an individual who is employed when first interviewed is considered left-censored because the transition from unemployment to employment (the failure event) happened prior to the beginning of the follow-up period.

4.2   Truncation

Truncation is often confused with censoring because it also gives rise to incomplete observations over time. In statistics, truncation usually refers to complete ignorance of the event of interest and of the covariates over part of the distribution.

In survival-data applications, it is difficult to be so ignorant when the event is an absorbing one, such as death. That we observe you today means that you did not die yesterday. Therefore, for such absorbing events, truncation is defined as a period over which the subject was not observed but is, a posteriori, known not to have failed. Truncation causes the statistical difficulty that, had the subject failed, he or she would never have been observed.

If multiple failures for the same subject are possible, truncation is defined as not knowing how many (or even if) events occurred during a given period.

However, we will restrict ourselves here to absorbing failure events, such as death.

4.2.1   Left truncation (delayed entry)

In left truncation, the period of ignorance extends from on or before the onset of risk to some time after the onset of risk. Consider the following subject represented below: for a while, the subject was not observed, but then the subject came under observation. Left truncation usually arises because we encounter a subject who came at risk some time ago.

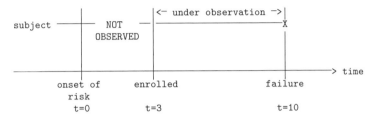

Can we include this subject in our study? Yes, but we must account for the fact that, had the subject failed earlier, we never would have encountered this subject. The subject's subsequent survival can be analyzed, but we do not want to make too much of the fact that the subject survived until the point we encountered him. Consider an 89-year-old smoker who just arrives to enroll in our study. Does he supply whopping evidence that smoking lengthens life? No, because, had he died, he would have never enrolled in our study; the only way this 89-year-old made it to our study was by surviving all 89 prior years.

In parametric models, left truncation is easily dealt with. The contribution to the overall likelihood of subject i (when failure time is measured exactly and without truncation) is

$$L_i(\boldsymbol{\beta}|t_i, \mathbf{x}_i) = S(t_i|\mathbf{x}_i, \boldsymbol{\beta})h(t_i|\mathbf{x}_i, \boldsymbol{\beta})$$

If subject i, who is ultimately observed to fail at time t_i, arrived late to our study (say, at time t_{0i}), then we simply need to add one more condition to subject i's contribution to the likelihood: that he had already survived until time t_{0i}, which has probability $S(t_{0i}|\mathbf{x}_i, \boldsymbol{\beta})$, thus

$$L_i(\boldsymbol{\beta}|t_i, \mathbf{x}_i) = \{S(t_i|\mathbf{x}_i, \boldsymbol{\beta})/S(t_{0i}|\mathbf{x}_i, \boldsymbol{\beta})\}h(t_i|\mathbf{x}_i, \boldsymbol{\beta}) \tag{4.2}$$

because $S(t_i|\mathbf{x}_i, \boldsymbol{\beta})/S(t_{0i}|\mathbf{x}_i, \boldsymbol{\beta})$ is the probability of surviving to t_i, given survival up to t_{0i}.

In fact, (4.2) applies even to cases where there is no left truncation and the subject enrolled at time $t_{0i} = 0$, because $S(0|\mathbf{x}_i, \boldsymbol{\beta}) = 1$.

Left truncation is easily handled in semiparametric models, too. Then one simply must omit the subject from all individual binary-outcome analyses during the truncation period because the subject could not possibly have failed during that period and still manage to be around when the period ended. We cannot treat the subject as being at risk of failure during the truncation period since the only reason we observe him at all is because he survived the period.

4.2.2   Interval truncation (gaps)

Interval truncation is just a variation on left truncation. Here think of a subject who disappears for a while but then reports back to the study, causing a gap in our follow-up. The statistical issue is that, had the subject died, he or she never could have reported back. Throughout this text, and throughout the Stata documentation, we treat the terms "gaps" and "interval truncation" as synonyms.

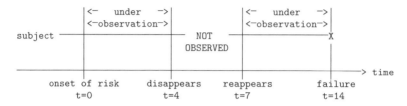

For parametric models, assume that subject i is observed up until time t_{1i}, goes unobserved until time t_{2i}, and then is observed up until time t_{3i}, at which point the subject dies. Then the probability of observing those parts that we did observe is

$$S(t_{1i}|\mathbf{x}_i,\boldsymbol{\beta})\{S(t_{3i}|\mathbf{x}_i,\boldsymbol{\beta})/S(t_{2i}|\mathbf{x}_i,\boldsymbol{\beta})\}h(t_{3i}|\mathbf{x}_i,\boldsymbol{\beta})$$

Thus this is subject i's contribution to the overall likelihood, denoted by L_i.

In semiparametric models, this problem is handled the same way as left truncation. One simply omits the subject from all individual binary-outcome analyses during the truncation period because the subject could not possibly have failed at those times.

4.2.3   Right truncation

Right truncation is indistinguishable from right censoring, which was discussed in section 4.1.1. There is a point beyond which the subject is not observed, and because time may extend all the way to infinity, failure is certain to occur eventually.

5 Recording survival data

5.1 The desired format

Given the possibility of censoring and truncation, the simple way we have shown for the recording of survival data, using one observation per subject and recording just time of failure,

```
time      x
   1      3
   5      2
   9      4
  20      9
  22     -4
```

is simply not adequate. Instead, we want to record (and think about) survival data as follows:

```
id     t0     t1     outcome      x
 1      0      1           1      3
 2      0      5           1      2
 3      0      9           1      4
 4      0     20           1      9
 5      0     22           1     -4
```

In this format, we make explicit the beginning and ending times of each record, using variables t0 and t1—recording the analysis times during which the subject was under observation—and we record the outcome at the end of the span in variable outcome, where outcome==1 denotes a failure and outcome==0 denotes nonfailure (censoring). You can call the variables what you wish, but these three variables concisely describe the information.

This method will allow us to record observations that are right-censored,

```
id     t0     t1     outcome      x
 1      0      1           1      3
 2      0      5           1      2
 3      0      9           1      4
 4      0     20           1      9
 5      0     22           0     -4     <- id 5 did not fail yet
```

that have left truncation,

```
id     t0     t1     outcome     x
 1      0      1           1     3
 2      0      5           1     2
 3      3      9           1     4      <- id 3 entered late, at t=3
 4      0     20           1     9
 5      0     22           0    -4
```

or that have interval truncation,

```
id     t0     t1     outcome     x
 1      0      1           1     3
 2      0      5           1     2
 3      3      9           1     4
 4      0      9           0     9      <- id 4 was unobserved
 4     11     20           1     9      <- between times 9 and 11
 5      0     22           0    -4
```

Do you see in this last example how the subject with `id==4` was not observed between times 9 and 11? The first record for the subject records the span 0–9, and the second records the span 11–20. Because there is no record for the span 9–11, the subject was unobserved (or perhaps just not at risk).

Each record in the data records a span during which the subject was under observation, and it includes an outcome variable recording whether a failure was observed at the end of the span. Subjects may have multiple records, as does subject 4 above, and so subjects can have delayed entry, multiple gaps, and truly complicated histories. For example, look at subject 5 below:

```
id     t0     t1     outcome     x
 1      0      1           1     3
 2      0      5           1     2
 3      3      9           1     4
 4      0      9           0     9
 4     11     20           1     9
 5      2      4           0    -4      <- subj. enrolled late, disappeared
 5      6      8           0    -4      <- then reappeared for a while
 5     10     15           0    -4      <- then reappeared for a while again
 5     17     21           0    -4      <- finally reappeared and was censored
```

Stata can understand data presented in this manner. Each of its statistical commands will do the statistically appropriate thing with these data, which is really not as amazing as it looks. If you examine the developments of the previous chapter, you will find that they follow a pattern:

For parametric analysis, for each span,

1. The likelihood contribution is $S(t_1|x_i, \boldsymbol{\beta})/S(t_0|x_i, \boldsymbol{\beta})$ for the observational period.

2. The likelihood contribution is multiplied by $h(t_1|x_i, \boldsymbol{\beta})$ if the period ends in a failure event.

For semiparametric analysis,

1. Individual studies are performed at all `t1`s for which `outcome` is equal to 1.
2. Each observation enters the individual study at time t if for that subject `t0` $< t \le$ `t1`.

Moreover, these individual time-span observations record the independent pieces of information of your data, meaning that records can be omitted from analysis without bias, even if that means including only pieces of a subject's history in an analysis. There is, of course, a loss of efficiency, but there is no bias. Let us explain.

First, understand that there is no difference in recording

```
id    t0    t1    outcome    x
 2     0     5          1    2    (subject observed between 0 and 5)
```

or

```
id    t0    t1    outcome    x
 2     0     2          0    2    (subject observed between 0 and 2)
 2     2     5          1    2    (subject observed between 2 and 5)
```

or

```
id    t0    t1    outcome    x
 2     0     2          0    2    (subject observed between 0 and 2)
 2     2     4          0    2    (subject observed between 2 and 4)
 2     4     5          1    2    (subject observed between 4 and 5)
```

All three of these variations record the same information: subject 2 was under continuous observation between analysis times 0 and 5 and, at $t = 5$, exhibited a failure. Why would you want to split this information into pieces? Perhaps times 0, 2, and 4 were the times the subject reported to the diagnostic center—or filled in a survey—and at those times you gathered information about characteristics that change with time:

```
id    t0    t1    outcome    x    exercise
 2     0     2          0    2           0
 2     2     4          0    2           1
 2     4     5          1    2           0
```

You could, if you wished, restrict your analysis to the subsample `exercise==0`. That would make it appear as if this subject had a gap between times 2 and 4. The subject was really not outside of observation, but he or she would still be interval truncated for this analysis. Or you could restrict your analysis to the subsample `exercise==1`, in which case this subject would appear to enter late into the sample and be right-censored.

You could even include `exercise` among the explanatory variables in your model, stating that the hazard shifted according to the current value of this variable.

Finally, this format allows recording multiple-failure data. Below the subject fails at times 4 and 7, and even after the second failure, we observe the subject a little while longer:

id	t0	t1	outcome	x	exercise
14	0	2	0	2	0
14	2	4	1	2	1
14	4	5	0	2	1
14	5	7	1	2	0
14	7	9	0	2	0

In our theoretical discussions, we seldom mentioned failure events that could reoccur, but it is an easy enough generalization. Assuming the failure events are of the same kind, such as first heart attack, the main issue is how you want to treat second and subsequent failure events. Is the hazard of failure at times 4 and 7 equal to $h(4)$ and $h(7)$, respectively, meaning the clock keeps ticking; or is it $h(4)$ and $h(3)$, meaning the clock gets reset; or is it $h_1(4)$ and $h_2(3)$, meaning they are on different clocks altogether?

5.2   Other formats

Unless you have the dubious pleasure of having entered the data yourself, the data probably do not come in the ever-so-desirable duration format. Instead of seeing

id	t0	t1	outcome	x	exercise
2	0	2	0	2	0
2	2	4	0	2	1
2	4	5	1	2	0

you have something that looks like

id	t	event	x	exercise
2	0	enrolled	2	0
2	2	checkup		1
2	4	checkup		0
2	5	failed		

or like

id	x	t1	event1	ex1	t2	event2	ex2	t3	event3	ex3	...
2	2	0	enrolled	0	2	checkup	1	4	checkup	1	...

These formats are called snapshot or transaction data. The first example is called "long-form snapshot data" and the second, "wide-form snapshot data". Each observation records a time and the measurements taken at that time. As long as we are being realistic about this, the data probably do not record analysis time but rather calendar time, so in the long form, the data might appear as

id	date	event	x	exercise
2	20jan2000	enrolled	2	0
2	22jan2000	checkup		1
2	24jan2000	checkup		0
2	25jan2000	failed		

and would appear similarly in the wide form. The event variable is seldom called that. Instead, it is often called a record type, especially when the data are in the long form, and the variable is often coded numerically. Record type 1 means an enrollment, 2 a checkup, and 9 a failure, for instance.

id	date	rectype	x	exercise
2	20jan2000	1	2	0
2	22jan2000	2		1
2	24jan2000	2		0
2	25jan2000	9		

Anyway, the problem that you must address—long before you get to use your first survival analysis command—is converting these data into the duration format.

For now, ignore the issue of date versus analysis time—that the "time" variable in the data might not correspond to analysis time. Let's focus on what must happen to convert these data into duration data. We can worry about defining a reasonable analysis-time variable later. Also for now we consider only the problem of long-form snapshot data because if we can solve that problem we can solve the other; converting wide-form data into long-form data is easy in Stata by using the reshape command; see [D] **reshape**.

Let's consider the following set of transactions:

id	date	rectype	x	exercise
2	20jan2000	1	2	0
2	22jan2000	2		1
2	24jan2000	2		0
2	25jan2000	9		

We will form a record from each pair of these records that contains a beginning date, an ending date, and the appropriate values. In choosing these appropriate values, we need to realize that snapshot datasets really contain two kinds of variables:

1. *Enduring or characteristic variables*
   These variables state characteristics that are valid for longer than an instant or that you are at least willing to treat that way. A subject's gender, age rounded to years, and blood pressure at the time of hospitalization are examples. In the examples above, we will assume that x and exercise are enduring variables.

2. *Instantaneous or event variables*
   These variables state events that occurred at an instant in time, such as having a heart attack, going bankrupt, finding a job, or other failure-type events. In the example above, rectype is an instantaneous variable because it records, among other things, whether the subject failed. In some other dataset, rectype might not record that (it might just record whether the record was a result of a visit or a telephone interview), and some other variable might record whether the subject failed at that instant. Then both of those variables would be instantaneous variables. The interview process does not continue after the date, nor does the act of failing.

Our goal is to take pairs of records, such as the first two in our snapshot dataset,

id	date	rectype	x	exercise
2	20jan2000	1	2	0
2	22jan2000	2		1

and form one time-span record from them:

id	date0	date1	rectype	x	exercise
2	20jan2000	22jan2000	?	?	?

But what is the value of each variable over this time span? For the enduring variables, we will obtain their values from the first observation of the pair:

id	date0	date1	rectype	x	exercise
2	20jan2000	22jan2000	?	2	0

For the instantaneous variables, we will obtain the value from the second observation of the pair:

id	date0	date1	rectype	x	exercise
2	20jan2000	22jan2000	2	2	0

In duration data, we want the values of the variables over a particular span of time, and we want events that occurred at the *end of the period*. So we are going to ignore the values of the transient variables in the first-of-pair observation and get those values from the second-of-pair observation. The net result of this will be to discard the values of the instantaneous variables recorded in the first observation of the snapshot dataset. It is dangerous to discard any data because you never know when you are going to need them. To get that back, when we look at our snapshot data,

id	date	rectype	x	exercise
2	20jan2000	1	2	0
2	22jan2000	2		1
2	24jan2000	2		0
2	25jan2000	9		

we will pretend that we see

id	date	rectype	x	exercise	
2	.	.	.	.	<- new
2	20jan2000	1	2	0	
2	22jan2000	2		1	
2	24jan2000	2		0	
2	25jan2000	9			

and convert those records. The result, then, of carrying out this procedure is

id	date0	date1	rectype	x	exercise
2	.	20jan2000	1	.	.
2	20jan2000	22jan2000	2	2	0
2	22jan2000	24jan2002	2	.	1
2	24jan2002	25jan2000	9	.	0

Thus the interpretation of a single record is as follows:

1. A record spans the period `date0` to `date1`.

2. The values of the enduring variables (here `x` and `exercise`) are the values they had on `date0`.

3. The values of the instantaneous variables (here `rectype`) are the values they had on `date1`.

Now you may be concerned about the missing values for `x`. Would it not be better if the enduring variables had their values carried down from one observation to the next? Yes, but do not try to solve all the problems at one time. We could solve this problem later by typing

```
. sort id date1
. by id: replace x = x[_n-1] if x>=.
```

The hard part is transforming

id	date	rectype	x	exercise
2	20jan2000	1	2	0
2	22jan2000	2		1
2	24jan2000	2		0
2	25jan2000	9		

into

id	date0	date1	rectype	x	exercise
2	.	20jan2000	1	.	.
2	20jan2000	22jan2000	2	2	0
2	22jan2000	24jan2002	2	.	1
2	24jan2002	25jan2000	9	.	0

but we do not have to do this manually. Stata has one command, **snapspan**, that will do that for us. Its syntax is

> **snapspan** *idvar time_var instantaneous_vars*, **generate**(*new_begin_date*)

and so, to convert this dataset, all we need to type is

```
. snapspan id date rectype, gen(date0)
. rename date date1
```

snapspan requires that we specify an ID variable *idvar*, a time variable *time_var*, and one or more variables we wish to treat as instantaneous *instantaneous_vars*. Any variables in our data that are left unspecified are treated as enduring variables.

That leaves just the problem of defining analysis time, t, for the date and time variables already in our data. That also is easy, but we have to get a little ahead of ourselves.

Once we have the data in duration form,

id	date0	date1	rectype	x	exercise
2	.	20jan2000	1	.	.
2	20jan2000	22jan2000	2	2	0
2	22jan2000	24jan2002	2	.	1
2	24jan2002	25jan2000	9	.	0

we use a command called `stset` to tell Stata about the data, and in that process, we can tell `stset` how to form the analysis-time variable for us. Basically, given the above, we might type

```
. stset date1, id(id) time0(date0) origin(time date0) failure(rectype==9)
```

and why we would do that is covered in the next chapter.

5.3   Example: Wide-form snapshot data

Let's consider the following wide-form dataset:

```
. use http://www.stata-press.com/data/cggm/wide1

. describe

Contains data from http://www.stata-press.com/data/cggm/wide1.dta
  obs:           3
  vars:          14                              28 Jan 2008 14:42
  size:          180 (99.9% of memory free)
```

	storage	display	value	
variable name	type	format	label	variable label
id	float	%9.0g		Subject ID
sex	float	%9.0g		Sex (1=female)
date1	float	%d		1st interview
event1	float	%9.0g		
x1	float	%9.0g		
date2	float	%d		2nd interview
event2	float	%9.0g		
x2	float	%9.0g		
date3	float	%d		3rd interview
event3	float	%9.0g		
x3	float	%9.0g		
date4	float	%d		4th interview
event4	float	%9.0g		
x4	float	%9.0g		

```
Sorted by:
```

. list

1.	id 1	sex 0	date1 20jan2000	event1 1	x1 5	date2 22jan2000	event2 9	x2 3

	date3 .	event3 .	x3 .	date4 .	event4 .	x4 .

2.	id 2	sex 1	date1 14feb2000	event1 1	x1 8	date2 18feb2000	event2 2	x2 5

	date3 22feb2000	event3 2	x3 5	date4 .	event4 .	x4 .

3.	id 3	sex 0	date1 11nov1999	event1 1	x1 2	date2 14nov1999	event2 2	x2 2

	date3 18nov1999	event3 2	x3 3	date4 22nov1999	event4 9	x4 3

To convert the data into the desired duration format, we must first **reshape** the data and then issue a **snapspan** command:

```
. reshape long date event x, i(id)
(note: j = 1 2 3 4)

Data                                wide    ->    long
-----------------------------------------------------------------
Number of obs.                         3    ->      12
Number of variables                   14    ->       6
j variable (4 values)                       ->      _j
xij variables:
                    date1 date2 ... date4   ->    date
                  event1 event2 ... event4  ->    event
                        x1 x2 ... x4        ->    x
-----------------------------------------------------------------

. drop if missing(date)
(3 observations deleted)
. snapspan id date event, gen(date0) replace
```

(*Continued on next page*)

```
. list, noobs sepby(id)
```

id	_j	sex	date	event	x	date0
1	.	.	20jan2000	1	.	.
1	1	0	22jan2000	9	5	20jan2000
2	.	.	14feb2000	1	.	.
2	1	1	18feb2000	2	8	14feb2000
2	2	1	22feb2000	2	5	18feb2000
3	.	.	11nov1999	1	.	.
3	1	0	14nov1999	2	2	11nov1999
3	2	0	18nov1999	2	2	14nov1999
3	3	0	22nov1999	9	3	18nov1999

We left out the drop if missing(date) on our first try, and when we did, we got

```
. snapspan id date event, gen(date0) replace
3 observations have date==.
either fix them or drop them
by typing drop if date==.
r(459);
```

So, we followed the instructions and dropped the observations with missing date. snapspan refuses to work with records for which it does not know the date. In our original wide dataset, we had missing dates when we had no information. We also specify the replace option to snapspan, which just permits Stata to change the data in memory without requiring that the data be saved first (after reshape and drop here).

6   Using stset

We left off in the last chapter having agreed that we would record our survival data in duration format:

```
id      date0      date1    event     x  exercise
 1   20jan2000  21jan2000       9     3         1
 2   15dec1999  20dec1999       6     2         0
 3   04jan2000  13jan2000       4     4         1
 4   31jan2000  08feb2000       3     9         1
 4   10feb2000  19feb2000       9     9         0
 5   12jan2000  14jan2000       3    -4         0
 5   16jan2000  18jan2000       3    -4         1
 5   20jan2000  25jan2000       3    -4         1
 5   27jan2000  01feb2000       9    -4         0
```

Each record above documents a span of time for a subject. We have made our example realistic in that, rather than recording analysis time measured from the instant of the onset of risk, our data record calendar dates. Our dataset contains two types of variables, instantaneous and enduring, and we know which are which.

Here `event` is an instantaneous variable. It records the event that happened at the end of the span, i.e., `date1`. Our `event` variable is coded so that `event==9` means failure, and the other codes mean something else.

The other two variables, `x` and `exercise`, are enduring variables, meaning that they hold their given values over the span indicated, or at least we are willing to treat them that way.

The first step is to `stset` the data. `stset`'s simplest syntax is

```
. stset time
```

but that is for those with a simple dataset such as

```
time      x
   1      3
   5      2
   9      4
  20      9
  22     -4
```

Here every observation records a failure time, and that is it; every observation is assumed to fail, and `time` is already defined as or assumed to be analysis time. `stset`'s second simplest syntax is

```
. stset time, failure(failed)
```

and that is for datasets like

```
time    failed     x
   1         1     3
   5         1     2
   9         1     4
  20         1     9
  22         0    -4
```

Every observation records a failure or censoring time, and the variable `failed` tells us which.

An example of `stset`'s more complete syntax is

```
. stset date1, id(id) time0(date0) origin(time date0) failure(event==9)
```

and that is for those of us who have more complicated survival datasets such as

```
id      date0      date1      event     x  exercise
 1   20jan2000  21jan2000         9     3         1
 2   15dec1999  20dec1999         6     2         0
 3   04jan2000  13jan2000         4     4         1
 4   31jan2000  08feb2000         3     9         1
 4   10feb2000  19feb2000         9     9         0
 5   12jan2000  14jan2000         3    -4         0
 5   16jan2000  18jan2000         3    -4         1
 5   20jan2000  25jan2000         3    -4         1
 5   27jan2000  01feb2000         9    -4         0
```

6.1   A short lesson on dates

In what follows, we give an example in which the data contain the dates over which subjects are at risk. Our data might record, for instance, that a patient became at risk starting on 31jan2000 and failed on 08feb2000 although perhaps our data records this in the format 1/31/2000 and 2/8/2000, or January 1, 2000, and February 8, 2000. What follows is little more than a tutorial on Stata's date functions. Stata has time functions as well, so in another example, the patient might be at risk starting at 08feb2000 08:22:14 and fail at 08feb2000 13:10:15. Times are not much different from dates, so the example that follows will be for the slightly simpler data case. See [D] **dates and times** for all the details for both dates and times.

In Stata, dates are recorded as integers representing the number of days from January 1, 1960. That is, 0 means January 1, 1960; 1 means January 2, 1960; and so on. In addition, -1 means December 31, 1959; -2 means December 30, 1959; and so on. Consider the following dataset:

```
id      date0      date1      event      x   exercise
1       14629      14630          9       3          1
2       14593      14598          6       2          0
3       14613      14622          4       4          1
4       14640      14648          3       9          1
4       14650      14659          9       9          0
5       14621      14623          3      -4          0
5       14625      14627          3      -4          1
5       14629      14634          3      -4          1
5       14636      14641          9      -4          0
```

Subject 1 came at risk between days 14,629 and 14,630, meaning the subject was at risk for 1 day. Dates stored in this format are convenient for this reason; we can subtract them to get the number of days between dates:

```
. generate span = date1 - date0
```

Numbers like 14,629 and 14,630 in this dataset refer to 14,629 and 14,630 days after 01jan1960, which turn out to be 20jan2000 and 21jan2000. Stata can display numbers in this format as dates if we put a date format on them:

```
. format date0 date1 %td
```

With that, now when we list the data, we see

```
id      date0      date1      event      x   exercise   span
1    20jan2000  21jan2000          9       3          1      1
2    15dec1999  20dec1999          6       2          0      5
3    04jan2000  13jan2000          4       4          1      9
4    31jan2000  08feb2000          3       9          1      8
4    10feb2000  19feb2000          9       9          0      9
5    12jan2000  14jan2000          3      -4          0      2
5    16jan2000  18jan2000          3      -4          1      2
5    20jan2000  25jan2000          3      -4          1      5
5    27jan2000  01feb2000          9      -4          0      5
```

The %td format changes how the variables are displayed, not their underlying contents. The values for date0 and date1 for the first subject are still 14,629 and 14,630, and we can still calculate things like date1 − date0 if we wish.

You will seldom receive datasets that contain numbers like 14,629 and 14,630. Rather, you will receive datasets that contain 20jan2000 and 21jan2000, or 1/20/00 and 1/21/00, or "January 20, 2000" and "January 21, 2000", and those dates will be stored as strings. The way to deal with such data is to (1) read the data into string variables in its original form, (2) convert the strings into numeric variables (Stata date variables) using Stata's date() function, and finally (3) format the resulting numeric variable with a %td format so that the resulting numeric variables are still readable to you.

1. The first step, reading the dates into a string variable, can itself be difficult. If the date recorded in the original is all run together—such as 1/21/98, 1/21/1998, 21/1/98, 21/1/1998, 21-1-1998, 21jan1998, 21January1998—then it is easy enough.

Whatever method you use to read your data—`infile, infix`, etc.—Stata likes
to keep things together.

Problems arise when the original date is recorded with blanks (spaces), such as 21
jan 1998 or January 21, 1998. Fortunately, most datasets on computers seldom
come in this format, but when they do, Stata wants to break the pieces into
separate variables unless the whole thing is in quotes.

When the original data does contain spaces, it is usually easier to accept that
Stata wants to break the pieces into separate variables, so you end up perhaps
with day, month, and year string variables. You can assemble them into one string
by typing

    . gen date = string(day) + " " + month + " " + string(year)

If `month` is instead a numeric variable (1, 2, ..., 12), put a `string()` function
around it to convert it into a string.

2. The second step, converting the string variable containing the date into a Stata
   date variable, is easy. Type

    . gen mydate = date(date, "DMY")

   noting that `DMY` is capitalized.

   The second argument to `date()` specifies the order of the day, month, and year
   in the string `date`. The argument `DMY` means day followed by month followed by
   year. Thus if your string date is instead in the order month, day, year, type `MDY`.

   This method works, assuming your year variable is a full, four-digit year. If your
   years are two-digit years (98 meaning 1998, 02 meaning 2002), type

    . gen mydate = date(date, "DMY", 2040)

   The last argument gives Stata a rule for converting two-digit years to four-digit
   years. Specifying 2040 says that, in making the translation, the largest year to be
   produced is 2040. So, year 39 would be interpreted as 2039 and year 40 as 2040,
   but year 41 would be interpreted as 1941. Specify a reasonable value, given your
   data, for this third argument.

3. The third step is just a matter of putting a date format on your new date variable.
   Type

    . format mydate %td

   If later you want to see the date in its numeric form, change the format back to a
   numeric format:

    . format mydate %8.0g

You can then switch back and forth at will.

6.2   Purposes of the stset command

The first purpose of the `stset` command is to tell Stata about the structure of your survival data so that you do not have to remember and repeat that information every time you issue a survival command. This reduces the likelihood of mistakes.

The second purpose of the `stset` command is to burn considerable computer time by performing checks to verify that what you claim as true indeed makes sense. You would complain about performance if these checks were done every time you issued a survival analysis command, and because of that, most software packages simply never check at all. Stata checks when you `stset` your dataset, and it checks again anytime you type `stset` without arguments.

The third purpose of `stset` is to allow you to describe complicated rules for when an observation is included and excluded, what defines onset of risk and failure, and how analysis time is defined for the time and date variables in your data.

6.3   Syntax of the stset command

Given everything the `stset` command can do, the syntax is truly impressive. `stset` is a wonderful command once you know it, but it can be a hurdle that may look insurmountable to new users—one only need to look at [ST] **stset** to see how daunting this command can appear.

The basic syntax of `stset` is

> `stset` *time_of_failure_var*

or

> `stset` *time_of_failure_or_censoring_var*, `failure(`*one_if_failure_var*`)`

Everything else is a detail. The time variable immediately following `stset` specifies when the failure or censoring occurs. The optional *one_if_failure_var* contains 1 if the observation contains a failure and 0 otherwise.

▷ **Example**

1. All observations record failure times:

       failtime   x
              1   3
              5   2
              9   4
             20   9
             22   -4

    The `stset` command is

        . stset failtime

    because `failtime` is the name of the variable in this dataset that contains the failure time.

2. Some observations are censored:

```
lasttime     x     failed
       1     3          1
       5     2          1
       9     4          1
      20     9          1
      22    -4          0
```

The `stset` command is

```
. stset lasttime, failure(failed)
```

because `failed` is the name of the variable in this dataset that contains 1 when variable `lasttime` records a failure time and 0 when `lasttime` records a censoring time.

◁

6.3.1 Specifying analysis time

Recall from section 3.3 our discussion of analysis time. It is 0 at the onset of risk, so it is probably shifted from calendar time by a different amount for each subject— analysis time might be time from diagnosis or time from beginning of unemployment. The duration might not even be a time measure but instead a time-intensity measure such as degree-days, or in rare instances, analysis time might have nothing whatsoever to do with time—it might be distance from the end of a bolt of cloth.

In any case, before any analysis can begin, you need to define analysis time:

$$t = somefunction(\text{time/date variables in your data})$$

If you are lucky, all the time/date variables in your data are already in terms of analysis time, but if they are not, you have two alternatives:

1. Define t yourself and then `stset` your data for t.
2. Specify `stset`'s `origin()` and possibly `scale()` options so that `stset` can calculate t based on the time/date variables in your data.

We prefer the latter because `stset` has a lot of error checking built into it.

From now on, we will write *time* to mean time as you have it recorded in your data and t to mean analysis time. Let *time* be a variable recorded in the units of time for your data. We obtain t from

$$t = \frac{time - origin}{scale}$$

The default definitions for *origin* and *scale* are 0 and 1, respectively; that is, by default, *time* = t.

By specifying `stset`'s `origin()` and `scale()` options, you can change the definitions of *origin* and *scale*. For instance, in your data, *time* might be calendar date, and you might want to define `origin()` as the date of diagnosis, thus making time from diagnosis analysis time and making diagnosis itself denote the onset of risk.

Recall that $t = 0$ is the time of the onset of risk and that Stata treats this seriously: all the survival analysis commands ignore data before $t = 0$. They do this automatically, irrevocably, and without emphasis. Mostly, this approach is desirable, but sometimes it surprises a user. Consider the user for whom calendar time is the appropriate definition of t. Now remember how Stata records dates: $t = 0$ corresponds to January 1, 1960. This means that

1. Without specifying `origin()`, records before January 1, 1960, are ignored, even if the user is interested only in estimating semiparametric models for which the exact timing of the onset of risk is otherwise irrelevant. If the user has data going back before January 1, 1960, `origin()` should be specified to be some date before then.

2. Risk begins accumulating on January 1, 1960. If the user is fitting a parametric model, he or she may want to think carefully about this, especially if the earliest date in the data is much later than this.

Defining t to be the calendar date is indeed rare, and we emphasize that such a definition usually involves an explicit definition of `origin()`. If t is already defined to be the analysis time (with a meaningful origin), then `origin()` need not be specified.

The `scale()` option is usually of no analytical significance. Most people use `scale()` to put t in more readable units. They define a reasonable `origin()` so that $t = time - origin$ provides an analytically sufficient definition but then decide they would rather have time recorded in months than in days, and so they define `scale(30)`, meaning $t = (time - origin)/30$. `scale()` can also be used to define time-intensity measures.

Anyway, focusing on the ever-so-important `origin()` option, here are some valid examples:

- `origin(time 20)`
  This sets *origin* = 20. It is a rather odd thing to do because it specifies that *time* is almost analysis time, but it needs 20 subtracted from it; the onset of risk occurred at *time* = 20.

- `origin(time td(15feb1999))`
  This sets *origin* = 14, 290, the integer value for the date February 15, 1999. Matching on calendar date is appropriate for date-formatted data, and here the onset of risk occurred on February 15, 1999. Writing `td(15feb1999)` is just how to write date constants in Stata and is preferable to having to figure out that February 15, 1999, translates to the integer 14,290.

- `origin(time bdate)`
  This sets *origin* = `bdate`, where `bdate` is a variable in the dataset, which we assume records the birthdate of subjects. Analysis time is now age, and onset of risk is now birth.

There is a little something extra to think about here: if the data contain multiple records for some subjects, which value of `bdate` should be used for the onset of risk? Well, your first reaction is probably that one would hope that `bdate` was the same across all the records for the same subject. So would we, but perhaps `bdate` is missing in some records. Perhaps `bdate` is recorded only on the first record of every subject or only on the last. It would not matter. Stata would find the nonmissing `bdate` and use it.

If `bdate` took on different values in different records within a subject, Stata would find and use the earliest value of `bdate`. Here that is not so useful, and you would probably prefer that Stata flag this as an error. In other cases, however, it is useful. For instance:

- `origin(time diagdate)`
  This sets *origin* = `diagdate`, where `diagdate` is a variable in the dataset. If subjects have multiple records, Stata will use the earliest (smallest) nonmissing value of `diagdate`, meaning that if a subject was diagnosed twice, it would be the first diagnosis date that mattered.

  In any case, analysis time is now time from (earliest) diagnosis, and (earliest) diagnosis corresponds to the onset of risk.

- `origin(time min(diagdate,d2date))`
  You can specify any expression for *origin*, and the one we just chose is the minimum value recorded in the two variables `diagdate` and `d2date`. Perhaps `diagdate` is the actual date of diagnosis and `d2date` is the judgment by a professional, for a few patients, as to when diagnosis would have occurred had the patients shown up earlier at the health center.

  In any case, analysis time is now the earliest time of either the earliest diagnosis or imputed diagnosis, and (possibly imputed) diagnosis is the onset of risk.

- `origin(event==3)`
  This is something new: `event` is a variable in the data. We are instructing Stata to thumb through the records for each patient and find the earliest date at which the variable `event` took on the value 3. In doing this, variable `event` is assumed to be an instantaneous variable, meaning that it occurs at the instant of the end of the time span, so it is the end-of-span time that is important.

  This construct is useful in multiple-record situations. We have multiple records on each subject, and we have a variable that records what happened on various dates. In these situations, it is common for our failure event to be one value of the variable and for our onset-of-risk event to be another value of the same variable.

  In any case, analysis time is now the time from `event==3`, and the occurrence of `event==3` is the onset of risk. If `event==3` never occurs, then the subject is automatically excluded from the analysis. If `event==3` occurs more than once for a subject, it is the earliest occurrence that matters.

  One final note: `event` is the name of a variable. If your event variable were named `code`, then you would type `origin(code==3)`. Distinguish this from the

word `time` in, for instance, `origin(time diagdate)`. There you always type the
word `time` and follow that with the name of your particular variable. (Actually,
if you omit the word `time`, Stata will probably still understand you, and if it does
not, Stata will complain.)

- `origin(event==3 4 7)`
  This is just a variation on the above. You may specify more than one number
  following the double-equals sign, and in fact, you may specify what Stata calls
  a "numlist". See `help numlist` in Stata for more information. Here Stata will
  select the earliest occurrence of any of these three events.

- `origin(event==3 4 7 time stdate)`
  This sets onset of risk to be the earliest of (1) the earliest time that variable `event`
  takes on the value 3, 4, or 7, and (2) the earliest time recorded in variable `stdate`.

- `origin(min)`
  Do not specify this for analysis. `origin(min)` is for playing a trick on Stata to
  get the data `stset`, which is useful for performing some data-management tasks.
  We will discuss this later.

6.3.2 Variables defined by stset

After you `stset` your data, you will find four new variables named _t0, _t, _d, and _st
in your dataset.

- _t0 and _t
  These two variables record the time span in analysis time, t, units for each record.
  Each record starts at _t0 and concludes at _t.

- _d
  This variable records the outcome at the end of the span and contains 1 if the
  time span ends in a failure and 0 if it does not.

- _st
  This variable records whether this observation is relevant (is to be used) in the
  current analysis. For each observation, the variable contains 1 if the observation
  is to be used and 0 if it is to be ignored. For example, if we specify `origin(time`
  `td(19feb1999))`, then observations dated before February 19, 1999, will have _st
  set to 0.

At a technical level, the purpose of `stset` is to set these variables, and then the other
`st` commands—the analysis commands—do their calculations using them and ignoring
your original time/date and event variables. If you have _t0, _t, _d, and _st set correctly,
it does not matter which `stset` command you typed. If they are set incorrectly, it does
not matter how much you think the `stset` command should have set them differently.

▷ **Example**

We now load the duration-format dataset considered at the beginning of this chapter and stset it.

```
. use http://www.stata-press.com/data/cggm/stset_ex1
. list, noobs sepby(id)
```

id	date0	date1	event	x	exercise
1	20jan2000	21jan2000	9	3	1
2	15dec1999	20dec1999	6	2	0
3	04jan2000	13jan2000	4	4	1
4	31jan2000	08feb2000	3	9	1
4	10feb2000	19feb2000	9	9	0
5	12jan2000	14jan2000	3	-4	0
5	16jan2000	18jan2000	3	-4	1
5	20jan2000	25jan2000	3	-4	1
5	27jan2000	01feb2000	9	-4	0

```
. stset date1, origin(time date0) id(id) failure(event==9) time0(date0)
  (output omitted )
. list id date0 date1 _t0 _t _d _st, noobs sepby(id)
```

id	date0	date1	_t0	_t	_d	_st
1	20jan2000	21jan2000	0	1	1	1
2	15dec1999	20dec1999	0	5	0	1
3	04jan2000	13jan2000	0	9	0	1
4	31jan2000	08feb2000	0	8	0	1
4	10feb2000	19feb2000	10	19	1	1
5	12jan2000	14jan2000	0	2	0	1
5	16jan2000	18jan2000	4	6	0	1
5	20jan2000	25jan2000	8	13	0	1
5	27jan2000	01feb2000	15	20	1	1

We have used options that we have not yet explained just to make stset work. However, the only thing we want you to focus on right now is _t0 and _t and to remember that we typed

```
. stset date1, origin(time date0) ...
```

The variables _t0 and _t record the span of each record in *analysis time* units; the original date0 and date1 record the same thing in *time* units.

◁

6.3.3 Specifying what constitutes failure

In your data, either you already have a variable that marks the failures—contains 1 if failure and 0 otherwise—or you have a variable that contains various codes such that, when the code is a certain value, it means failure for this analysis.

stset's failure() option specifies the failure event.

1. *Simple example.* The variable failed contains zeros and ones, where failed==1 means failure:

time	failed	x
1	1	3
5	1	2
9	1	4
20	1	9
22	0	-4

   The failure() option for this is

   ```
   . stset time, failure(failed)
   ```

2. *Another example.* The variable failed contains various codes: 0 means nonfailure, and the other codes indicate that failure occurred and the reason:

time	failed	x
1	1	3
5	7	2
9	2	4
20	8	9
22	0	-4

   If you want to set this dataset so that all failures regardless of reason are treated as failures, then the failure() option would be

   ```
   . stset time, failure(failed)
   ```

3. *A variation on the previous example.* Using the same dataset, you want to set it so that failure only for reason 9 is treated as a failure, and other values are treated as nonfailures (censorings). Here the failure() option would be

   ```
   . stset time, failure(failed==9)
   ```

4. *A surprising example.* The variable failed contains zeros, ones, and missing values.

time	failed	x
1	1	3
5	.	2
9	.	4
20	8	9
22	0	-4

   This is the same dataset as used in the above example, except that variable failed contains a missing value in the second and third observations. Ordinarily, you would expect Stata to ignore observations with missing values, but failure() is an exception. By default, if the failure variable contains missing, it is treated as if it contains 0.

If you wanted to treat all failures as failures, regardless of whether the reasons were known, you would type

```
. gen fail = cond(failed!=0,1,0)
. stset time, failure(fail)
```

5. *A more complicated example.* In this analysis, you have multiple records per subject. Variable event contains various codes, and event==9 means failure:

```
id       date0       date1       event    x   exercise
1     20jan2000   21jan2000         9     3       1
2     15dec1999   20dec1999         6     2       0
3     04jan2000   13jan2000         4     4       1
4     31jan2000   08feb2000         3     9       1
4     10feb2000   19feb2000         9     9       0
5     12jan2000   14jan2000         3    -4       0
5     16jan2000   18jan2000         3    -4       1
5     20jan2000   25jan2000         3    -4       1
5     27jan2000   01feb2000         9    -4       0
```

The failure() option for this is

```
. stset date1, failure(event==9) ...
```

6. *Another more complicated example.* In this analysis, the variable event contains various codes; event equal to 9, 10, or 11 means failure:

```
id       date0       date1       event    x   exercise
1     20jan2000   21jan2000         9     3       1
2     15dec1999   20dec1999         6     2       0
3     04jan2000   13jan2000         4     4       1
4     31jan2000   08feb2000         3     9       1
4     10feb2000   19feb2000        11     9       0
5     12jan2000   14jan2000         3    -4       0
5     16jan2000   18jan2000         3    -4       1
5     20jan2000   25jan2000         3    -4       1
5     27jan2000   01feb2000        10    -4       0
```

The failure() option for this is

```
. stset date1, failure(event==9 10 11) ...
```

That is, you can specify a numlist following the double-equals sign such as failure(event==9 10 11), or equivalently, failure(event==9/11).

After stsetting your data, the new variable _d in your dataset marks the failure event, and it contains 1 if failure and 0 otherwise. Looking back at the example at the end of section 6.3.2, note how the values of _d were set. After stsetting your data, you can always check that _d is as you expect it to be.

6.3.4 Specifying when subjects exit from the analysis

When does a subject exit from the analysis, and at what point are his or her records irrelevant?

A subject exits (1) when the subject's data run out or (2) upon first failure. For some analyses, we may wish subjects to exit earlier or later.

1. If we are analyzing a cancer treatment and a subject has a heart attack, we may wish to treat the subject's data as censored at that point and all later records as irrelevant to this particular analysis.

2. If we are analyzing an event for which repeated failure is possible, such as heart attacks, we may wish the subject to continue in the analysis even after having a first heart attack.

Consider the data for the following subject:

```
. use http://www.stata-press.com/data/cggm/id12
. list
```

	id	begin	end	event	x
1.	12	20jan2000	21jan2000	3	3
2.	12	21jan2000	25jan2000	8	3
3.	12	25jan2000	30jan2000	4	3
4.	12	30jan2000	31jan2000	8	3

First, pretend that the failure event under analysis is `event==9`, which, for this subject, never occurs. Then all the data for this subject would be used, and the corresponding _d values would be 0:

```
. stset end, failure(event==9) origin(time begin) id(id) time0(begin)
(output omitted )
. list id begin end _t0 _t _d _st, noobs
```

id	begin	end	_t0	_t	_d	_st
12	20jan2000	21jan2000	0	1	0	1
12	21jan2000	25jan2000	1	5	0	1
12	25jan2000	30jan2000	5	10	0	1
12	30jan2000	31jan2000	10	11	0	1

We used options `id()` and `time0()` without explanation, but we promise we will explain them in just a bit. In any case, their use is not important to the current discussion.

Now instead pretend that the failure event is `event==8`. By default, Stata would interpret the data as

```
. stset end, failure(event==8) origin(time begin) id(id) time0(begin)
(output omitted)
. list id begin end _t0 _t _d _st, noobs
```

id	begin	end	_t0	_t	_d	_st
12	20jan2000	21jan2000	0	1	0	1
12	21jan2000	25jan2000	1	5	1	1
12	25jan2000	30jan2000	.	.	.	0
12	30jan2000	31jan2000	.	.	.	0

Variable `_st` is how `stset` marks whether an observation is being used in the analysis. Here only the first two observations for this subject are relevant because, by default, data are ignored after the first failure. If `event==8` marks a heart attack, for instance, Stata would ignore data on subjects after their first heart attack.

The `exit()` option is how one controls when subjects exit. Examples of `exit()` include the following:

- `exit(failure)`
  This is just the name for how Stata works by default. When do subjects exit from the analysis? They exit when they first fail, even if there are more data following that failure. Of course, subjects who never fail exit when they run out of data.

- `exit(event==4)`
  If you specify the `exit()` option, you take complete responsibility for specifying the exit-from-analysis rules. `exit(event==4)` says that subjects exit when the variable `event` takes on value 4, and that is the only reason except for, of course, running out of data.

  If you coded `failure(event==4) exit(event==4)`, that would be the same as coding `failure(event==4) exit(failure)`, which would be the same as omitting the `exit()` option altogether. Subjects would exit upon failure.

  If you coded `failure(event==8) exit(event==4)`, subjects would not exit upon failure unless it just so happened that their data ended when `event` was equal to 8. Multiple failures per subject would be possible because, other than running out of data, subjects would be removed only when `event==4`. Subjects would be dropped from the analysis the first time `event==4`, even if that was before the first failure.

- `exit(event==4 8)`
  Now subjects exit when `event` first equals either 4 or 8.

  If you coded `failure(event==8) exit(event==4 8)`, you are saying that subjects exit upon failure and that they may exit before that when `event` equals 4.

- `exit(time lastdate)`
  This is another example that allows for multiple failures of the same subject. Here each subject exits as of the earliest date recorded in variable `lastdate`, regardless of the number of failures, if any, or subjects exit when they run out of data. `lastdate`, it is assumed, is recorded in units of *time* (not analysis time *t*).

- `exit(time .)`
  This also allows for multiple failures of the same subject. It is a variation of the above. It is used to indicate that each subject should exit only when he or she runs out of data, regardless of the number of failures, if any.

- `exit(time td(20jan2000))`
  This is just a variation on the previous example, and here the exit-from-analysis date is a fixed date regardless of the number of failures. This would be an odd thing to do.

- `exit(event==4 8 time td(20jan2000))`
  This example is not so odd. Subjects exit from the analysis at the earlier date of (1) the earliest date at which `event` 4 or 8 occurs, and (2) January 20, 2000.

  Consider coding

  ```
  failure(event==8) exit(event==4 8 time td(20jan2000))
  ```

  You would be saying that subjects exit upon failure, that they exit before that if and when `event` 4 occurs, and that anyone still left around is removed from the analysis as of January 20, 2000, perhaps because that is the last date at which you have complete data.

You can check that you have specified `exit()` correctly by examining the variables `_d` and `_st` in your data; `_d` is 1 at failure and 0 otherwise, and `_st` is 1 when an observation is used and 0 otherwise:

```
. stset end, failure(event==8) exit(event==4) origin(time begin) id(id)
> time0(begin)
(output omitted)
. list id begin end event x _t0 _t _d _st, noobs
```

id	begin	end	event	x	_t0	_t	_d	_st
12	20jan2000	21jan2000	3	3	0	1	0	1
12	21jan2000	25jan2000	8	3	1	5	1	1
12	25jan2000	30jan2000	4	3	5	10	0	1
12	30jan2000	31jan2000	8	3	.	.	.	0

```
. stset end, failure(event==8) exit(event==4 8) origin(time begin) id(id)
> time0(begin)
(output omitted)
```

```
. list id begin end event x _t0 _t _d _st, noobs
```

id	begin	end	event	x	_t0	_t	_d	_st
12	20jan2000	21jan2000	3	3	0	1	0	1
12	21jan2000	25jan2000	8	3	1	5	1	1
12	25jan2000	30jan2000	4	3	.	.	.	0
12	30jan2000	31jan2000	8	3	.	.	.	0

There is nothing stopping us from `stset`ting the data, looking, and then `stset`ting again if we do not like the result. Specifying

```
failure(event==8) exit(event==4 8)
```

would make more sense in the above example if, in addition to subject 12, we had another subject for which `event` 4 preceded `event` 8.

6.3.5 Specifying when subjects enter the analysis

When do subjects enter the analysis? We want subjects to enter at the onset of risk or, if they are not under observation at that point, after that. That is Stata's default rule, but Stata has to assume that "under observation" corresponds to the presence of data.

Stata's default answer is that subjects enter at analysis time $t = 0$ (as specified by `origin()`), or if their earliest records in the data are after that, subjects enter then.

Some datasets, however, contain records reporting values before the subject was really under observation. The records are historical; they were added to the data after the subject enrolled, and had the subject failed during that early period, the subject would never have been around to enroll in our study. Consider the following data:

id	begin	end	event	x
27	.	11jan2000	2	.
27	11jan2000	15jan2000	10	3
27	15jan2000	21jan2000	8	3
27	21jan2000	30jan2000	9	3

Here pretend that `event==2` is the onset of risk but `event==10` is enrollment in our study. Subject 27 enrolled in our study on January 15 but came at risk before that—on January 11, a fact we determined when the subject enrolled in our study. Another subject might have the events reversed,

id	begin	end	event	x
27	.	11jan2000	10	.
27	11jan2000	15jan2000	2	3
27	15jan2000	21jan2000	8	3
27	21jan2000	30jan2000	9	3

and yet another might have the events coincident (indicated, perhaps, by `event==12`):

```
id      begin         end   event    x
29         .   11jan2000      12    .
29  11jan2000   21jan2000       8    3
29  21jan2000   30jan2000       9    3
```

Option `enter()` specifies when subjects enter. This option works the same as `exit()`, only the meaning is the opposite. Some examples are

- `enter(event==2)`
  Subjects enter when time `event` is 2 or $t = 0$, whichever is later. Specifying `enter(event==2)` does not necessarily cause all subjects to enter at the point `event` equals 2 if, in the data, `event` takes on the value 2 prior to $t = 0$ for some subjects. Those subjects would still enter the analysis at $t = 0$.

- `enter(event== 2 12)`
  This example temporarily defines t' as the earliest t (time in analysis-time units) that `event==2` or `event==12` is observed for each subject. Subjects enter at the later of t' or $t = 0$. For example, if t' happens to correspond to a date earlier than that specified in `origin()`, then the onset of risk is just taken to be $t = 0$. The result is no different than if you typed

  ```
  . gen ev_2_12 = (event==2) | (event==12)
  . stset ..., ... enter(ev_2_12 == 1)
  ```

- `enter(time intvdate)`
  `intvdate` contains the date of interview recorded in *time* (not analysis time t) units. For each subject, `stset` finds the earliest time given in `intvdate` and then enforces the rule that the subject cannot enter before then.

- `enter(event==2 12 time intvdate)`
  This is a typical compound specifier. For each set of records that share a common `id()` variable, `stset` finds the earliest time at which events 2 or 12 occurred. It then finds the earliest time of `intvdate` and takes the later of those two times. Subjects cannot enter before then.

Specifying `enter()` affects `_st`. It does not affect how analysis time is measured—only `origin()` and `scale()` do that. Below `event==2` is the onset of risk, `event==10` is enrollment in our study, and `event==12` is simultaneous enrollment and onset of risk. Remember, instantaneous variables such as `event` are relevant at the end of the time span, which is variable `end` in these data:

(Continued on next page)

```
. use http://www.stata-press.com/data/cggm/id27_29
. list, noobs sepby(id)
```

id	begin	end	event	x
27	.	11jan2000	2	.
27	11jan2000	15jan2000	10	3
27	15jan2000	21jan2000	8	3
27	21jan2000	30jan2000	9	3
28	.	11jan2000	10	.
28	11jan2000	15jan2000	2	3
28	15jan2000	21jan2000	8	3
28	21jan2000	30jan2000	9	3
29	.	11jan2000	12	.
29	11jan2000	21jan2000	8	3
29	21jan2000	30jan2000	9	3

```
. stset end, origin(event==2 12) enter(event==10 12) failure(event==9)
> time0(begin) id(id)
  (output omitted )
. list id begin end event x _t0 _t _d _st, noobs sepby(id)
```

id	begin	end	event	x	_t0	_t	_d	_st
27	.	11jan2000	2	.	.	.	.	0
27	11jan2000	15jan2000	10	3	.	.	.	0
27	15jan2000	21jan2000	8	3	4	10	0	1
27	21jan2000	30jan2000	9	3	10	19	1	1
28	.	11jan2000	10	.	.	.	.	0
28	11jan2000	15jan2000	2	3	.	.	.	0
28	15jan2000	21jan2000	8	3	0	6	0	1
28	21jan2000	30jan2000	9	3	6	15	1	1
29	.	11jan2000	12	.	.	.	.	0
29	11jan2000	21jan2000	8	3	0	10	0	1
29	21jan2000	30jan2000	9	3	10	19	1	1

In studying these results, look particularly at subject 28:

id	begin	end	event	x	_t0	_t	_d	_st
28	.	11jan2000	10	.	.	.	.	0
28	11jan2000	15jan2000	2	3	.	.	.	0
28	15jan2000	21jan2000	8	3	0	6	0	1
28	21jan2000	30jan2000	9	3	6	15	1	1

We specified origin(event==2 12) enter(event==10 12), so the subject entered our study on January 11 (when event==10) but did not come at risk until January 15 (when event==2). So how is it that this subject's second record—11jan2000 to 15jan2000— has _st==0 when the record occurred while under observation? Because analysis time t was negative prior to January 15 (when event became 2), and subjects cannot be in the analysis prior to $t = 0$.

6.3.6   Specifying the subject-ID variable

If there are multiple records per subject in your data, as there have been in many of our examples, you must specify a subject-ID variable using the id() option to stset. We have been doing that all along but without explaining.

If you do not specify the id() option, each record is assumed to reflect a different subject. If you do specify id(*varname*), subjects with equal values of the specified variable *varname* are assumed to be the same subject.

It never hurts to specify an ID variable, even in single-record data, because for various reasons you may later want to create multiple records for each subject.

For multiple-record data, when you specify an ID variable, Stata verifies that no records overlap:

```
. use http://www.stata-press.com/data/cggm/id101
. list, noobs
```

id	begin	end	event	x
101	20jan2000	21jan2000	3	3
101	21jan2000	26jan2000	8	3
101	25jan2000	30jan2000	4	3
101	30jan2000	31jan2000	8	3

```
. stset end, failure(event==9) origin(time begin) id(id) time0(begin)
                id:  id
     failure event:  event == 9
obs. time interval:  (begin, end]
 exit on or before:  failure
    t for analysis:  (time-origin)
            origin:  time begin
```

```
    4  total obs.
    1  overlapping records (end[_n-1]>begin)                PROBABLE ERROR

    3  obs. remaining, representing
    1  subject
    0  failures in single failure-per-subject data
    7  total analysis time at risk, at risk from t =          0
                           earliest observed entry t =        0
                               last observed exit t =        11
```

Notice the PROBABLE ERROR that stset flagged. What did stset do with these two records? It kept the earlier one and ignored the later one:

(Continued on next page)

```
. list id begin end event x _t0 _t _d _st, noobs
```

id	begin	end	event	x	_t0	_t	_d	_st
101	20jan2000	21jan2000	3	3	0	1	0	1
101	21jan2000	26jan2000	8	3	1	6	0	1
101	25jan2000	30jan2000	4	3	.	.	.	0
101	30jan2000	31jan2000	8	3	10	11	0	1

stset will not complain when, rather than an overlap, there is a gap. Note the gap between the second and third records:

```
. use http://www.stata-press.com/data/cggm/id102
. list, noobs
```

id	begin	end	event	x
102	20jan2000	21jan2000	3	3
102	21jan2000	25jan2000	8	3
102	27jan2000	30jan2000	4	3
102	30jan2000	31jan2000	8	3

```
. stset end, failure(event==9) origin(time begin) id(id) time0(begin)
                 id:  id
      failure event:  event == 9
obs. time interval:  (begin, end]
exit on or before:   failure
    t for analysis:  (time-origin)
            origin:  time begin
```

```
        4  total obs.
        0  exclusions

        4  obs. remaining, representing
        1  subject
        0  failures in single failure-per-subject data
        9  total analysis time at risk, at risk from t =            0
                              earliest observed entry t =           0
                                last observed exit t =             11
. list id begin end event x _t0 _t _d _st, noobs
```

id	begin	end	event	x	_t0	_t	_d	_st
102	20jan2000	21jan2000	3	3	0	1	0	1
102	21jan2000	25jan2000	8	3	1	5	0	1
102	27jan2000	30jan2000	4	3	7	10	0	1
102	30jan2000	31jan2000	8	3	10	11	0	1

stset did not even mention the gap because interval truncation is a valid ingredient of any statistical analysis we would perform with these data. Stata does, however, have the command stdescribe, which describes datasets that have been stset and, in particular, will tell you if you have any gaps. stdescribe is covered in more detail in section 7.3.

6.3.7   Specifying the begin-of-span variable

Another option we have been using without explanation is time0(). The last stset command we illustrated was

```
. stset end, failure(event==9) origin(time begin) id(id) time0(begin)
```

and, in fact, the time0() option has appeared in most of our examples. time0() is how you specify the beginning of the span, and if you omit this option Stata will determine the beginning of the span for you. For this reason, you must specify time0() if you have time gaps in your data and you do not want stset to assume that you do not.

Rather than having the data

```
id       begin       end     event     x    exercise
12   20jan2000  21jan2000       3       3        1
12   21jan2000  25jan2000       8       3        0
12   25jan2000  30jan2000       4       3        1
12   30jan2000  31jan2000       8       3        1
```

which has no gaps, you might have the same data recorded as

```
id    enrolled       date     event     x    exercise
12   20jan2000  21jan2000       3       3        1
12              25jan2000       8       3        0
12              30jan2000       4       3        1
12              31jan2000       8       3        1
```

or even

```
id        date     event     x   exercise
12   20jan2000       1       .        .
12   21jan2000       3       3        1
12   25jan2000       8       3        0
12   30jan2000       4       3        1
12   31jan2000       8       3        1
```

In this last example, we added an extra record at the top. In any case, all these datasets report the same thing: this subject enrolled in the study on January 20, 2000, and we also have observations for January 21, January 25, January 30, and January 31.

We much prefer the first way we showed these data,

```
id       begin       end     event     x    exercise
12   20jan2000  21jan2000       3       3        1
12   21jan2000  25jan2000       8       3        0
12   25jan2000  30jan2000       4       3        1
12   30jan2000  31jan2000       8       3        1
```

because it clarifies that these are time-span records and because gaps are represented naturally. The interpretation of these records is

```
            time span                         enduring variables
        --------------------                  ------------
  id        begin         end      event      x    exercise
  12    20jan2000   21jan2000          3       3           1
  12    21jan2000   25jan2000          8       3           0
  12    25jan2000   30jan2000          4       3           1
  12    30jan2000   31jan2000          8       3           1
                     ---------      -----
                     end time    instantaneous
                                  variable(s)
```

The enduring variables are relevant over the entire span, and the instantaneous variables are relevant only at the instant of the end of the span. Instantaneous variables may or may not be relevant to any statistical models fit to these data but are relevant when stsetting the data. In general, event and failure variables are instantaneous, and variables that record characteristics are enduring.

The second way we showed of recording these data changes nothing; we are just omitting the begin-of-span variable:

```
  id    enrolled        date      event      x    exercise
  12    20jan2000   21jan2000          3       3           1
  12                25jan2000          8       3           0
  12                30jan2000          4       3           1
  12                31jan2000          8       3           1
```

Consider the second record here. Over what period is exercise==0? Between January 21 and January 25—not January 25 to January 30 (during which it is 1). The time span for a record is the period from the record before it to this record, except for the first record, in which case it is from enrolled in this dataset.

We can stset this dataset, and the only issue is setting analysis time appropriately:

- stset date, id(id) origin(time enrolled) ...
  if the date enrolled corresponds to the onset of risk

- stset date, id(id) origin(event==3) ...
  if the occurrence of event==3 marks the onset of risk

- stset date, id(id) origin(event==8) ...
  if the occurrence of event==8 marks the onset of risk

We may omit the time0() option, and when we do that, stset obtains the time span by comparing adjacent records and assuming no gaps:

```
. use http://www.stata-press.com/data/cggm/id12b
. stset date, failure(event==8) origin(time enrolled) id(id)
                    id:  id
         failure event:  event == 8
   obs. time interval:  (date[_n-1], date]
   exit on or before:  failure
      t for analysis:  (time-origin)
              origin:  time enrolled
```

```
    4  total obs.
    2  obs. begin on or after (first) failure
```

```
    2  obs. remaining, representing
    1  subject
    1  failure in single failure-per-subject data
    5  total analysis time at risk, at risk from t =          0
                      earliest observed entry t =          0
                       last observed exit t =          5
```

```
. list id enrolled date event x exercise _t0 _t _d _st, noobs
```

id	enrolled	date	event	x	exercise	_t0	_t	_d	_st
12	20jan2000	21jan2000	3	3	1	0	1	0	1
12	.	25jan2000	8	3	0	1	5	1	1
12	.	30jan2000	4	3	1	.	.	.	0
12	.	31jan2000	8	3	1	.	.	.	0

The third way we showed that these data might be recorded is by adding an extra observation at the top:

```
id      date    event    x   exerszd
12   20jan2000     1     .      .
12   21jan2000     3     3      1
12   25jan2000     8     3      0
12   30jan2000     4     3      1
12   31jan2000     8     3      1
```

Here event==1 marks enrollment. The best way to think about these data is

```
            time span                  enduring variables
        --------------------              -----------
id                     date    event    x   exerszd
12   -infinity   20jan2000       1     .      .
12   20jan2000   21jan2000       3     3      1
12   21jan2000   25jan2000       8     3      0
12   25jan2000   30jan2000       4     3      1
12   31jan2000   31jan2000       8     3      1
                 ---------     -----
                 end time    instantaneous
                             variable(s)
```

Records, just as before, are paired. Variable date records the end of a time span, and the beginning of a time span is obtained from the record before. For the first record, just pretend the beginning of the span is the beginning of time. The enduring variables are missing over this span, and all we have is an event, which occurs at the end of each span. To stset this dataset, you would type

- `stset date, id(id) origin(event==1) ...`
  if the date of enrollment corresponds to the onset of risk

- `stset date, id(id) origin(event==3) ...`
  if the occurrence of `event==3` marks the onset of risk

- `stset date, id(id) origin(event==8) ...`
  if the occurrence of `event==8` marks the onset of risk

6.3.8   Convenience options

The previous sections have covered all the important options—options that affect the definition of failure,

> `failure()`:   `failure(`*varname*[`==`*numlist*]`)`

options that affect the definition of analysis time,

> `origin()`:   `origin(`[*varname*`==`*numlist*] `time` *exp* `|` `min)`

> `scale()`:   `scale(#)`

options that affect when the subject is under observation,

> `enter()`:   `enter(`[*varname*`==`*numlist*] `time` *exp*`)`

> `exit()`:   `exit(failure` `|` [*varname*`==`*numlist*] `time` *exp*`)`

and options that provide the details that Stata needs to know,

> `id()`:   `id(`*varname*`)`

> `time0()`:   `time0(`*varname*`)`

That leaves the convenience options: `if()`, `ever()`, `never()`, `before()`, and `after()`. Each of these options takes any valid Stata expression as an argument and simply provides a convenient way to select records.

- `if()`
  `if()` is like the standard Stata syntax `if` *exp* except that it is preferred in cases where you do not specify the `time0()` option. When you have recorded in your data only the ending times of each span, Stata derives the span by pairing records and using the ending times. The standard `if` *exp* in Stata's syntax removes records from consideration, meaning that if you have

patno	mytime	x1	x2	event
3	7	20	5	14
3	9	22	5	23
3	11	21	5	29

and you type `stset mytime if x1!=22, ...`, it will appear to Stata as if the dataset is missing the middle record, and thus Stata will incorrectly determine that the last record spans the period 7–11.

If you type `stset mytime, if(x1!=22) ...`, Stata will correctly determine that the third record spans the period 9–11 because Stata will leave the middle record in while it determines the beginning and ending times, and only then will it remove the record from consideration.

The problem that `if()` solves arises only when you do not specify the `time0()` option because, with `time0()`, Stata knows exactly when each record begins and ends.

- `ever()`
  The subject is eligible for inclusion in the analysis only if the stated expression is ever true, whether in the past, even before $t = 0$, or in the future—even after the failure event or `exit()`.

- `never()`
  The subject is eligible for inclusion in the analysis only if the stated expression is never true, whether in the past, even before $t = 0$, or in the future—even after the failure event or `exit()`.

- `before()`
  Only the subject's records that occur before the first time the expression is true are eligible for inclusion in the analysis. The expression is assumed to be relevant at the end of time spans (based on instantaneous variables).

- `after()`
  Only the subject's records that occur after the first time the expression is true are eligible for inclusion in the analysis. The expression is assumed to be relevant at the end of time spans (based on instantaneous variables).

Be careful using `before()` and `after()` with time variables rather than event variables. There is no danger in using something like `before(event==7)` or `after(event==8)`. However, consider something such as

```
before(end==td(28jan2000))
```

with the following data:

```
id      begin        end     event    x    exercise
12    20jan2000   21jan2000      3     3        1
12    21jan2000   25jan2000      8     3        0
12    25jan2000   30jan2000      4     3        1
12    30jan2000   31jan2000      8     3        1
```

In this dataset, the variable `end` never takes on the value 28jan2000. Specifying `before(end>=td(28jan2000))` does not solve the problem either. That would select the first two observations but omit the span January 25–January 28.

Coding things such as `before(event==4)`, on the other hand, is perfectly safe.

7 After stset

After you have `stset` your data, there are five things you should do before starting your formal analysis:

1. Look at `stset`'s output.

2. List at least a few of your data, examining _t0, _t, _d, and _st to be sure that they appear as you want them.

3. Type `stdescribe`. `stdescribe` provides a brief description of your data, so if there are problems, perhaps you will spot them.

4. Type `stvary` if you have multiple-record (meaning multiple records per subject) data. It reports whether variables within subjects vary over time and on their pattern of missing values.

5. If you discover problems in step 4, use `stfill` and `streset` to fix them.

7.1 Look at stset's output

In most of the examples in chapter 6, we omitted the output from `stset` for the sake of brevity. For instance, in one of our examples, we showed

```
. use http://www.stata-press.com/data/cggm/id27_29
. list, noobs sepby(id)
```

id	begin	end	event	x
27	.	11jan2000	2	.
27	11jan2000	15jan2000	10	3
27	15jan2000	21jan2000	8	3
27	21jan2000	30jan2000	9	3
28	.	11jan2000	10	.
28	11jan2000	15jan2000	2	3
28	15jan2000	21jan2000	8	3
28	21jan2000	30jan2000	9	3
29	.	11jan2000	12	.
29	11jan2000	21jan2000	8	3
29	21jan2000	30jan2000	9	3

```
. stset end, origin(event==2 12) enter(event==10 12) failure(event==9)
> time0(begin) id(id)
```
(output omitted)

Let us now examine what stset actually displayed:

```
. stset end, origin(event==2 12) enter(event==10 12) failure(event==9)
> time0(begin) id(id)

                 id:  id
      failure event:  event == 9
  obs. time interval:  (begin, end]
   enter on or after:  event==10 12
   exit on or before:  failure
      t for analysis:  (time-origin)
              origin:  event==2 12

    11  total obs.
     3  entry time missing (begin>=.)                  PROBABLE ERROR
     1  obs. end on or before enter()
     1  obs. end on or before origin()

     6  obs. remaining, representing
     3  subjects
     3  failures in single failure-per-subject data
    49  total analysis time at risk, at risk from t =        0
                          earliest observed entry t =        0
                            last observed exit t =          19
```

Notice the PROBABLE ERROR message? Here it was in fact not an error, but we did not want to explain that at the time. Now we will explain and note that most probable errors are in fact errors, and the source of probable errors is worth exploring.

There are two parts to stset's output. The first simply repeats in a more readable format what you specified in the stset command and fills in any defaults:

```
                 id:  id
      failure event:  event == 9
  obs. time interval:  (begin, end]
   enter on or after:  event==10 12
   exit on or before:  failure
      t for analysis:  (time-origin)
              origin:  event==2 12
```

This output is fairly self-explanatory. One subtle item of interest is (begin, end], where Stata is reporting from where it obtained the time span for each record. For this Stata uses the interval notation ( ], which mathematically means an interval for which the left endpoint is excluded and the right endpoint is included. The interval starts just after begin and continues through end, where begin and end are two variables in our data.

The notation has statistical meaning. Say that subject 1 fails at $t = 5$ and subject 2 is right-censored at the same time. When Stata's semiparametric analyses commands perform the individual binary-outcome analyses for each failure time, would Stata include or exclude subject 2 from the analysis for $t = 5$? Stata *would* include subject 2 because the duration is interpreted as being up to and including the time listed.

The second part of `stset`'s output summarizes the results of applying the above definitions to this particular dataset:

```
11  total obs.
 3  entry time missing (begin>=.)                          PROBABLE ERROR
 1  obs. end on or before enter()
 1  obs. end on or before origin()
 _____

 6  obs. remaining, representing
 3  subjects
 3  failures in single failure-per-subject data
49  total analysis time at risk, at risk from t =          0
                         earliest observed entry t =       0
                              last observed exit t =      19
```

This second part splits into two pieces, separated by the solid horizontal line. The first piece contains some odd things you should know about, and the second piece reports the characteristics of the dataset that was just set.

Among the odd things you should know about, some may be flagged as probable errors. This is Stata's way of telling you that it thinks something is wrong and that you need to understand why this occurred so you can safely assert that Stata is wrong. The messages not flagged as probable errors are facts that Stata considers common enough. If pressed, you should be able to explain why each of those arose too, but you can start out by assuming that there is a good reason and wait for the data to prove you wrong.

For probable errors, start with a presumption of guilt on your part; for the rest, start with the presumption of innocence.

In fact, what is being reported in the first piece of the output is a complete accounting for the records in your data. For this example, Stata reports that

1. We started with 11 records in the dataset.

2. Based on what you specified, we decided to exclude five of the records as follows:

   a. Three were omitted because the variable `begin` contained missing values, and Stata finds this odd.

   b. One was omitted because "`obs. end on or before enter()`", meaning the time span of the record was before the subject could enter into the analysis given the rule `enter(event==10 12)`, which you specified.

   c. One was omitted because "`obs. end on or before origin()`", meaning the time span of the record was before $t = 0$ given the rule `origin(event==2 12)`, which you specified.

Here `begin` contained missing values, but that was okay because we never tried to use the enduring variables from that open-ended time span, and in fact, all the enduring

variable x contained missing values for these records. As a result, we noted the PROBABLE
ERROR message and investigated to find that there was indeed no error.

The warnings that stset might issue include (in order from the least serious to the
most serious)

```
obs. end on or before enter()
obs. end on or before origin()

entry time missing                         PROBABLE ERROR
entry on or after exit (etime>t)           PROBABLE ERROR

multiple records at same instant (t[_n-1]==t)  PROBABLE ERROR
overlapping records (t[_n-1]>entry time)       PROBABLE ERROR


ignored because patid missing
weights invalid                            PROBABLE ERROR
event time missing                         PROBABLE ERROR
```

7.2   List some of your data

stset does not change any existing data. All it does is define the new variables _t0,
_t, _d, and _st, which incorporate the information contained in the data with what you
specify in stset. As such, you should check these variables on at least a small part of
your data to make sure that the results were as you intended.

- _t0 and _t record the time span in analysis-time units.
- _d records the outcome (failure or censoring) at the end of each time span.
- _st records whether the observation is relevant to the current analysis.

These are the variables that matter in the sense that all other survival analysis commands (the st family) work with these variables rather than with your original variables.

This last fact means that if you change your data, you need to stset it again so
that these variables can change along with the data. In fact, stset is smart enough to
remember which syntax and options you had previously used; thus, all you would have
to do is type stset with no arguments.

Often you will have trimmed and cleaned your data to the point where all observations should be relevant to the analysis. Since _st is an indicator variable that marks
inclusion into subsequent analyses, verify that _st==1 in all observations. Either type
assert _st==1, or type summarize _st and verify the mean of _st is one.

If you expect records to be excluded, on the other hand, look at a few of the records
for which _st==0 to verify that the correct observations were excluded.

❏ **Technical note**

In survival models, the "response" is the triple (t_0, t, d), where t_0 marks the beginning of a time span, t marks the end, and d indicates failure or censoring. As such, when we `stset` our data, all we are doing is generating the "response" variables based on the information in our data and generating an indicator variable (`_st`) that determines whether our response makes sense given what we know about our data. Thus, by using `stset`, we guarantee that no matter what model we choose to fit, we are using the same response.

❏

7.3   Use stdescribe

`stdescribe` presents a brief description of the dataset you have just set. Continuing our example from the previous section,

```
. stdescribe
             failure _d:  event == 9
       analysis time _t:  (end-origin)
                 origin:  event==2 12
       enter on or after:  event==10 12
                     id:  id
```

			per subject		
Category	total	mean	min	median	max
no. of subjects	3				
no. of records	6	2	2	2	2
(first) entry time		1.333333	0	0	4
(final) exit time		17.66667	15	19	19
subjects with gap	0				
time on gap if gap	0	.	.	.	.
time at risk	49	16.33333	15	15	19
failures	3	1	1	1	1

Of particular interest is the line beginning "`subjects with gap`". `stset` will flag overlaps but will not even mention gaps, because gaps are not errors. `stdescribe`, on the other hand, will detect gaps and will even provide summary statistics on the lengths of the gaps.

The rest of the output can be thought of as that from a specialized **summarize** command for survival data.

(Continued on next page)

❑ **Technical note**

Notice the first part of the output from `stdescribe`, the part that reads

```
. stdescribe
          failure _d:  event == 9
    analysis time _t:  (end-origin)
             origin:  event==2 12
    enter on or after:  event==10 12
                 id:  id
```

That portion of the output is not really being produced by `stdescribe` but by the `st` system itself. It will appear in the output of every `st` command—to remind you how you have `stset` your data.

You may get tired of repeatedly seeing that reminder and, if so, type

```
. stset, noshow
```

Later, if you want to get it back again, type

```
. stset, show
```

                                                                                        ❑

7.4   Use stvary

Here is the result of running `stvary` on this analysis:

```
. stvary
          failure _d:  event == 9
    analysis time _t:  (end-origin)
             origin:  event==2 12
    enter on or after:  event==10 12
                 id:  id
```

	subjects for whom the variable is				
			never	always	sometimes
variable	constant	varying	missing	missing	missing
x	3	0	3	0	0

With a larger dataset with more covariates, we might see something like

```
. stvary
          failure _d:  event == 9
    analysis time _t:  (end-origin)
             origin:  event==2 12
    enter on or after:  event==10 12
                 id:  id
```

	subjects for whom the variable is				
			never	always	sometimes
variable	constant	varying	missing	missing	missing
sex	500	0	20	0	480
bp	80	400	100	20	380
x	497	3	65	0	435

You want to become practiced at reading this output because it will uncover problems in your data.

Let us start with the variable `sex`. Within subject, it never varies—the `varying` column is 0. That is good—no one changes their sex in the middle of the study. Also good is that there is not one subject in the data for which the variable is "always missing". For 480 subjects, however, the value of `sex` is sometimes missing, and only for 20 is it recorded in every observation. This is a problem that we will want to fix. In the 480 "sometimes missing" subjects, we are seeing something like

```
id      sex
42        .
42        1
42        .
```

For 480 subjects, the variable `sex` is at least filled in once and is at least missing once. This would be easy enough to fix by typing

```
. sort id end
. quietly by id: replace sex = sex[_n-1] if missing(sex)
. quietly by id: replace sex = sex[_N]
```

but for now we will focus instead on just making a list of the problems. In section 7.5, we will see that there is actually an easier way to fix these types of problems.

Variable `bp` (blood pressure) is shown to be constant for 80 subjects and varying for 400. Varying means that it takes on different, nonmissing values at different times. For 20 subjects, according to `stvary`, the variable is always missing.

Some examples of how `stvary` categorizes constant, varying, and always missing are

```
id      bp                              id      bp
56       .    <- always missing         60       .    <- constant
56       .                              60      180
56       .                              60       .

57      180   <- constant               61      180   <- varying
57      180                             61      190
57      180                             61      200

58      180   <- constant               62      180   <- varying
58       .                              62      190
58      180                             62      180

59      180   <- constant               63      180   <- varying
59       .                              63       .
59       .                              63      190
```

Does `bp` have problems? It does if we want to use it as an explanatory variable in our models because any observations for which `bp` is missing would be dropped from the analysis. Perhaps that is necessary given that we are uncertain as to the value of `bp` when it is missing. On the other hand, we may be willing to fill in `bp` values from the last time they were known so that

```
id        bp        bp     <- becomes
58       180       180
58         .       180
58       180       180

59       180       180
59         .       180
59         .       180

60         .         .
60       180       180
60         .       180

63       180       180
63         .       180
63       190       190
```

which is easy enough to do for cases where bp is not always missing,

```
. sort id end
. quietly by id: replace bp = bp[_n-1] if missing(bp)
```

and in fact, we show an automatic way to do this in section 7.5.

Returning to the stvary output, variable x in our data also looks problematic, because it is constant for 497 subjects and varies for only three. This looks suspicious.

1. Should x really be constant for everyone, and so the three subjects for which it varies represent coding errors?

2. We recall from above that stvary has a strange concept of constant—if we observe the variable just once and all the other values are missing, then stvary categorizes that result as constant. Ergo, could all 497 "constant" subjects in our data have x observed just once? Some 435 could be like that because 435 are "sometimes missing". We should look, and if we find that to be true, we should wonder about the reasonableness of filling in subsequent values.

7.5   Perhaps use stfill

stfill is an alternative to doing things like

```
. sort id end
. quietly by id: replace sex = sex[_n-1] if missing(sex)
. quietly by id: replace sex = sex[_N]
```

and

```
. sort id end
. quietly by id: replace bp = bp[_n-1] if missing(bp)
```

The basic syntax of `stfill` is

stfill *varlist*, {baseline|forward}

The `baseline` option will fill in all the records with the value first observed for each subject, even if the current value is nonmissing. This has the effect of making the variables in *varlist* constant within each subject. The `forward` option, alternatively, will fill in missing values by carrying forward prior values of variables.

For example, if you type

```
. stfill sex bp, forward
```

`stfill` will fix the variables `sex` and `bp` from our previous example—almost. `stfill` will not backfill, or fill in earlier times with values from later times, because backfilling is harder to justify.

There is another issue regarding the use of `stfill`, and this issue applies to other `st` data-management commands. When you `stset` the data, the result may be that only a subset of the available records is marked as valid for future analysis. For example, suppose you have the following recorded in your dataset:

```
id     date0      date1     event  ...
71  01jan2000  04jan2000       2
71  04jan2000  07jan1999       3
71  07jan2000  13jan2000       6
71  13jan2000  18jan2000       6
71  18feb2000  24feb2000       9
71  24jan2000  28jan2000       3
71  28jan2000  31jan2000       3
```

Then you `stset` the data with

```
. stset date1, id(id) failure(event==9) origin(event==3)
```

which, for this subject, results in

```
id     date0      date1     event  _t0    _t    _d   _st
71  01jan2000  04jan2000       2     .     .     .     0
71  04jan2000  07jan1999       3     .     .     .     0
71  07jan2000  13jan2000       6     0     6     0     1
71  13jan2000  18jan2000       6     6    11     0     1
71  18feb2000  24feb2000       9    11    17     1     1
71  24jan2000  28jan2000       3     .     .     .     0
71  28jan2000  31jan2000       3     .     .     .     0
```

Thus we have seven records for this subject, but for our analysis we are using only the middle three.

We call this whole collection of data on the patient a *history*. We refer to the records before the onset of risk as the *past history*, the records being used as the *current history*, and the records after failure or censoring as the *future history*.

Now if we are going to fill values forward, it would be a shame not to use the information we have in the past. If we are going to go to the effort of cleaning up data, it would be a shame not to clean all of it, including the future. However, all the `st` commands—including the `st` data-management commands—restrict themselves to the current history. They pretend the other data do not exist.

There is a trick, however, to make Stata recognize the other records for data management. If you type

```
. streset, past
```

Stata will temporarily set the dataset so that it includes the past and the current history, and when you later type

```
. streset
```

Stata will make things just as they were, with only the current history set. If you type

```
. streset, future
```

Stata will temporarily set the dataset so that it includes the current and the future history, and of course, typing `streset` by itself will restore things to be as they were. If you type

```
. streset, past future
```

Stata will temporarily set the dataset to include the full history.

What do these commands really do? They concoct an absurd definition of analysis time and failure so that the observations temporarily appear as if they are in the analysis. These definitions are not appropriate for analysis, but that does not matter because Stata knows when it is using these artificial definitions and will not let you do anything inappropriate in using the `st` analysis commands.

So, returning to our example, the best way to fix `sex` and `bp` is

```
. stfill sex, baseline
. streset, past future
. stfill bp, forward
. streset
```

We `stfill sex, baseline` before including the past history, just to make sure that the value used coincides with the onset of risk.

7.6   Example: Hip fracture data

Pretend that a study was performed to quantify the benefit of a new inflatable device to protect elderly persons from hip fractures resulting from falls. The device is worn around the hips at all times. It is hypothesized that the device will reduce the incidence of hip fractures in this population. Forty-eight women over the age of 60, with no previous histories of hip trauma, were recruited for this study. Of these 48 women, 28 were randomly given the device and instructed on how to wear it. The remaining 20 women were not provided with the device and were used as study control subjects. All 48 women were monitored closely, and blood calcium levels were drawn approximately every 5 months. The time to hip fracture or censoring was recorded in months. It was decided at study onset that, if a woman was ever hospitalized during follow-up, she would not be considered at risk of falling and fracturing her hip. This creates gaps in the data.

The dataset below is real, so feel free to use it. The data, however, are fictional.

```
. use http://www.stata-press.com/data/cggm/hip
(hip fracture study)

. describe

Contains data from http://www.stata-press.com/data/cggm/hip.dta
  obs:            106                          hip fracture study
  vars:             7                          30 Jan 2008 11:58
  size:         1,484 (99.9% of memory free)

              storage  display    value
variable name   type   format     label      variable label

id              byte   %4.0g                  patient ID
time0           byte   %5.0g                  begin of span
time1           byte   %5.0g                  end of span
fracture        byte   %8.0g                  fracture event
protect         byte   %8.0g                  wears device
age             byte   %4.0g                  age at enrollment
calcium         float  %8.0g                  blood calcium level

Sorted by:

. summarize

    Variable |       Obs        Mean    Std. Dev.       Min        Max

          id |       106    28.21698    13.09599          1         48
       time0 |       106    4.792453    5.631065          0         15
       time1 |       106     11.5283    8.481024          1         39
    fracture |       106    .2924528    .4570502          0          1
     protect |        48    .5833333    .4982238          0          1

         age |        48      70.875    5.659205         62         82
     calcium |       106    10.10849    1.407355       7.25      12.32

. sort id time0

. by id: list time0-calcium
```

```
-> id = 1

     | time0   time1   fracture   protect   age   calcium |
  1. |    0       1          1         0    76      9.35   |
```

```
-> id = 2
```

	time0	time1	fracture	protect	age	calcium
1.	0	1	1	0	80	7.8

(output omitted)

```
-> id = 17
```

	time0	time1	fracture	protect	age	calcium
1.	0	8	0	0	66	11.48
2.	8	15	1	.	.	10.79

```
-> id = 18
```

	time0	time1	fracture	protect	age	calcium
1.	0	5	0	0	64	11.58
2.	15	17	1	.	.	11.59

(output omitted)

```
-> id = 47
```

	time0	time1	fracture	protect	age	calcium
1.	0	5	0	1	63	12.18
2.	5	15	0	.	.	11.64
3.	15	35	0	.	.	11.79

```
-> id = 48
```

	time0	time1	fracture	protect	age	calcium
1.	0	5	0	1	67	11.21
2.	5	15	0	.	.	11.43
3.	15	39	0	.	.	11.29

Here time is already recorded in analysis-time units, which just means we will not have to bother with the `origin()` option when we type `stset`.

Our data do, however, have multiple observations per subject to accommodate the time-varying covariate `calcium`, and we will assume that the value of this variable is fixed over the interval spanned by each record.

`age` records the age of each participant at the time of enrollment in the study. Glancing at our data, you will notice that `age` appears to be coded only in the first record for each subject. All the records are like that. If we later copy this value of age down (i.e., propagate age values from past to future observations), we will be treating age as fixed.

In any case, the stset command for this dataset is

```
. stset time1, id(id) time0(time0) failure(fracture)
                id:  id
     failure event:  fracture != 0 & fracture < .
 obs. time interval:  (time0, time1]
  exit on or before:  failure

      106  total obs.
        0  exclusions

      106  obs. remaining, representing
       48  subjects
       31  failures in single failure-per-subject data
      714  total analysis time at risk, at risk from t =        0
                             earliest observed entry t =        0
                                  last observed exit t =       39
```

Let us now go through the data verification process described earlier in this chapter. We begin by looking at the above output; examining _t0, _t, _d, and _st; confirming that stdescribe does not reveal any surprises; confirming that stvary makes a similarly unsurprising report; and finally using stfill to fix any problems uncovered by stvary.

First, we look at _t0, _t, _d, and _st in the beginning, middle, and end of the data.

```
. list id time0 time1 frac _t0 _t _d _st if id<=3
```

	id	time0	time1	fracture	_t0	_t	_d	_st
1.	1	0	1	1	0	1	1	1
2.	2	0	1	1	0	1	1	1
3.	3	0	2	1	0	2	1	1

```
. list id time0 time1 frac _t0 _t _d _st if 16<=id & id<=18, sepby(id)
```

	id	time0	time1	fracture	_t0	_t	_d	_st
23.	16	0	5	0	0	5	0	1
24.	16	5	12	1	5	12	1	1
25.	17	0	8	0	0	8	0	1
26.	17	8	15	1	8	15	1	1
27.	18	0	5	0	0	5	0	1
28.	18	15	17	1	15	17	1	1

```
. list id time0 time1 frac _t0 _t _d _st if id>=47, sepby(id)
```

	id	time0	time1	fracture	_t0	_t	_d	_st
101.	47	0	5	0	0	5	0	1
102.	47	5	15	0	5	15	0	1
103.	47	15	35	0	15	35	0	1
104.	48	0	5	0	0	5	0	1
105.	48	5	15	0	5	15	0	1
106.	48	15	39	0	15	39	0	1

That looks good. Let's see what **stdescribe** has to say:

```
. stdescribe
           failure _d:  fracture
     analysis time _t:  time1
                   id:  id
```

			per subject		
Category	total	mean	min	median	max
no. of subjects	48				
no. of records	106	2.208333	1	2	3
(first) entry time		0	0	0	0
(final) exit time		15.5	1	12.5	39
subjects with gap	3				
time on gap if gap	30	10	10	10	10
time at risk	714	14.875	1	11.5	39
failures	31	.6458333	0	1	1

Starting with the first two lines, **stdescribe** reports that there are 48 subjects in our data with 106 records; the average number of records per subject is just over two, with three being the maximum number of records for any one subject. From this, we see that the **st** system correctly recognizes that there are multiple records per subject. In cases where there is only one observation per subject, the reported totals in the first and second line would be equal, and the mean, min, and max number of records per subject would all be equal to 1.

In lines 3 and 4, **stdescribe** reports that everyone entered at time 0—there is no delayed entry—and that the average exit time was 15.5 months, with a minimum of 1 month and a maximum of 39. Be careful when interpreting this reported average exit time. This is just the average of the follow-up times; it is *not* the average survival time because some of our subjects are censored. When there are no censored observations, the average exit time reported by **stdescribe** does equal the average survival time.

In lines 5 and 6, **stdescribe** reports that there are three subjects with gaps, each 10 months long. This is a strange finding. In most datasets with time gaps, the gaps vary in length. We were immediately suspicious of these results and wanted to verify them. One way to identify these three subjects is to make use of the fact that, when there are no gaps between consecutive observations for a subject, the ending time of the first record equals the beginning time of the next record. So, we will sort the data by subject and time, and then we will generate a dummy variable, gap = 1, for those observations with gaps. Then we can list the observations.

```
. sort id _t0 _t
. quietly by id: gen gap=1 if _t0 != _t[_n-1] & _n>1
. list id if gap==1
```

	id
28.	18
55.	30
63.	33

```
. list id time0 time1 _t0 _t if id==18 | id==30 | id==33, sepby(id)
```

	id	time0	time1	_t0	_t
27.	18	0	5	0	5
28.	18	15	17	15	17
54.	30	0	5	0	5
55.	30	15	19	15	19
62.	33	0	5	0	5
63.	33	15	23	15	23

All appears well; each of these records had a gap lasting 10 months, yet perhaps we still might want to check that the data were entered correctly.

Returning to the stdescribe output, on line 7 we observe that subjects were at risk of failure for a total of 714 months. This is simply the sum of the time spanned by the records, calculated by stdescribe by calculating the length of the interval represented by each record (_t0, _t] and then summing these lengths conditional on _st==1.

Finally, in line 8, stdescribe reports that there were 31 failures, or 31 hip fractures, in our data. The maximum number of per-subject failures is one, indicating that we have single failure-per-subject data, and the minimum number of failures is zero, indicating the presence of censored observations. Of course, we can also see that there are censored observations when we compare the total number of failures and the total number of observations on line 2. As we may expect, for a dataset with multiple failures per observation, stdescribe will provide in line 8 summary statistics indicating the existence of multiple failures-per-subject data.

(Continued on next page)

So, all looks fine based on `stdescribe`. Next let's try `stvary`:

```
. capture drop gap
. stvary
        failure _d:  fracture
   analysis time _t:  time1
               id:  id

           subjects for whom the variable is
                                              never     always   sometimes
      variable |  constant    varying        missing    missing   missing

       protect |     48          0               8         0         40
           age |     48          0               8         0         40
       calcium |      8         40              48         0          0
```

By default, `stvary` reports on all variables in the dataset, omitting the variables used or created by `stset`. (`stvary` optionally allows variable lists, so we can specify those variables that we wish to examine.)

The variable `protect` records if the subject is in the experimental or the control group, `age` records the subject's age at enrollment in the study, and `calcium` records the subject's blood calcium concentration. Recall that this last characteristic was examined approximately every 5 months, so the variable varies over time.

Looking at `stvary`'s output, what catches our eye is the many "sometimes missings". Well, we already knew that `protect` and `age` did not have the values filled in on subsequent records and that we would have to fix that. Let us now follow our own advice and use `streset` and `stfill` to fix this problem. Here `streset` is unnecessary because there are no observations for which `_st==0`, but it never hurts to be cautious:

```
. streset, past future
  (output omitted)
. stfill age protect, forward
        failure _d:  fracture
   analysis time _t:  (time1-origin)
            origin:  min
   exit on or before:  time .
               id:  id
replace missing values with previously observed values:
           age:  58 real changes made
       protect:  58 real changes made
. streset
  (output omitted)
```

Now `stvary` reports

```
. stvary
        failure _d:  fracture
  analysis time _t:  time1
              id:  id
        subjects for whom the variable is
                                            never    always   sometimes
     variable |  constant    varying        missing  missing  missing
    ----------+--------------------------------------------------------
      protect |     48          0              48        0        0
          age |     48          0              48        0        0
      calcium |      8         40              48        0        0
```

Satisfied, we can save these data as `hip2.dta`.

```
. save hip2
file hip2.dta saved
```

When you `stset` a dataset and `save` it, Stata remembers how the data were set the next time you use the data.

8 Nonparametric analysis

The last two chapters served as a tutorial on `stset`. Once you `stset` your data, you can use any `st` survival command, and the nice thing is that you do not have to continually restate the definitions of analysis time, failure, and rules for inclusion.

As previously discussed in chapter 1, the analysis of survival data can take one of three forms—nonparametric, semiparametric, and parametric—all depending on what we are willing to assume about the form of the survivor function and about how the survival experience is affected by covariates.

Nonparametric analysis follows the philosophy of letting the dataset speak for itself and makes no assumption about the functional form of the survivor function (and thus no assumption about, for example, the hazard, cumulative hazard). The effects of covariates are not modeled either—the comparison of the survival experience is done at a qualitative level across the values of the covariates.

Most of Stata's nonparametric survival analysis is performed via the `sts` command, which calculates estimates, saves estimates as data, draws graphs, and performs tests, among other things; see [ST] **sts**.

8.1 Inadequacies of standard univariate methods

Before we proceed, however, we must discuss briefly the reasons that the typical preliminary data analysis tools do not translate well into the survival analysis paradigm. For example, the most basic of analyses would be one that analyzed the mean time to failure or median time to failure. Let us use the hip-fracture dataset, which we `stset` at the end of chapter 7:

```
. use http://www.stata-press.com/data/cggm/hip2
(hip fracture study)
. list id _t0 _t fracture protect age calcium if 20<=id & id<=22, sepby(id)
```

	id	_t0	_t	fracture	protect	age	calcium
32.	20	0	5	0	0	67	11.19
33.	20	5	15	0	0	67	10.68
34.	20	15	23	1	0	67	10.46
35.	21	0	5	0	1	82	8.97
36.	21	5	6	1	1	82	7.25
37.	22	0	5	0	1	80	7.98
38.	22	5	6	0	1	80	9.65

Putting aside for now the possible effects of the covariates, if we were interested in estimating the population mean time to failure, we might be tempted to use the standard tools such as

```
. ci _t
```

Variable	Obs	Mean	Std. Err.	[95% Conf. Interval]
_t	106	11.5283	.8237498	9.894958 13.16165

We might quickly realize that this is not what we want because there are multiple records for each individual. We could just consider those values of _t corresponding to the last record for each individual,

```
. sort id _t
. quietly by id: gen last = _n==_N
. ci _t if last
```

Variable	Obs	Mean	Std. Err.	[95% Conf. Interval]
_t	48	15.5	1.480368	12.52188 18.47812

and we now have a mean based on 48 observations (one for each subject). This will not serve, however, because _t does not always correspond to failure time—some times in our data are censored, meaning that the failure time in these cases is known only to be greater than _t. As such, the estimate of the mean is biased downward.

Dropping the censored observations and redoing the analysis will not help. Consider an extreme case of a dataset with just one censored observation and assume the observation is censored at time 0.1, long before the first failure. For all you know, had that subject not been censored, the failure might have occurred long after the last failure in the data and so had a large effect on the mean. Wherever the censored observation is located in the data, we can repeat that argument, and so, in the presence of censoring, obtaining estimates of the mean survival time calculated in the standard way is simply not possible.

Estimates of the median survival time are similarly not possible to obtain using standard nonsurvival tools. The standard way of calculating the median is to order the observations and report the middle one as the median. In the presence of censoring, that ordering is impossible to ascertain. (The modern way of calculating the median is to turn to the calculation of survival probabilities and find the point at which the survival probability is 0.5. See section 8.5.)

Thus even the most simple analysis—never mind the more complicated regression models—will break down when applied to survival data. Also there are even more issues related to survival data—truncation, for example—that would only further complicate the estimation.

Instead, survival analysis is a field of its own. Given the nature of the role that time plays in the analysis, much focus is given to the functions that characterize the distribution of the survival time: the hazard function, the cumulative hazard function, and the survivor function being the most common ways to describe the distribution. Much of survival analysis is concerned with the estimation of and inference for these functions of time.

8.2 The Kaplan–Meier estimator

8.2.1 Calculation

The estimator of Kaplan and Meier (1958) is a nonparametric estimate of the survivor function $S(t)$, which is the probability of survival past time t or, equivalently, the probability of failing after t. For a dataset with observed failure times, $t_1, \ldots, t_k$, where k is the number of distinct failure times observed in the data, the Kaplan–Meier estimate [also known as the *product limit* estimate of $S(t)$] at any time t is given by

$$\widehat{S}(t) = \prod_{j|t_j \leq t} \left(\frac{n_j - d_j}{n_j} \right) \tag{8.1}$$

where n_j is the number of individuals at risk at time t_j and d_j is the number of failures at time t_j. The product is over all observed failure times less than or equal to t.

How does this estimator work? Consider the hypothetical dataset of subjects given in the usual format,

```
id    t    failed
 1    2         1
 2    4         1
 3    4         1
 4    5         0
 5    7         1
 6    8         0
```

and form a table that summarizes what happens at each time in our data (whether a failure time or a censored time):

t	No. at risk	No. failed	No. censored
2	6	1	0
4	5	2	0
5	3	0	1
7	2	1	0
8	1	0	1

At $t = 2$, the earliest time in our data, all six subjects were at risk, but at that instant, only one failed (id==1). At the next time, $t = 4$, five subjects were at risk, but at that instant, two failed. At $t = 5$, three subjects were left, and no one failed, but one subject was censored. This left us with two subjects at $t = 7$, of which one failed. Finally, at $t = 8$, we had one subject left at risk, and this subject was censored at that time.

Now we ask the following:

- What is the probability of survival beyond $t = 2$, the earliest time in our data? Because five of the six subjects survived beyond this point, the estimate is 5/6.

- What is the probability of survival beyond $t = 4$ given survival right up to $t = 4$? Because we had five subjects at risk at $t = 4$, and two failed, we estimate this probability to be 3/5.

- What is the probability of survival beyond $t = 5$ given survival right up to $t = 5$? Because three subjects were at risk, and no one failed, the probability estimate is $3/3 = 1$.

and so on. We can now augment our table with these component probabilities (calling them p):

t	No. at risk	No. failed	No. censored	p
2	6	1	0	5/6
4	5	2	0	3/5
5	3	0	1	1
7	2	1	0	1/2
8	1	0	1	1

- The first value of p, 5/6, is the probability of survival beyond $t = 2$.

- The second value, 3/5, is the (conditional) probability of survival beyond $t = 4$ given survival up until $t = 4$, which in these data is the same as survival beyond $t = 2$. Thus, unconditionally, the probability of survival beyond $t = 4$ is $(5/6)(3/5) = 1/2$.

- The third value, 1, is the conditional probability of survival beyond $t = 5$ given survival up until $t = 5$, which in these data is the same as survival beyond $t = 4$. Unconditionally, the probability of survival beyond $t = 5$ is thus equal to $(1/2)(1) = 1/2$.

Thus the Kaplan–Meier estimate is the running product of the values of p that we have previously calculated, and we can add it to our table.

t	No. at risk	No. failed	No. censored	p	$\widehat{S}(t)$
2	6	1	0	5/6	5/6
4	5	2	0	3/5	1/2
5	3	0	1	1	1/2
7	2	1	0	1/2	1/4
8	1	0	1	1	1/4

Because the Kaplan–Meier estimate in (8.1) operates only on observed failure times (and not at censoring times), the net effect is simply to ignore the cases where $p = 1$ in calculating our product; ignoring these changes nothing.

In Stata, the Kaplan–Meier estimate is obtained using the `sts list` command, which gives a table similar to the one we constructed:

```
. clear
. input id time failed
            id       time     failed
  1. 1 2 1
  2. 2 4 1
  3. 3 4 1
  4. 4 5 0
  5. 5 7 1
  6. 6 8 0
  7. end
. stset time, fail(failed)
  (output omitted )
. sts list
          failure _d:  failed
    analysis time _t:  time
            Beg.            Net      Survivor     Std.
    Time   Total   Fail   Lost      Function     Error      [95% Conf. Int.]

       2      6      1      0        0.8333      0.1521      0.2731    0.9747
       4      5      2      0        0.5000      0.2041      0.1109    0.8037
       5      3      0      1        0.5000      0.2041      0.1109    0.8037
       7      2      1      0        0.2500      0.2041      0.0123    0.6459
       8      1      0      1        0.2500      0.2041      0.0123    0.6459
```

The column "Beg. Total" is what we called "No. at risk" in our table; the column "Fail" is "No. failed"; and the column "Net lost" is related to our "No. censored" column but is modified to handle delayed entry (see sec. 8.2.3).

The standard error reported for the Kaplan–Meier estimate is that given by Green-wood's (1926) formula:

$$\widehat{\mathrm{Var}}\{\widehat{S}(t)\} = \widehat{S}^2(t) \sum_{j|t_j \leq t} \frac{d_j}{n_j(n_j - d_j)} \tag{8.2}$$

These standard errors, however, are not used for confidence intervals. Instead, the asymptotic variance of $\ln\{-\ln \widehat{S}(t)\}$,

$$\widehat{\sigma}^2(t) = \frac{\sum \frac{d_j}{n_j(n_j - d_j)}}{\left\{ \sum \ln\left(\frac{n_j - d_j}{d_j}\right) \right\}^2}$$

is used, where the sums are calculated over $j|t_j \leq t$ (Kalbfleisch and Prentice 2002, 18). The confidence bounds are then calculated as $\widehat{S}(t)$ raised to the power $\exp\{\pm z_{\alpha/2}\widehat{\sigma}(t)\}$, where $z_{\alpha/2}$ is the $(1 - \alpha/2)$ quantile of the standard normal distribution.

8.2.2 Censoring

When censoring occurs at some time other than an observed failure time, for a different subject the effect is simply that the censored subjects are dropped from the "No. at risk" total without processing the censored subject as having failed. However, when some subjects are censored at the same time that others fail, we need to be a bit careful about how we order the censorings and failures. When we went through the calculations of the Kaplan–Meier estimate in section 8.2.1, we did so without explaining this point, yet be assured that we were following some convention.

The Stata convention for handling a censoring that happens at the same time as a failure is to assume that the failure occurred before the censoring, and in fact, all Stata's st commands follow this rule. In chapter 7, we defined a time span based on the stset variables _t0 and _t to be the interval $(t_0, t]$, which is open at the left endpoint and closed at the right endpoint. Therefore, if we apply this definition of a time span, then any record shown to be censored at the end of this span can be thought of as instead being censored at some time $t + \epsilon$ for an arbitrarily small ϵ. The subject can fail at time t, but if the failure is censored, then Stata assumes that the censoring took place just a little bit later; thus failures occur before censorings.

This is how Stata handles this issue, but there is nothing wrong with the convention that handles censorings as occurring before failures when they appear to happen concurrently. One can force Stata to look at things this way by subtracting a small number from the time variable in your data for those records that are censored, and most of the time the number may be chosen small enough as to not otherwise affect the analysis.

❑ **Technical note**

If you force Stata to treat censorings as occurring before failures, be sure to modify the time variable in your data and not the _t variable that stset has created. In general, manually changing the values of the stset variables _t0, _t, _d, and _st is dangerous, because these variables have relations to your variables, and some of the data-management st commands exploit that relationship.

Thus instead of something such as

```
. replace _t = _t - 0.0001 if _d == 0
```

use

```
. replace time = time - 0.0001 if failed == 0
. stset time, failure(failed)
```

Better yet, use

```
. replace time = time - 0.0001 if failed == 0
. stset
```

because stset will remember the details of how you previously set your data and will apply these same settings to the modified data.

❑

8.2.3 Left truncation (delayed entry)

Left truncation refers to subjects who do not come under observation until after they are at risk. By the time you begin observing this subject, they have already survived for some time, and you are observing them only because they did not fail during that time.

At one level, such observations cause no problems with the Kaplan–Meier calculation. In (8.1), n_j is the number of subjects at risk (eligible to fail), and this number needs to take into account that subjects are not at risk of failing until they come under observation. When they enter, we simply increase n_j to reflect this fact.

For example, if you have the following data (subject 6 enters at $t_0 = 4$ and is censored at $t = 7$),

```
id     t0     t1    failed
 1      0      2       1
 2      0      4       1
 3      0      4       1
 4      0      5       0
 5      0      7       1
 6      4      7       0
 7      0      8       0
```

then the "risk-group" table is

t	No. at risk	No. failed	No. censored	No. added
2	6	1	0	0
4	5	2	0	1
5	4	0	1	0
7	3	1	1	0
8	1	0	1	0

and now it is just a matter of making the Kaplan–Meier calculations based on how many are in the "No. at risk" and "No. failed" columns. We will let Stata do the work:

```
. clear

. input id time0 time1 failed

              id      time0      time1     failed
  1.  1       0         2          1
  2.  2       0         4          1
  3.  3       0         4          1
  4.  4       0         5          0
  5.  5       0         7          1
  6.  6       4         7          0
  7.  7       0         8          0
  8. end

. stset time1, fail(failed) time0(time0)

(output omitted )

. sts list

         failure _d:  failed
   analysis time _t:  time1
```

	Beg.		Net	Survivor	Std.		
Time	Total	Fail	Lost	Function	Error	[95% Conf.	Int.]
2	6	1	0	0.8333	0.1521	0.2731	0.9747
4	5	2	-1	0.5000	0.2041	0.1109	0.8037
5	4	0	1	0.5000	0.2041	0.1109	0.8037
7	3	1	1	0.3333	0.1925	0.0461	0.6756
8	1	0	1	0.3333	0.1925	0.0461	0.6756

Notice how Stata listed the delayed entry at $t = 4$: "Net Lost" is -1. To conserve columns, rather than listing censorings and entries separately, Stata combines them into one column containing censorings-minus-entries and labels that column as "Net Lost".

There is a level at which delayed entries cause considerable problems. In these entries' presence, the Kaplan–Meier procedure for calculating the survivor curve can yield absurd results. This happens when some late arrivals to the study enter after everyone before them has failed.

Consider the following output from `sts list` for such a dataset:

```
. sts list

        failure _d:  failed
  analysis time _t:  time1

              Beg.           Net      Survivor    Std.
   Time      Total   Fail   Lost     Function    Error     [95% Conf. Int.]

      2         6      1       0       0.8333    0.1521    0.2731    0.9747
      4         5      2      -1       0.5000    0.2041    0.1109    0.8037
      5         4      0       1       0.5000    0.2041    0.1109    0.8037
      7         3      1       1       0.3333    0.1925    0.0461    0.6756
      8         1      1       0       0.0000       .          .         .
      9         0      0      -3       0.0000       .          .         .
     10         3      1       0       0.0000       .          .         .
     11         2      1       1       0.0000       .          .         .
```

We constructed these data to include three more subjects to enter at $t = 9$, after everyone who was previously at risk had failed. At $t = 8$, $\widehat{S}(t)$ has reached zero, never to return. Why does this happen? Note the product form of (8.1). Once a product term of zero (which occurs at $t = 8$) has been introduced, the product is zero, and further multiplication by anything nonzero is pointless. This is a shortcoming of the Kaplan–Meier method, and in section 8.3 we show that there is an alternative.

❏ **Technical note**

There is one other issue about the Kaplan–Meier estimator regarding delayed entry. When the earliest entry into the study occurs after $t = 0$, one may still calculate the Kaplan–Meier estimation, but the interpretation changes. Rather than estimating $S(t)$, you are now estimating $S(t|t_{\min})$, the probability of surviving past time t given survival to time $t_{\min}$, where $t_{\min}$ is the earliest entry time.

❏

8.2.4   Interval truncation (gaps)

Interval truncation is really no different from censoring followed by delayed entry. The subject disappears from the risk groups for a while and then reenters. The only issue is making sure that our "No. at risk" calculations reflect this fact, but Stata is up to that.

As with delayed entry, if a subject with a gap reenters after a final failure—meaning that a prior Kaplan–Meier estimate of $S(t)$ is zero—then all subsequent estimates of $S(t)$ will also be zero regardless of future activity.

8.2.5   Relationship to the empirical distribution function

The cumulative distribution function is defined as $F(t) = 1 - S(t)$, and in fact, by specifying the `failure` option, you can ask `sts list` to list the estimate of $F(t)$, which is obtained as 1 minus the Kaplan–Meier estimate:

```
. clear

. input id time0 time1 failed
              id        time0        time1      failed
     1.  1       0       2             1
     2.  2       0       4             1
     3.  3       0       4             1
     4.  4       0       5             0
     5.  5       0       7             1
     6.  6       4       7             0
     7.  7       0       8             0
     8. end

. stset time1, fail(failed) time0(time0)
```

(output omitted)

```
. sts list, failure

       failure _d:  failed
 analysis time _t:  time1
```

Time	Beg. Total	Fail	Net Lost	Failure Function	Std. Error	[95% Conf. Int.]	
2	6	1	0	0.1667	0.1521	0.0253	0.7269
4	5	2	-1	0.5000	0.2041	0.1963	0.8891
5	4	0	1	0.5000	0.2041	0.1963	0.8891
7	3	1	1	0.6667	0.1925	0.3244	0.9539
8	1	0	1	0.6667	0.1925	0.3244	0.9539

For standard nonsurvival datasets, the *empirical distribution function* (edf) is defined to be

$$\widehat{F}_{\text{edf}}(t) = \sum_{j|t_j \leq t} n^{-1}$$

where we have $j = 1, \ldots, n$ observations. That is, $\widehat{F}_{\text{edf}}(t)$ is a step function that increases by $1/n$ at each observation in the data. Of course, $\widehat{F}_{\text{edf}}(t)$ has no mechanism to account for censoring, truncation, and gaps, but when none of these exist, it can be shown that

$$\widehat{S}(t) = 1 - \widehat{F}_{\text{edf}}(t)$$

where $\widehat{S}(t)$ is the Kaplan–Meier estimate. To demonstrate, consider the following simple dataset, which has no censoring or truncation:

```
. clear

. input t
              t
     1.  1
     2.  4
     3.  4
     4.  5
     5. end

. stset t
```

(output omitted)

```
. sts list, failure
        failure _d:  1 (meaning all fail)
  analysis time _t:  t
```

Time	Beg. Total	Fail	Net Lost	Failure Function	Std. Error	[95% Conf. Int.]	
1	4	1	0	0.2500	0.2165	0.0395	0.8721
4	3	2	0	0.7500	0.2165	0.3347	0.9911
5	1	1	0	1.0000	.	.	.

This reproduces $\widehat{F}_{\text{edf}}(t)$, which is a nice property of the Kaplan–Meier estimator. Despite its sophistication in dealing with the complexities caused by censoring and truncation, it reduces to the standard methodology when these complexities do not exist.

8.2.6 Other uses of sts list

The `sts list` command lists the Kaplan–Meier survivor function. Let us use our hip-fracture dataset (the version already `stset`):

```
. use http://www.stata-press.com/data/cggm/hip2, clear
(hip fracture study)

. sts list
        failure _d:  fracture
  analysis time _t:  time1
              id:  id
```

Time	Beg. Total	Fail	Net Lost	Survivor Function	Std. Error	[95% Conf. Int.]	
1	48	2	0	0.9583	0.0288	0.8435	0.9894
2	46	1	0	0.9375	0.0349	0.8186	0.9794
3	45	1	0	0.9167	0.0399	0.7930	0.9679
4	44	2	0	0.8750	0.0477	0.7427	0.9418
(output omitted)							
13	21	1	0	0.5384	0.0774	0.3767	0.6752
15	20	1	-2	0.5114	0.0781	0.3507	0.6511
16	21	1	0	0.4871	0.0781	0.3285	0.6283
(output omitted)							
35	2	0	1	0.1822	0.0760	0.0638	0.3487
39	1	0	1	0.1822	0.0760	0.0638	0.3487

`sts list` can also produce less-detailed output. For instance, we can ask to see five equally spaced survival times in our data by specifying the `at()` option:

(Continued on next page)

```
. sts list, at(5)

        failure _d:  fracture
   analysis time _t:  time1
                id:  id

                  Beg.              Survivor     Std.
      Time       Total    Fail      Function     Error      [95% Conf. Int.]

         1         48       2        0.9583      0.0288      0.8435    0.9894
        13         21      18        0.5384      0.0774      0.3767    0.6752
        25         10       9        0.2776      0.0749      0.1443    0.4282
        37          2       2        0.1822      0.0760      0.0638    0.3487
        49          1       0           .           .          .         .
```

Note: survivor function is calculated over full data and evaluated at
       indicated times; it is not calculated from aggregates shown at left.

sts list will also list side-by-side comparisons of the estimated survivor function. Our
hip-fracture data has two study groups: one was assigned to wear an experimental protective device (protect==1) and a control group did not wear the device (protect==0).

```
. sts list, by(protect) compare

        failure _d:  fracture
   analysis time _t:  time1
                id:  id

                          Survivor Function
   protect                    0           1

   time       1           0.9000      1.0000
              5           0.6000      1.0000
              9           0.4364      0.8829
             13           0.1870      0.7942
             17           0.0831      0.7501
             21           0.0831      0.7501
             25             .         0.5193
             29             .         0.4544
             33             .         0.3408
             37             .         0.3408
             41             .            .
```

sts list has options that allow you to control these lists to get the desired output; see
[ST] sts for more details.

8.2.7 Graphing the Kaplan–Meier estimate

sts graph graphs (among other things) the Kaplan–Meier estimate. Typed without
arguments, sts graph graphs the overall (estimated) survivor function for your data.

```
. use http://www.stata-press.com/data/cggm/hip2
(hip fracture study)
. sts graph

        failure _d:  fracture
  analysis time _t:  time1
               id:  id
```

This produces the graph shown in figure 8.1.

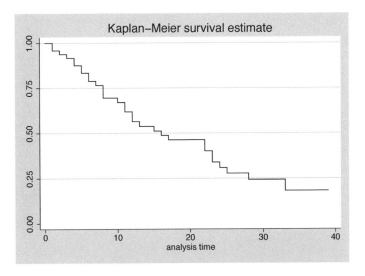

Figure 8.1. Kaplan–Meier estimate for hip-fracture data

sts graph has many options, including most of the options available with twoway. These options are designed to enhance the visual aspect of the graph and to show various aspects of the data.

sts graph with the by() option can plot multiple survivor curves. We can type the following to compare the survivor curves of the treatment group (protect==1) versus the control group (protect==0),

```
. sts graph, by(protect)
```

which produces the graph given in figure 8.2.

(Continued on next page)

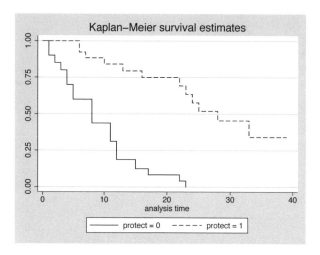

Figure 8.2. Kaplan–Meier estimates for treatment versus control

As hoped, we see that the treatment group has a better survival experience than the control group.

`sts graph` also has options to indicate the number of censored observations, delayed entries, or the number of subjects at risk. Specifying `censored(number)`, for example, displays tick marks approximately at the times when censoring occurred and, above each tick mark, displays the number of censored observations,

```
. sts graph, censored(number)
```

which produces the graph given in figure 8.3.

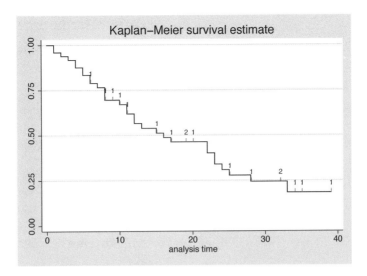

Figure 8.3. Kaplan–Meier estimate with the number of censored observations

We could also combine `censored(number)` with `by(protect)` if we wish.

Option `enter` is an alternative to `censored()`. It attempts to show more information than `censored()` by including the number who enter with the censoring information. The number entered is displayed below the curve, and the number censored is displayed above the curve. Unlike `censored()`, these numbers are centered along the flat portions of the curve, not where the censorings or entries exactly occurred.

When we specify option `lost`, the numbers displayed are censored minus entered, which is, say, the effective number lost. These numbers are displayed above the curve, centered along the flat portions.

Options `enter` and `lost` are most useful with delayed-entry data.

`sts graph` with the `risktable` option adds a table showing the number at risk beneath the survivor curve. Typing

```
. sts graph, by(protect) risktable
```

produces the graph given in figure 8.4.

(*Continued on next page*)

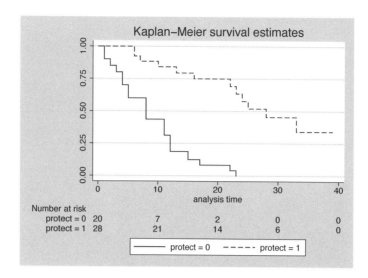

Figure 8.4. Kaplan–Meier estimates with a number-at-risk table

By default, the number at risk is shown for each time reported on the x axis. This may be changed by specifying time points as a *numlist* in `risktable(`*numlist*`)`. The `risktable()` option also offers many suboptions allowing you to customize the look of your at-risk table and to add additional information to the table. For example, we can label the groups in the table and include the number of failure events as follows:

```
. sts graph, by(protect) risktable(, order(1 "Control" 2 "Treatment") failevents)
> legend(label(1 "Control") label(2 "Treatment"))
```

This produces the graph given in figure 8.5. We also used the `legend()` option to label group categories in the legend of the graph.

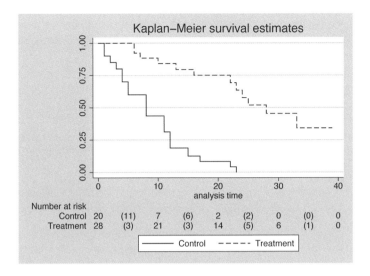

Figure 8.5. Kaplan–Meier estimates with a customized number-at-risk table

The `protect==0` group is now labeled `Control`, and the `protect==1` group is labeled `Treatment`. The number of failures are reported in parentheses between the displayed at-risk times. You can read more about the `risktable()` option in [ST] **sts graph**.

`sts graph` has other useful options for plotting Kaplan–Meier curves, including an option to add confidence bands and the `noorigin` option to specify that we wish the plotted survivor curve to begin at the first exit time instead of at $t = 0$. Read [ST] **sts graph** in the Stata manual to explore the behavior of these and other options.

8.3   The Nelson–Aalen estimator

The cumulative hazard function is defined as

$$H(t) = \int_0^t h(u)du$$

where $h()$ is the hazard function. In chapter 2, we discussed the count-data interpretation of $H(t)$, namely that it may be interpreted as the number of expected failures in $(0, t)$ for a subject if failure were a repeatable process.

Here we obtain a calculation formula for the empirical cumulative hazard function just as, in section 8.2, we gave a calculation formula for the empirical survivor function. One way would be to use the theoretical relationship between $H(t)$ and $S(t)$,

$$H(t) = -\ln\{S(t)\}$$

where for $S(t)$ we could use the Kaplan–Meier estimator. There is, however, another nonparametric method for estimating $H(t)$ that has better small-sample properties. The estimator is from Nelson (1972) and Aalen (1978),

$$\widehat{H}(t) = \sum_{j|t_j \leq t} \frac{d_j}{n_j}$$

where n_j is the number at risk at time t_j, d_j is the number of failures at time t_j, and the sum is over all distinct failure times less than or equal to t. Thus, given some data,

```
id     t   failed
 1     2      1
 2     4      1
 3     4      1
 4     5      0
 5     7      1
 6     8      0
```

we can write the risk table

t	n_j	d_j	No. censored
2	6	1	0
4	5	2	0
5	3	0	1
7	2	1	0
8	1	0	1

We calculate the number of failures per subject at each observed time, $e_j = d_j/n_j$, and then sum these to form $\widehat{H}(t)$:

t	n_j	d_j	No. censored	e_j	$\widehat{H}(t)$
2	6	1	0	0.1667	0.1667
4	5	2	0	0.4000	0.5667
5	3	0	1	0.0000	0.5667
7	2	1	0	0.5000	1.0667
8	1	0	1	0.0000	1.0667

$\widehat{H}(t)$ is the Nelson–Aalen estimator of the cumulative hazard, and **sts list** with the **cumhaz** option will make this calculation:

```
. clear
. input id time failed

            id          time        failed
  1.  1     2           1
  2.  2     4           1
  3.  3     4           1
  4.  4     5           0
  5.  5     7           1
  6.  6     8           0
  7. end

. stset time, fail(failed)

  (output omitted)

. sts list, cumhaz

          failure _d:  failed
    analysis time _t:  time
```

Time	Beg. Total	Fail	Net Lost	Nelson–Aalen Cum. Haz.	Std. Error	[95% Conf. Int.]	
2	6	1	0	0.1667	0.1667	0.0235	1.1832
4	5	2	0	0.5667	0.3283	0.1820	1.7639
5	3	0	1	0.5667	0.3283	0.1820	1.7639
7	2	1	0	1.0667	0.5981	0.3554	3.2015
8	1	0	1	1.0667	0.5981	0.3554	3.2015

The standard errors reported above are based on the variance calculation (Aalen 1978),

$$\widehat{\mathrm{Var}}\{\widehat{H}(t)\} = \sum_{j|t_j \leq t} \frac{d_j}{n_j^2}$$

and the confidence intervals reported are $\widehat{H}(t) \exp\{\pm z_{\alpha/2}\widehat{\phi}(t)\}$, where

$$\widehat{\phi}^2(t) = \frac{\widehat{\mathrm{Var}}\{\widehat{H}(t)\}}{\{\widehat{H}(t)\}^2}$$

estimates the asymptotic variance of $\ln \widehat{H}(t)$ and $z_{\alpha/2}$ is the $(1 - \alpha/2)$ quantile of the normal distribution.

The cumhaz option is also valid for sts graph, in which case you are plotting the Nelson–Aalen cumulative hazard instead of the Kaplan–Meier survivor function. Using the hip-fracture data, we can compare the Nelson–Aalen curves for treatment versus control,

```
. use http://www.stata-press.com/data/cggm/hip2, clear
(hip fracture study)
. sts graph, by(protect) cumhaz
```

which produces the curves shown in figure 8.6. Naturally, we see that the cumulative hazard is greater for the control group.

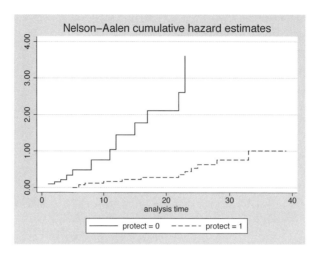

Figure 8.6. Nelson–Aalen curves for treatment versus control

Theoretically, the survivor and cumulative hazard functions are related by

$$S(t) = \exp\{-H(t)\}$$

or, if you prefer, $H(t) = -\ln\{S(t)\}$. We can use these relations to convert one estimate to the other. In small samples, the Kaplan–Meier product-limit estimator is superior when estimating the survivor function, and the Nelson–Aalen estimator is superior when estimating the cumulative hazard function. For the survivor function and the cumulative hazard function, both the Kaplan–Meier estimator and the Nelson–Aalen estimator are consistent estimates of each, and the statistics are asymptotically equivalent (Klein and Moeschberger 2003, 104). That is, in very large samples, it does not matter how you estimate the survivor function, whether by Kaplan–Meier or by transforming the Nelson–Aalen.

We can compare the survivor curves estimated both ways. The command `sts generate` will prove useful here. `sts generate` will create variables containing the Kaplan–Meier or Nelson–Aalen estimates, depending on which you request. If we specify the Kaplan–Meier, we can then convert this estimate into an estimate of the cumulative hazard, based on the Kaplan–Meier. If we specify the Nelson–Aalen, we can then convert this estimate into an estimate of the survivor function, based on the Nelson–Aalen.

With the hip-fracture data still in memory,

```
. sts generate kmS = s          /* obtain K-M survivor estimate */
. sts generate naH = na         /* obtain N-A cumulative hazard estimate */
. gen naS = exp(-naH)           /* calculate N-A survivor estimate */
. gen kmH = -log(kmS)           /* calculate K-M cumulative hazard estimate */
```

```
. label var kmS "K-M"
. label var naS "N-A"
. label var kmH "K-M"
. label var naH "N-A"
```

First, we graph the comparison of the survivor functions, producing figure 8.7:

```
. line kmS naS _t, c(J J) sort
```

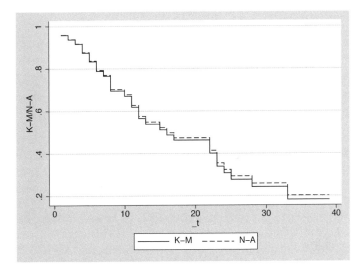

Figure 8.7. Estimated survivor functions. K–M = Kaplan–Meier; N–A = Nelson–Aalen.

The top curve is the survivor function estimated by transforming the Nelson–Aalen estimator, and the bottom one is the Kaplan–Meier. These results are not unusual. A Taylor expansion shows that the Nelson–Aalen estimator of the survivor function is always greater than or equal to that from the Kaplan–Meier estimator. See, for example, Appendix 1 of Hosmer, Lemeshow, and May (2008) for details.

Next we graph the comparison of the cumulative hazard functions, producing figure 8.8:

```
. line naH kmH _t, c(J J) sort
```

(*Continued on next page*)

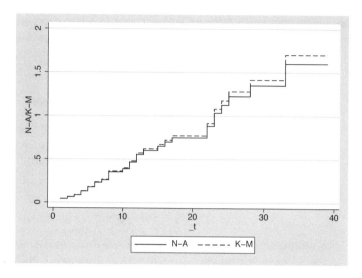

Figure 8.8. Estimated cumulative hazard functions. N–A = Nelson–Aalen; K–M = Kaplan–Meier.

   The top curve is the cumulative hazard obtained by transforming the Kaplan–Meier survivor function, and the bottom one is the Nelson–Aalen curve. By analogy to the above discussion, the Kaplan–Meier version of the cumulative hazard is always greater than or equal to that from the Nelson–Aalen estimator.

   The above graphs are typical for most datasets in that they demonstrate that most of the time, the Kaplan–Meier and Nelson–Aalen estimators are similar once the transformation has been made so that they both estimate the same things. The Nelson–Aalen estimator, however, does offer another advantage. Recall in section 8.2.3 where we discussed a shortcoming of the Kaplan–Meier estimator, namely, that once the estimator is zero, it remains zero regardless of future activity. The Nelson–Aalen estimator does not suffer from this problem because it will increase with every failure event, even after the Kaplan–Meier estimate would have reached zero. When this is at issue, the Nelson–Aalen estimator is preferred.

❏ **Technical note**

   What led to the Kaplan–Meier estimator falling to and staying at zero in section 8.2.3 was that (1) there was a gap during which no one in the data was at risk, and (2) before the gap there was no one in the group at risk who was known to have survived longer than when the gap started. That is, there was one group of people at risk, and that group evaporated before anyone from the second group came under observation. In the first group, the empirically calculated survivor function had indeed fallen to zero, meaning that at the last time anything was known about the group, all that were left were known to fail.

The Nelson–Aalen estimator deals better with data like that, but understand that if your data contain a gap during which no one is at risk, then your data are absolutely silent about what the risks might be during that period.

If there is such a gap, the Nelson–Aalen estimator produces a consistent estimate of

$$H^*(t) = \int_0^t h(u)I(u)du$$

where $I(u)$ is an indicator function equal to 0 over those periods where no subject is at risk, and 1 otherwise.

Hence, up until the gap, the Nelson–Aalen curve is an estimate of $H(t)$, but after the gap, it is an estimate of $H(t|\text{no risk during the gap})$.

Few datasets actually have such gaps. Gaps in subject histories are not a problem; it is only when all the gaps in all the subjects add up in such a way that no one is at risk during a period that the dataset itself will have such a gap. When datasets do have such gaps, you want to avoid the Kaplan–Meier estimate and use the Nelson–Aalen, and still you need to think carefully as you interpret results.

For instance, let us imagine that you do a short study on smoking and monitor subjects for only 2 years. The second-oldest person in your data was 40 years old at enrollment, meaning that he was 42 at the end of the study. The oldest person was 80 at enrollment. You would not want to draw the conclusion that the risk of death due to smoking is 0 between ages 42 and 80. You observed no one during that period and so have no information on what the risk might be.

❑

8.4   Estimating the hazard function

`sts graph` can also be used to plot an estimate of the hazard function, $h(t)$. Because the hazard is the derivative of the cumulative hazard, $H(t)$, it would seem straightforward to estimate the hazard itself. However, examination of figure 8.6 and the subsequent graphs reveals that the estimated cumulative hazards available to us are step functions and thus cannot be directly differentiated. That is not to say that it is not straightforward to take figure 8.6 and picture in our minds what the derivative of the cumulative hazard would look like; for the control group, it would be fairly linear (because the cumulative hazard is parabolic), and for the treatment group, the derivative would start off as constant for some time (because the cumulative hazard is initially linear) and then increase.

We can estimate the hazard by taking the steps of the Nelson–Aalen cumulative hazard and smoothing them with a kernel smoother. More precisely, for each observed death time, t_j, if we define the estimated *hazard contribution* to be

$$\Delta\widehat{H}(t_j) = \widehat{H}(t_j) - \widehat{H}(t_{j-1})$$

we can obtain these hazard contributions using `sts generate` *newvar* `= h`. Then we can estimate $h(t)$ with

$$\widehat{h}(t) = b^{-1} \sum_{j=1}^{D} K_t \left(\frac{t - t_j}{b} \right) \Delta \widehat{H}(t_j)$$

for some kernel function $K_t()$ and bandwidth b; the summation is over the D times at which failure occurs (Klein and Moeschberger 2003, 167; Muller and Wang 1994).

This whole process can be automated by specifying option `hazard` to `sts graph`. Using our hip-fracture data, we can graph the estimated hazards for both the treatment and control groups as follows:

```
. use http://www.stata-press.com/data/cggm/hip2, clear
(hip fracture study)
. sts graph, hazard by(protect) kernel(gaussian) width(4 5)
```

This produces figure 8.9. The graph agrees with our informal analysis of the Nelson–Aalen cumulative hazards. In applying the kernel smoother, we specified a Gaussian (normal) kernel function and bandwidths of four for the control group (`protect==0`) and five for the treatment group (`protect==1`) although suitable defaults would have been provided had we not specified these; see [R] **kdensity** for a list of available kernel functions and their definitions.

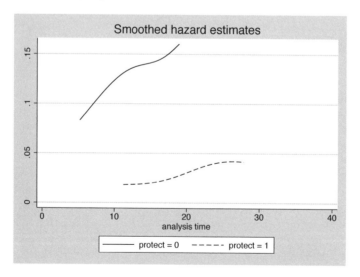

Figure 8.9. Smoothed hazard functions

In practice, you will often find that plotting ranges of smoothed hazard curves are narrower than those of their cumulative-hazard and survivor counterparts. Kernel smoothing requires averaging values over a moving window of data. Near the endpoints of the plotting range, these windows contain insufficient data for accurate estimation, and the resulting estimators are said to contain *boundary bias*. The standard approach Stata takes in dealing with this problem is simply to restrict the plotting range so that hazard estimates exhibiting boundary bias are not displayed. An alternate approach— one that results in wider plotting ranges—is to use a boundary kernel; see the technical note below for details.

❑ **Technical note**

We should clarify the notation for the kernel function $K_t()$ we used in the above formula. The conventional kernel estimator uses a symmetric kernel function $K()$. Applying this estimator directly to obtain a hazard function smoother results in biased estimates in the boundary regions near the endpoints.

Consider the left and right *boundary regions*, $B_L = \{t : t_{\min} \le t < b\}$ and $B_R = \{t : t_{\max} - b < t \le t_{\max}\}$, respectively, where $t_{\min}$ and $t_{\max}$ are the minimum and maximum observed failure times. Using a symmetric kernel in these regions leads to biased estimates because there are no failures observed before time $t_{\min}$ and after time $t_{\max}$; that is, the support of the kernel exceeds the available range of the data. A more appropriate choice is to use in the boundary region some asymmetric kernel function, referred to as a *boundary kernel*, $K_{\text{bnd}}()$. This method consists of using a symmetric kernel for time points in the interior region, $K_t() = K()$, and a respective boundary kernel for time points in the boundary regions, $K_t() = K_{\text{bnd}}()$. The method of boundary kernels is described more thoroughly in Gray (1990), Muller and Wang (1994), and Klein and Moeschberger (2003, 167) to name a few.

`sts graph, hazard` uses the boundary adjustments suggested in Muller and Wang (1994) with the `epan2`, `biweight`, and `rectangular` kernels. For other kernels (see [R] **kdensity**), no boundary adjustment is made. Instead, the default graphing range is constrained to be the range $[t_{\min} + b, t_{\max} - b]$, following the advice of Breslow and Day (1987). You can also request that no boundary-bias adjustment be made at all by specifying option `noboundary`.

We change the kernel to `epan2` in the above to obtain figure 8.10.

```
. sts graph, hazard by(protect) kernel(epan2) width(6 7)
        failure _d:  fracture
   analysis time _t:  time1
               id:  id
```

We also specified larger bandwidths in `width(6 7)`. The modified (boundary) `epan2` kernel is used to correct for the left- and right-boundary bias. Notice the wider plotting range compared with the Gaussian example.

(Continued on next page)

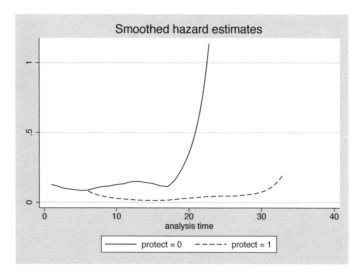

Figure 8.10. Smoothed hazard functions with the modified `epan2` kernel for the left and right boundaries

There are few subjects remaining at risk past analysis time 15 in the control group (`protect=0`). This leads to a poor estimate of the hazard function in that region. Hess, Serachitopol, and Brown (1999) found that the kernel-based hazard function estimators tend to perform poorly if there are fewer than 10 subjects at risk. There are 11 subjects at risk at time 8 and 7 subjects at risk at time 11 in the control group. It would be reasonable to limit the graphing range to, say, time 10 (option `tmax(10)`) when graphing the hazard function for the control group.

❏

One interesting feature of smoothed hazard plots is that you can assess the assumption of proportional hazards (the importance of which will be discussed in chapter 9 on the Cox model) by plotting the estimated hazards on a log scale.

```
. sts graph, hazard by(protect) kernel(gaussian) width(4 5) yscale(log)
```

By examining figure 8.11, we find the lines to be somewhat parallel, meaning that the proportionality assumption is violated only slightly. When hazards are proportional, the proportionality can be exploited by using a Cox model to assess the effects of treatment more efficiently; see chapter 9.

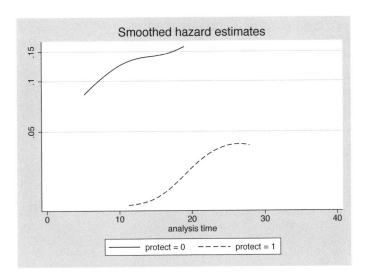

Figure 8.11. Smoothed hazard functions, log scale

8.5   Estimating mean and median survival times

As we mentioned in section 8.1, standard univariate methods used to estimate means and percentiles may not be appropriate with complex survival data. In section 2.4, we gave formal definitions of the mean and median survival times as functions of the survivor function. These relationships form the basis of the survival methods for estimating means and percentiles.

Recall the definition of the median survival time, $t_{50} = \tilde{\mu}_T$, as the time beyond which 50% of subjects are expected to survive, i.e., $S(\tilde{\mu}_T) = 0.5$. A natural way to estimate the median then is to use the Kaplan–Meier estimator $\hat{S}(t)$ in place of $S(t)$. Because this estimator is a step function, the nonparametric estimator of the median survival time is defined to be

$$\hat{t}_{50} = \min\{t_i | \hat{S}(t_i) \leq 0.5\}$$

More generally, the estimate of the pth percentile of survival times, t_p, is obtained as the smallest observed time t_i for which $\hat{S}(t_i) \leq 1 - p/100$ for any p between 0 and 100.

In Stata, you can use stci to obtain the median survival time provided for you by default or use the p(#) option with stci to obtain any other percentile. For our hip-fracture data, we can obtain median survival times for both the treatment and control group as follows:

```
. use http://www.stata-press.com/data/cggm/hip2
(hip fracture study)

. stci, by(protect)

        failure _d:  fracture
  analysis time _t:  time1
              id:  id
```

| | no. of | | | | |
protect	subjects	50%	Std. Err.	[95% Conf. Interval]	
0	20	8	1.251531	4	12
1	28	28	2.786697	22	.
total	48	16	4.753834	11	23

From the output, the estimated median in the control group is 8 months, and in the treatment group the median is 2.3 years (28 months). This agrees with the graphs of the survivor curves given in figure 8.2.

The large-sample standard errors reported above are based on the following formula given by Collett (2003, 35) and Klein and Moeschberger (2003, 122):

$$\widehat{\mathrm{SE}}\{\widehat{t}_p\} = \frac{\sqrt{\widehat{\mathrm{Var}}\{\widehat{S}(t_p)\}}}{\widehat{f}(t_p)}$$

where $\widehat{\mathrm{Var}}\{\widehat{S}(t_p)\}$ is the Greenwood pointwise standard error estimate (8.2) and $\widehat{f}(t_p)$ is the estimated density function at the pth percentile.

Confidence intervals, however, are not calculated based on these standard errors. For a given confidence level α, they are obtained by inverting the respective $100(1 - \alpha)\%$ confidence interval for $S(t_p)$ based on a $\ln\{-\ln S(t)\}$ transformation, as given in section 8.2.1. That is, the confidence interval for the pth percentile t_p is defined by the pair (t_L, t_U) such that $P\{L(t_L) \leq S(t_p) \leq U(t_U)\} = 1 - \alpha$ where $L(\cdot)$ and $U(\cdot)$ are the upper and the lower pointwise confidence limits of $S(t)$; see section 8.2.1. Computationally, $\widehat{t}_L$ and $\widehat{t}_U$ are estimated as the smallest observed times at which the upper and lower confidence limits for $S(t)$ are less than or equal to $1 - p/100$. For a review of other methods of obtaining confidence intervals see, for example, Collett (2003), Klein and Moeschberger (2003), and Andersen and Keiding (2006).

In the above example, the 95% confidence interval for the median survival time of the `protect==0` group is $(4, 12)$ (months) and of the `protect==1` group is $(22, .)$. A missing value (dot) for the upper confidence limit for the `protect==1` group indicates that this limit could not be determined. For this group, the estimated upper confidence limit of the survivor function never falls below 0.5:

```
. sts list if protect==1, at(5)
          failure _d:  fracture
    analysis time _t:  time1
                  id:  id
                Beg.                    Survivor      Std.
      Time     Total      Fail         Function      Error      [95% Conf. Int.]

         5        28         0           1.0000         .            .          .
        16        18         6           0.7501      0.0891      0.5242     0.8798
        27        10         4           0.5193      0.1141      0.2824     0.7120
        38         2         2           0.3408      0.1318      0.1135     0.5871
        49         1         0              .           .            .          .
```

Note: survivor function is calculated over full data and evaluated at
       indicated times; it is not calculated from aggregates shown at left.

Although the median survival time is commonly used to estimate the location of the survival distribution because it tends to be right skewed, the mean of the distribution may also be of interest in some applications.

The mean μ_T is defined as an integral from zero to infinity of the survivor function $S(t)$. Similar to the median estimation, a natural way of estimating the mean then is to plug in the Kaplan–Meier estimator $\widehat{S}(t)$ for $S(t)$ in the integral expression. The nonparametric estimator of the mean survival time is defined as follows:

$$\widehat{\mu}_T = \int_0^{t_{\max}} \widehat{S}(t)dt$$

where $t_{\max}$ is the maximum observed failure time. The integral above is restricted to the range $[0, t_{\max}]$ because the Kaplan–Meier estimator is not defined beyond the largest observed failure time. Therefore, the mean estimated by using the above formula is often referred to as a *restricted mean*. A restricted mean $\widehat{\mu}_T$ will underestimate the true mean μ_T if the last observed analysis time is censored.

The standard error for the estimated restricted mean is given by Klein and Moeschberger (2003, 118) and Collett (2003, 340) as follows:

$$\widehat{\mathrm{SE}}\{\widehat{\mu}_T\} = \sum_{i=1}^{n} \widehat{A}_i \sqrt{\frac{d_i}{R_i(R_i - d_i)}}$$

where the sum is over all distinct failure times, $\widehat{A}_i$ is the estimated area under the Kaplan–Meier product-limit survivor curve from time t_i to $t_{\max}$, R_i is the number of subjects at risk at time t_i, and d_i is the number of failures at time t_i.

The $100(1 - \alpha)\%$ confidence interval for the estimated restricted mean is computed as $\widehat{\mu}_T \pm z_{\alpha/2}\widehat{\mathrm{SE}}\{\widehat{\mu}_T\}$, where $z_{\alpha/2}$ is the $(1 - \alpha/2)$ quantile of the standard normal distribution.

Continuing the hip-fracture example, we obtain group-specific restricted means by specifying option `rmean` with `stci`:

```
. stci, by(protect) rmean
        failure _d:  fracture
  analysis time _t:  time1
              id:  id
```

protect	no. of subjects	restricted mean	Std. Err.	[95% Conf. Interval]	
0	20	8.938312	1.39057	6.21285	11.6638
1	28	26.75578(*)	2.503475	21.8491	31.6625
total	48	18.89901(*)	2.006556	14.9662	22.8318

```
(*) largest observed analysis time is censored, mean is underestimated
```

The estimated mean survival time of the control group (8.94 months) is smaller than the estimated mean survival time of the experimental group (26.76 months). The respective 95% confidence intervals (6.21, 11.66) and (21.85, 31.66) do not overlap, suggesting that the treatment-group patients have a higher expected time-without-fracture. This agrees with the results obtained earlier for the median survival times. Here we have an estimate of the upper confidence limit for the treatment group because the confidence interval is based on the estimates of the restricted mean and its standard error.

Notice the note at the bottom of the table reported for the mean estimate of the treatment group and the overall mean. In this group, and in the combined sample, the last observed analysis time is censored; therefore, the restricted mean underestimates the true mean. Stata detects it and warns you about it.

Stata offers a way to alleviate this problem by computing an *extended mean*. The extended mean is computed by extending the Kaplan–Meier product-limit survivor curve to zero by using an exponentially fit curve and then computing the area under the entire curve (Klein and Moeschberger 2003, 100, 122). This approximation is ad hoc and must be evaluated with care. We recommend plotting the extended survivor function by using the graph option of stci.

For the above example,

```
. stci, by(protect) emean
        failure _d:  fracture
  analysis time _t:  time1
              id:  id
```

protect	no. of subjects	extended mean
0	20	8.938312(*)
1	28	39.10101
total	48	23.07244

```
(*) no extension needed
```

the extended mean of the treatment group is 39.10 months and is noticeably larger than the previous estimate of 26.76. The estimate of the overall mean increases from 18.90 to 23.07. The extended mean of the control group is the same as the restricted mean because the last observed time in this group ends in failure.

By examining the graphs of the extended overall and treatment-group survivor curves,

```
. stci, emean graph
```
(*output omitted*)

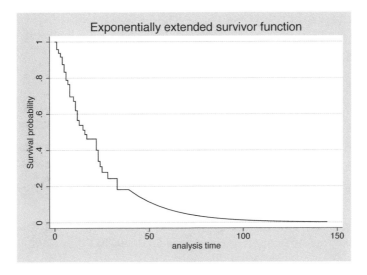

Figure 8.12. Exponentially extended Kaplan–Meier estimate

and

```
. stci if protect==1, emean graph
```
(*output omitted*)

(*Continued on next page*)

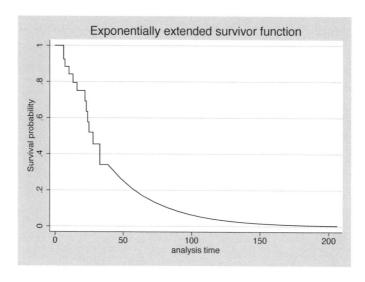

Figure 8.13. Exponentially extended Kaplan–Meier estimate treatment group

we conclude that the assumption of the exponential survivor function beyond the last observed failure time may be reasonable.

8.6   Tests of hypothesis

Earlier we discussed how to compare the survival experience between two (or more) groups using the by() option to sts list, sts graph, and stci. To form formal tests of hypothesis for the equality of survivor functions across groups, you can use the sts test command.

sts test will allow you to test the equality of survivor functions using one of several available nonparametric tests, namely, the log-rank (Mantel and Haenszel 1959), Wilcoxon (Breslow 1970; Gehan 1965), Tarone–Ware (1977), Peto–Peto–Prentice (Peto and Peto 1972; Prentice 1978), and generalized Fleming–Harrington (Harrington and Fleming 1982) tests.

All these tests are appropriate for testing the equality of survivor functions across two or more groups. These tests do not test the equality of the survivor functions at a specific time point. Instead, they are global tests in the sense that they compare the overall survivor functions. These tests work by comparing (at each failure time) the expected versus the observed number of failures for each group and then combining these comparisons over all observed failure times. The above tests differ only in respect to how they weight each of these individual comparisons that occur at each failure time when combining these comparisons to form one overall test statistic.

8.6.1   The log-rank test

Typing

```
. sts test varname, logrank
```

performs the log-rank test. *varname* should be a variable taking on different values for different groups, such as 1, 2, 3, ..., or any distinct set of values. In our hip-fracture data, we have the variable `protect==0` for the control subjects and `protect==1` for the treatment group, and so we can type

```
. use http://www.stata-press.com/data/cggm/hip2
(hip fracture study)
. sts test protect, logrank

         failure _d:  fracture
   analysis time _t:  time1
                id:  id

Log-rank test for equality of survivor functions

            |   Events        Events
 protect    | observed      expected
------------+----------------------------
 0          |       19          7.14
 1          |       12         23.86
------------+----------------------------
 Total      |       31         31.00
                chi2(1) =        29.17
                Pr>chi2 =       0.0000
```

Although the log-rank test is a rank test, it can be viewed as an extension of the familiar Mantel–Haenszel (1959) test applied to survival data. Let's say that we are interested in comparing the survival experience of r groups. Assume that, in all groups combined, there are k distinct failure times. Further assume that, at failure time t_j, there are n_j subjects at risk, of which d_j fail and $n_j - d_j$ survive. Then the log-rank test statistic is computed by constructing, at each of the k distinct failure times, an $r \times 2$ contingency table and then combining results from these k tables. To be clear, for each time t_j, we would have a table of the form given in table 8.1.

Table 8.1. $r \times 2$ contingency table for time t_j

Group	Failures at t_j	Survived at t_j	At risk at t_j
1	d_{1j}	$n_{1j} - d_{1j}$	n_{1j}
2	d_{2j}	$n_{2j} - d_{2j}$	n_{2j}
.	.	.	.
.	.	.	.
.	.	.	.
r	d_{rj}	$n_{rj} - d_{rj}$	n_{rj}
Total	d_j	$n_j - d_j$	n_j

The expected number of failures in group i at time t_j, under the null hypothesis of no difference in survival among the r groups, is $E_{ij} = n_{ij}d_j/n_j$. The chi-squared test statistic (distributed as χ^2 with $r-1$ degrees of freedom under the null) is calculated as a quadratic form $\mathbf{u}'\mathbf{V}^{-1}\mathbf{u}$ using the row vector

$$\mathbf{u}' = \sum_{j=1}^{k} W(t_j)(d_{1j} - E_{1j}, \ldots, d_{rj} - E_{rj}) \tag{8.3}$$

and the $r \times r$ variance matrix $\mathbf{V}$, where the individual elements are calculated by

$$V_{il} = \sum_{j=1}^{k} \frac{W^2(t_j)n_{ij}d_j(n_j - d_j)}{n_j(n_j - 1)}\left(\delta_{il} - \frac{n_{ij}}{n_j}\right) \tag{8.4}$$

where $i = 1, \ldots, r$; $l = 1, \ldots, r$; and $\delta_{il} = 1$ if $i = l$ and 0 otherwise.

The weight function $W(t_j)$ is what characterizes the different flavors of the tests computed by sts test and is defined as a positive function equal to zero when n_{ij} is zero. For the log-rank test, $W(t_j) = 1$ when n_{ij} is nonzero.

The test statistic is constructed by combining the information from the contingency tables obtained at every failure time, and consequently, the test takes into account the entire survival experience and not just a specific point in time.

Stata shows a summary of those k tables in the output. The above output included the "Events observed" category that refers to the number of failures observed—19 for the first group and 12 for the second—and "expected" refers to the number of events that would be expected if the two groups shared the same survivor function—7.14 for the first group and 23.86 for the second. Here the observed values are different enough from the expected so as to produce a highly significant chi-squared value. The log-rank test clearly rejects the null hypothesis that the survivor functions of the two groups are the same.

The relative survival experiences of the distinct groups may be characterized by the groups' hazard functions, and thus the null hypothesis of the tests computed by sts test may be expressed in the hazards. Namely, for sts test the null hypothesis is

$$H_o : h_1(t) = h_2(t) = \cdots = h_r(t)$$

This is the null hypothesis for all tests computed by sts test, and the different tests vary in power according to how H_o is violated. For example, the log-rank test is most powerful when the hazards are not equal but instead are proportional to one another.

8.6.2   The Wilcoxon test

Typing

```
sts test varname, wilcoxon
```

will perform the generalized Wilcoxon test of Breslow (1970) and Gehan (1965).

```
. use http://www.stata-press.com/data/cggm/hip2
(hip fracture study)

. sts test protect, wilcoxon

        failure _d:  fracture
  analysis time _t:  time1
              id:  id

Wilcoxon (Breslow) test for equality of survivor functions

          |  Events    Events    Sum of
  protect |  observed  expected   ranks
----------+---------------------------------
  0       |     19       7.14      374
  1       |     12      23.86     -374
----------+---------------------------------
  Total   |     31      31.00        0

              chi2(1) =    23.08
              Pr>chi2 =   0.0000
```

The Wilcoxon test is also a rank test and is constructed in the same way as the log-rank test, except that for this test we set $W(t_j) = n_j$ in (8.3) and (8.4). That is, the Wilcoxon test places more weight to tables at earlier failure times—when more subjects are at risk—than to tables for failures later in the distribution. The Wilcoxon test is preferred to the log-rank test when the hazard functions are thought to vary in ways other than proportionally. However, there is a drawback. Because of this weighting scheme, the Wilcoxon test can prove unreliable if the censoring patterns differ over the test groups.

8.6.3   Other tests

In addition to the log-rank and the generalized Wilcoxon, sts test can also perform other tests:

1. *The Tarone–Ware test*: sts test *varname*, tware
   Based on the work of Tarone and Ware (1977), this test is nearly identical to the Wilcoxon test, except that the weight function is $W(t_j) = \sqrt{n_j}$ instead of $W(t_j) = n_j$. As such, more weight is given to the earlier failure times when more subjects are at risk, but not as much as the Wilcoxon test. As a result, this test is less susceptible to problems should vast differences exist in the censoring patterns among the groups.

2. *The Peto–Peto–Prentice test*: sts test *varname*, peto
   Based on the work of Peto and Peto (1972) and Prentice (1978), this test uses as a

weight function an estimate of the overall survivor function; that is, $W(t_j) = \widetilde{S}(t_j)$, where $\widetilde{S}(t_j)$ is similar (but not exactly equal) to the Kaplan–Meier estimator. While more computationally intensive, this test is not susceptible to differences in censoring patterns among groups.

3. *The Fleming–Harrington test:* `sts test` *varname,* `fh(`*p q*`)`
   From Harrington and Fleming (1982), this test uses $W(t_j) = \{\widehat{S}(t_j)\}^p \{1 - \widehat{S}(t_j)\}^q$, where $\widehat{S}(t_j)$ is the Kaplan–Meier estimator, and p and q are chosen by the user so that the weighting scheme is customized. When $p > q$, more weight is given to earlier failures than to later ones; when $p < q$, the opposite is true. When $p = q = 0$, the test reduces to the standard log-rank test.

8.6.4 Stratified tests

The tests of equality of survivor functions performed by `sts test`, in all its incarnations, may be modified so that the tests are stratified. In a stratified test, we perform the test separately for different subgroups of the data and then combine the test results into one overall statistic. For instance, we might be testing that Group A had similar survival experiences as Group B, but then we might decide to stratify on sex, so we would test (1) that females in Group A had similar survival as did females in Group B and (2) that males in Group A had similar survival as did males in Group B. Why do this? Because perhaps we think the survival experiences of males and females differ. If Group A had more females than Group B and we just compared the two groups, then we could be misled into thinking there is a difference due to A and B.

For the stratified test, the calculations in (8.3) and (8.4) are formed over each stratum to form stratum-specific quantities $\mathbf{u}_s$ and $\mathbf{V}_s$, which are then summed over the strata to form $\mathbf{u} = \sum_s \mathbf{u}_s$ and $\mathbf{V} = \sum_s \mathbf{V}_s$. The test statistic is then calculated as $\mathbf{u}'\mathbf{V}\mathbf{u}$ and is redefined this way.

In our hip-fracture study, we know that age is an important factor associated with hip fractures. As age increases, so does the degree of osteoporosis, making bones more fragile and thus more susceptible to fractures. As a way of controlling for this, we might categorize each of the subjects into one of three age groups and then stratify on the age-group variable. In our data, age varies from 62 to 82, as we see by using `summarize`:

```
. use http://www.stata-press.com/data/cggm/hip2
(hip fracture study)

. summarize age
```

Variable	Obs	Mean	Std. Dev.	Min	Max
age	106	70.46226	5.467087	62	82

```
. gen agegrp = 1

. replace agegrp = 2 if age>65
(78 real changes made)

. replace agegrp = 3 if age>75
(20 real changes made)
```

```
. tabulate agegrp
    agegrp |      Freq.      Percent        Cum.
-----------+-----------------------------------
         1 |         28        26.42       26.42
         2 |         58        54.72       81.13
         3 |         20        18.87      100.00
-----------+-----------------------------------
     Total |        106       100.00
. sts test protect, logrank strata(agegrp)
         failure _d:  fracture
   analysis time _t:  time1
                 id:  id

Stratified log-rank test for equality of survivor functions
           |  Events          Events
   protect | observed     expected(*)
-----------+------------------------------
   0       |       19            7.15
   1       |       12           23.85
-----------+------------------------------
   Total   |       31           31.00
(*) sum over calculations within agegrp
              chi2(1) =       30.03
              Pr>chi2 =      0.0000
```

Even accounting for age, we still find a significant difference. The `detail` option will allow us to see the individual tables that go into the stratified results:

```
. sts test protect, logrank strata(agegrp) detail
         failure _d:  fracture
   analysis time _t:  time1
                 id:  id

Stratified log-rank test for equality of survivor functions
-> agegrp = 1
           |  Events          Events
   protect | observed        expected
-----------+------------------------------
   0       |        4            0.93
   1       |        1            4.07
-----------+------------------------------
   Total   |        5            5.00
              chi2(1) =       13.72
              Pr>chi2 =      0.0002
-> agegrp = 2
           |  Events          Events
   protect | observed        expected
-----------+------------------------------
   0       |       10            4.34
   1       |        8           13.66
-----------+------------------------------
   Total   |       18           18.00
              chi2(1) =       11.36
              Pr>chi2 =      0.0007
```

```
-> agegrp = 3
          │     Events          Events
 protect  │   observed        expected
──────────┼──────────────────────────────
 0        │          5            1.89
 1        │          3            6.11
──────────┼──────────────────────────────
 Total    │          8            8.00
               chi2(1) =           8.32
               Pr>chi2 =         0.0039
-> Total
          │     Events          Events
 protect  │   observed      expected(*)
──────────┼──────────────────────────────
 0        │         19            7.15
 1        │         12           23.85
──────────┼──────────────────────────────
 Total    │         31           31.00
(*) sum over calculations within agegrp
               chi2(1) =          30.03
               Pr>chi2 =         0.0000
```

The above illustrates how the test works. sts test ran separate log-rank tests for each of the age groups, with the results of each of those tests being the same as if we typed

```
. sts test protect if agegrp==1, logrank
. sts test protect if agegrp==2, logrank
. sts test protect if agegrp==3, logrank
```

Then sts test combined those three tests into one.

9 The Cox proportional hazards model

In section 3.2 on semiparametric models, we formulated an analysis of survival data where no parametric form of the survivor function is specified, yet the effects of the covariates are parameterized to alter the baseline survivor function (that for which all covariates are equal to zero) in a certain way. The Cox (1972) model, which assumes that the covariates multiplicatively shift the baseline hazard function, is by far the most popular because of its elegance and computational feasibility.

The Cox proportional hazards regression model (Cox 1972) asserts that the hazard rate for the jth subject in the data is

$$h(t|\mathbf{x}_j) = h_0(t) \exp(\mathbf{x}_j \boldsymbol{\beta}_x) \tag{9.1}$$

where the regression coefficients, $\boldsymbol{\beta}_x$, are to be estimated from the data.

The nice thing about this model is that $h_0(t)$, the baseline hazard, is given no particular parameterization and, in fact, can be left unestimated. The model makes no assumptions about the shape of the hazard over time—it could be constant, increasing, decreasing, increasing and then decreasing, decreasing and then increasing, or anything else you can imagine; what is assumed is that, whatever the general shape, it is the same for everyone. One subject's hazard is a multiplicative replica of another's; comparing subject j to subject m, the model states that

$$\frac{h(t|\mathbf{x}_j)}{h(t|\mathbf{x}_m)} = \frac{\exp(\mathbf{x}_j \boldsymbol{\beta}_x)}{\exp(\mathbf{x}_m \boldsymbol{\beta}_x)}$$

which is constant, assuming the covariates $\mathbf{x}_j$ and $\mathbf{x}_m$ do not change over time.

How exactly is this possible, given that a parametric regression model, in its likelihood calculations, contains terms using the hazard function and survivor function and that at first blush, a likelihood calculation involving the baseline hazard seems inevitable? Refer to section 3.2 where we gave a heuristic approach that explains why, by confining our analysis to only those times for which failure occurs and by conditioning on the fact that failures occurred only at those times, the baseline hazard drops out from the calculations. For a more technical treatment of how this happens, see Kalbfleisch and Prentice (2002, 130–133).

For now, however, just realize that estimation is still possible even after leaving the baseline hazard function unspecified, and this offers a considerable advantage when we cannot make reasonable assumptions about the shape of the hazard, for example, $h_0(t) = a$ (constant) or $h_0(t) = apt^{p-1}$ (Weibull). Compared with these parametric approaches, the advantage of the semiparametric Cox model is that we do not need to make assumptions about $h_0(t)$. If wrong, such assumptions could produce misleading results about $\boldsymbol{\beta}_x$. The cost is a loss in efficiency; if we knew the functional form of $h_0(t)$, we could do a better job of estimating $\boldsymbol{\beta}_x$.

9.1   Using stcox

Stata's stcox command fits Cox proportional hazards models. After stsetting your dataset or loading a dataset that has already been stset, you type stcox followed by the **x** (independent) variables. This syntax differs from most of Stata's other estimation commands in that you need not specify a response variable. For survival data, the response is the triple (t_0, t, d), which denotes the time span $(t_0, t]$ with failure/censoring indicator d, and Stata remembers these variables from when you stset your data.

Below we also specify the nohr option, for reasons explained later:

```
. use http://www.stata-press.com/data/cggm/hip2
(hip fracture study)

. stcox protect, nohr

         failure _d:  fracture
   analysis time _t:  time1
                id:  id

Iteration 0:    log likelihood = -98.571254
Iteration 1:    log likelihood = -86.655669
Iteration 2:    log likelihood = -86.370792
Iteration 3:    log likelihood =  -86.36904
Refining estimates:
Iteration 0:    log likelihood =  -86.36904

Cox regression -- Breslow method for ties

No. of subjects =          48               Number of obs   =        106
No. of failures =          31
Time at risk    =         714
                                            LR chi2(1)      =      24.40
Log likelihood  =   -86.36904               Prob > chi2     =     0.0000
```

_t _d	Coef.	Std. Err.	z	P>|z|	[95% Conf. Interval]
protect	-2.047599	.4404029	-4.65	0.000	-2.910773 -1.184426

These results report that for our hip-fracture data,

$$h(t|\text{protect}) = h_0(t) \exp(-2.047599 * \text{protect})$$

In our hip-fracture data, the variable protect equals 1 or 0; it equals 1 if the subject wears an inflatable hip-protection device and 0 otherwise. These results say that

$$h(t|\text{protect} == 0) = h_0(t);$$
$$h(t|\text{protect} == 1) = h_0(t) \exp(-2.047599)$$

so the ratio of these hazards (the *hazard ratio*) is $\exp(-2.047599) = 0.12904437$.

Whatever the hazard rate at a particular time for those who do not wear the device, the hazard at the same time for those who do wear the device is 0.129 times that hazard, which is substantially less. More exactly, these results report that if we constrain the hazard rate of those who wear the device to a multiplicative constant of the hazard of those who do not wear the device, then that multiplicative constant is estimated to be 0.129.

In obtaining these results, we made no assumption about the time profile of the hazard although we would guess, if forced, that the hazard probably increases. We did, however, assert that whatever the shape of the hazard for the group `protect==0`, it is that same shape for the group `protect==1` but multiplied by a constant.

9.1.1 The Cox model has no intercept

Stata reported no intercept for the model above, which makes the Cox model different from many of Stata's other estimation commands. The Cox model has no intercept because the intercept is subsumed into the baseline hazard $h_0(t)$, and mathematically speaking, the intercept is unidentifiable from the data. Pretend that we added an intercept to the model,

$$h(t|\mathbf{x}_j) = h_0(t) \exp(\beta_0 + \mathbf{x}_j \boldsymbol{\beta}_x)$$

thus

$$h(t|\mathbf{x}_j) = \{h_0(t) \exp(\beta_0)\} \exp(\mathbf{x}_j \boldsymbol{\beta}_x)$$

We would now call $\{h_0(t) \exp(\beta_0)\}$ our new baseline hazard. The value of β_0 is undefined because any value works as well as any other—it would merely change the definition of $h_0(t)$, which we do not define anyway.

9.1.2 Interpreting coefficients

Exponentiated individual coefficients have the interpretation of the ratio of the hazards for a 1-unit change in the corresponding covariate. For instance, if the coefficient on variable `age_in_years` in some model is 0.18, then a 1-year increase in age increases the hazard by 20% because $\exp(0.18) = 1.20$. If the coefficient on the variable `weight_in_kilos` is -0.2231, then a 1-kg increase in weight decreases the hazard by 20% because $\exp(-0.2231) = 0.8$. If the coefficient on variable `one_if_female` is 0.0488, then females face a hazard 5% greater than males because $\exp(0.0488) = 1.05$, and a 1-unit increase in `one_if_female` moves the subject from being male to female.

Let's see this more clearly: for a subject with covariates $x_1, x_2, \ldots, x_k$, the hazard rate under this model is

$$h(t|x_1, x_2, \ldots, x_k) = h_0(t) \exp(\beta_1 x_1 + \beta_2 x_2 + \cdots + \beta_k x_k)$$

For a subject with the same covariates except that x_2 is incremented by 1, the hazard would be

$$h(t|x_1, x_2 + 1, \ldots, x_k) = h_0(t) \exp\{\beta_1 x_1 + \beta_2(x_2 + 1) + \cdots + \beta_k x_k\}$$

and the ratio of the two hazards is thus $\exp(\beta_2)$.

In our `stcox` example, we fit a model using the covariate `protect`, and the hazard ratio for this variable was estimated to be $\exp(-2.0476) = 0.129$. A subject wearing the device is thus estimated to face a hazard rate that is only 12.9% of the hazard faced by a subject who does not wear the device.

This hazard-ratio interpretation is so useful that, in the previous example, we had to go out of our way and specify the `nohr` (no hazard ratio) option to keep `stcox` from exponentiating the coefficients for us. Here is what we would have seen by default without the `nohr` option:

```
. use http://www.stata-press.com/data/cggm/hip2
(hip fracture study)

. stcox protect

        failure _d:  fracture
   analysis time _t:  time1
              id:  id

Iteration 0:   log likelihood = -98.571254
Iteration 1:   log likelihood = -86.655669
Iteration 2:   log likelihood = -86.370792
Iteration 3:   log likelihood =  -86.36904
Refining estimates:
Iteration 0:   log likelihood =  -86.36904

Cox regression -- Breslow method for ties

No. of subjects =            48                 Number of obs   =        106
No. of failures =            31
Time at risk    =           714
                                                LR chi2(1)      =      24.40
Log likelihood  =    -86.36904                  Prob > chi2     =     0.0000
```

_t						
_d	Haz. Ratio	Std. Err.	z	P>|z|	[95% Conf. Interval]	
protect	.1290443	.0568315	-4.65	0.000	.0544336	.3059218

The only difference in results from typing `stcox protect, nohr` and `stcox protect` is that this time `stcox` reported hazard ratios—exponentiated coefficients—rather than the coefficients themselves. This is a difference only in how results are reported, not in the results themselves.

❑ **Technical note**

The reported standard error also changed. When we specified the `nohr` option, we obtained the estimate $\widehat{\beta}_x = -2.047599$ with the estimated standard error 0.4404029.

When we left off the `nohr` option, we instead obtained $\exp(\widehat{\beta}_x) = 0.1290443$ with the standard error 0.0568315, which is a result of fitting the model in this scale and was obtained by applying the *delta method* to the original standard-error estimate. The delta method obtains the standard error of a transformed variable by calculating the variance of the corresponding first-order Taylor expansion, which for the transform $\exp(\beta_x)$ amounts to multiplying the original standard error by $\exp(\widehat{\beta}_x)$. This trick of calculation yields identical results as does transforming the parameters prior to estimation and then reestimating.

The next two columns in the original output report the Wald test of the null hypothesis H_o: $\beta_x = 0$ versus the alternative H_a: $\beta_x \neq 0$, and the numbers reported in these columns remain unchanged. In the new output, the test corresponds to H_o: $\exp(\beta_x) = 1$ versus H_a: $\exp(\beta_x) \neq 1$. Hazard ratios equal to 1 correspond to coefficients equal to 0 because $\exp(0) = 1$. In the new output, the test corresponds to $\exp(\beta_x) = 1$, but what is reported is the test for $\beta_x = 0$. That is, the z-statistic is calculated using the original coefficient and its standard error, not the transformed coefficient and the transformed standard error. Were it calculated the other way, you would get a different, yet asymptotically equivalent, test. Confronted with this discrepancy, Stata leaves the results in the original metric because, in the original metric, they often have better small-sample properties. Tests based on estimates of linear predictors are often better left untransformed.

The confidence interval for the hazard ratio is based on the transformed (exponentiated) endpoints of the confidence interval for the original coefficient. An alternative, asymptotically equivalent method would have been to base the calculation directly on the hazard ratio and its standard error, but this confidence interval would not be guaranteed to be entirely positive even though we know a hazard ratio is always positive. For this reason, and because confidence intervals of the untransformed coefficients usually have better properties, Stata reports confidence intervals based on the "transform the endpoints" calculation.

❑

`stcox`, like all Stata estimation commands, will redisplay results when invoked without arguments. Typing `stcox` would redisplay the above hazard-ratio output. Typing `stcox, nohr` would redisplay results but would report coefficients rather than hazard ratios.

9.1.3   The effect of units on coefficients

The units in which you measure the covariates $\mathbf{x}$ make no substantive difference, but choosing the right units can ease interpretation.

If the covariate x_2 measures weight, it does not matter whether you measure that weight in kilograms or pounds—the coefficient will just change to reflect the change in units. If you fit the model using kilograms to obtain

$$h(t|\mathbf{x}) = h_0(t) \exp(\beta_1 x_1 + \beta_2 x_2 + \cdots + \beta_k x_k)$$

and now you wish to substitute $x_2^* = 2.2x_2$ (the same weight measured in pounds), then

$$h(t|\mathbf{x}) = h_0(t)\exp\{\beta_1 x_1 + (\beta_2/2.2)x_2^* + \cdots + \beta_k x_k\}$$

When estimating the coefficient on x_2^*, you are in effect estimating $\beta_2/2.2$, and results from `stcox` will reflect this in that the estimated coefficient for x_2^* will be the coefficient estimated for x_2 in the original model divided by 2.2. If the coefficient we estimated using kilograms was $\widehat{\beta}_2 = 0.4055$, say, then the coefficient we estimate using pounds would be $\widehat{\beta}_2/2.2 = 0.1843$, and this is what `stcox` would report. The models are, logistically speaking, the same. Weight would have an effect, and that effect in kilograms would be 0.4055*`kilograms`. That same effect in pounds would be 0.1843*`pounds`.

The effect on the reported hazard ratios is nonlinear. The estimated hazard ratio for a 1-kg increase in weight would be $\exp(0.4055) = 1.5$. The estimated hazard ratio for a 1-pound increase would be $\exp(0.1843) = 1.202$.

Changing the units of covariates to obtain hazard ratios reported in the desired units is a favorite trick among those familiar with proportional hazards models, and rather than remember the exact technique, it is popular to just let the software do it for you. For instance, if you are fitting a model that includes age,

```
. stcox protect age
```

and you want the hazard ratios reported for a 5-year increase in age rather than a 1-year increase, changing the units of age, type

```
. generate age5 = age/5
. stcox protect age5
```

If you do this with the hip-fracture dataset, in the first case you will get a reported hazard ratio of 1.110972, meaning that a 1-year increase in age is associated with an 11% increase in the hazard. In the second case, you will get a reported hazard ratio of 1.692448, meaning that a 5-year increase in age is associated with a 69% increase in the hazard, and note that $\ln(1.110972) = \ln(1.692448)/5$.

Changing units changes coefficients and exponentiated coefficients (hazard ratios) in the expected way. On the other hand, shifting the means of covariates changes nothing that Cox reports, although, as we will demonstrate later, shifting a covariate pays dividends when using `stcox` to estimate the baseline survivor or cumulative hazard function, because doing so effectively changes the definition of what is considered "baseline".

Nevertheless, coefficients and hazard ratios remain unchanged. Using the `hip2.dta` dataset, whether we measure age in years since birth or in years above 60 doesn't matter, and in fact,

```
. stcox protect age
. generate age60 = age - 60
. stcox protect age60
```

will yield identical displayed results from `stcox`. In both cases, the estimated hazard ratio for age would be 1.110972.

In most linear-in-the-parameters models, shifting from where the covariates are measured causes a corresponding change in the overall intercept of the model, but because that overall intercept is wrapped up in the baseline hazard in the Cox model, there is no change in reported results. For some constant shift c,

$$
\begin{aligned}
h(t|\mathbf{x}) &= h_0(t)\exp\{\beta_1 x_1 + \beta_2(x_2 - c) + \cdots + \beta_k x_k)\} \\
&= \{h_0(t)\exp(-\beta_2 c)\}\exp(\beta_1 x_1 + \beta_2 x_2 + \cdots + \beta_k x_k)
\end{aligned}
$$

so all we have really done is redefine the baseline hazard (something we do not need to estimate anyway).

9.1.4   Estimating the baseline cumulative hazard and survivor functions

In the Cox model given in (9.1), $h_0(t)$ is called the baseline hazard function and

$$
\exp(\mathbf{x}\boldsymbol{\beta}_x) = \exp(\beta_1 x_1 + \cdots + \beta_k x_k)
$$

is called the *relative hazard*. Thus $\mathbf{x}\boldsymbol{\beta}_x$ is referred to as the *log relative-hazard*, also known as the risk score.

From (9.1), $h_0(t)$ corresponds to the overall hazard when $\mathbf{x} = \mathbf{0}$ because then the relative hazard is 1.

Although the Cox model produces no direct estimate of the baseline hazard, estimates of functions related to $h_0(t)$ can be obtained after the fact, conditional on the estimates of $\boldsymbol{\beta}_x$ from the Cox model. One may obtain estimates of the baseline survivor function $S_0(t)$ corresponding to $h_0(t)$, the baseline cumulative hazard function $H_0(t)$, and the baseline *hazard contributions*, which may then be smoothed to estimate $h_0(t)$ itself.

We noted previously, when we fit the model of the relative hazard of hip fracture, that we suspected that the hazard was increasing over time although nothing in the Cox model would constrain the function to have that particular shape. We can verify our suspicion by asking `stcox` to return to us, along with the estimated Cox results, the estimate of the baseline cumulative hazard based upon them:

```
. use http://www.stata-press.com/data/cggm/hip2
(hip fracture study)
. stcox protect, basechazard(H0)
  (output omitted)
. line H0 _t, c(J) sort
```

(Continued on next page)

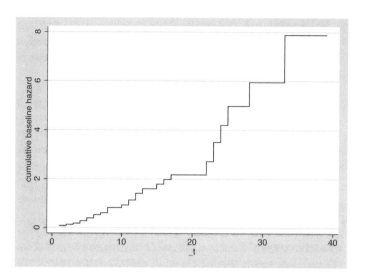

Figure 9.1. Estimated baseline cumulative hazard

We see that the cumulative hazard does appear to be increasing and at an increasing rate, meaning that the hazard itself is increasing (recall that the hazard is the derivative of the cumulative hazard).

Figure 9.1 is the cumulative hazard for a subject with all covariates equal to 0, which here means `protect==0`, the control group. In general, the (nonbaseline) cumulative hazard function in a Cox model is given by

$$
\begin{aligned}
H(t|\mathbf{x}) &= \int_0^t h(u|\mathbf{x})du \\
&= \exp(\mathbf{x}\boldsymbol{\beta}_x)\int_0^t h_0(u)du \\
&= \exp(\mathbf{x}\boldsymbol{\beta}_x)H_0(t)
\end{aligned}
$$

Thus the cumulative hazard for those who do wear the hip-protection device is $H(t) = 0.129H_0(t)$, and we can draw both cumulative hazards on one graph,

```
. gen H1 = H0 * 0.1290
. label variable H0 H0
. line H1 H0 _t, c(J J) sort
```

which produces figure 9.2.

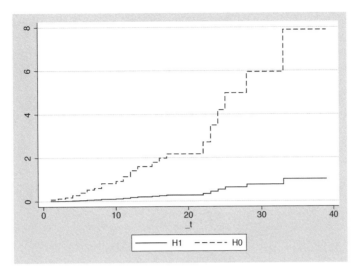

Figure 9.2. Estimated cumulative hazard: treatment versus controls

We can also retrieve the estimated survivor function when we fit the model by using the basesurv option to stcox,

```
. stcox protect, basesurv(S0)
  (output omitted )
. line S0 _t, c(J) sort
```

which produces figure 9.3.

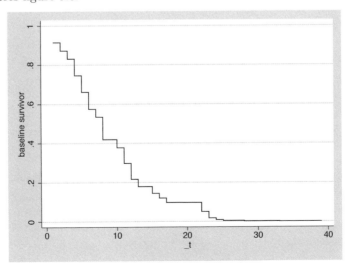

Figure 9.3. Estimated baseline survivor function

As with the cumulative hazard, the baseline survivor function $S_0(t)$ is the survivor function evaluated with all the covariates equal to zero. The formula for obtaining the value of the survivor function at other values of the covariates can be derived from first principles:

$$
\begin{aligned}
S(t|\mathbf{x}) &= \exp\{-H(t|\mathbf{x})\} \\
&= \exp\{-\exp(\mathbf{x}\boldsymbol{\beta}_x)H_0(t)\} \\
&= S_0(t)^{\exp(\mathbf{x}\boldsymbol{\beta}_x)}
\end{aligned}
$$

We can draw both survivor curves on one graph by typing

```
. gen S1 = S0 ^ 0.1290
. label variable S0 S0
. line S1 S0 _t, connect(J J) sort
```

which produces figure 9.4.

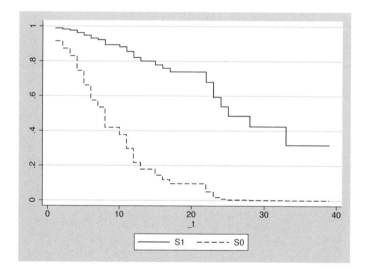

Figure 9.4. Estimated survivor: treatment versus controls

In drawing these graphs, we have been careful to ensure that the points were connected with horizontal lines by specifying `line`'s `connect()` option. We connected the points to emphasize that the estimated functions are really step functions, no different from the Nelson–Aalen and Kaplan–Meier estimators of these functions in models without covariates. These functions are estimates of empirically observed functions, and failures occur in the data only at specific times.

❏ **Technical note**

If you fit a Cox regression model with no covariates and retrieve an estimate of the baseline survivor function, you will get the Kaplan–Meier estimate. For example, with the hip-fracture data, typing

```
. sts gen S2 = s
. stcox, estimate basesurv(S1)
```

will produce variables S1 and S2 that are identical up to calculation precision. (To fit a Cox model with no covariates, we needed to specify the estimate option so that Stata knew we were not merely redisplaying results from the previous stcox fit.)

By the same token, if you fit a Cox regression model with no covariates and retrieve an estimate of the baseline cumulative hazard, you will get the Nelson–Aalen estimator.

For the details, we refer you to Kalbfleisch and Prentice (2002, 114–118). We will mention that the estimation of the baseline functions involves the estimation of quantities called *hazard contributions* at each failure time and that each hazard contribution is the increase in the estimated cumulative hazard at each failure time. Nominally, these calculations take into account the estimated regression parameters, so one can think of the estimated baseline survivor function from a Cox model as a covariate-adjusted Kaplan–Meier estimate—use the estimated β_x to put everyone on the same level by adjusting for the covariates, and then proceed with the Kaplan–Meier calculation.

In models with no covariates, these hazard contributions reduce to coincide with the calculations involved in the Kaplan–Meier and Nelson–Aalen curves.

❏

9.1.5 Estimating the baseline hazard function

We demonstrated how to use stcox to retrieve an estimate of the baseline survivor or baseline cumulative hazard function, $S_0(t)$ or $H_0(t)$, but estimates of $h_0(t)$ cannot be obtained directly from stcox. Because $h_0(t)$ is the derivative of $H_0(t)$, why not just take the derivative of the estimated $H_0(t)$ and use that as an estimate of $h_0(t)$? Or, because $h_0(t)$ is a function of the derivative of $S_0(t)$, why not follow a similar approach using the estimate of $S_0(t)$? The formal answer is that the derivative of these estimated functions is everywhere 0, except at the failure times, where it is undefined (these are step functions).

If you want an estimate of the baseline hazard itself, you will have to somehow smooth out the discontinuities in the rates of change associated with these step functions. One way to smooth the discontinuities is to use standard kernel-smoothing methodology, similar to what we did in section 8.4. Formally, if we define the baseline hazard contribution for each observed failure time, t_j, as $\widehat{h}_{t_j}$ (see the technical note below), we can estimate $h_0(t)$ using

$$\widehat{h}_0(t) = b^{-1} \sum_{j=1}^{D} K_t \left(\frac{t - t_j}{b} \right) \widehat{h}_{t_j}$$

for some kernel function $K_t()$ (see the discussion of boundary kernels in section 8.4) and bandwidth b; the summation is over the D times at which failure occurs.

We can estimate this in Stata by first specifying option `basehc(`*newvar*`)` to `stcox` to obtain the baseline hazard contributions and then by using `stcurve` to perform the smoothing.

```
. use http://www.stata-press.com/data/cggm/hip2, clear
(hip fracture study)
. stcox protect, basehc(h0)
  (output omitted )
. stcurve, hazard at(protect=0)
```

This produces figure 9.5, which is a graph of the estimated baseline hazard function (that is, the hazard for `protect==0`).

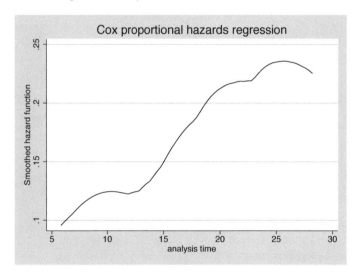

Figure 9.5. Estimated baseline hazard function

Using the same baseline hazard contributions, we can use `stcurve` to plot a comparison of the estimated hazards for treatments and controls, this time customizing the selection of kernel function and bandwidth:

```
. stcurve, hazard at1(protect=0) at2(protect=1) kernel(gaussian) width(4)
```

This produces figure 9.6. Comparing this graph with figure 8.9, we see the implications of the proportional-hazards assumption. The hazards depicted in figure 9.6 are indeed proportional, and if graphed on a log scale (we leave it to you as an exercise to try this), they would be parallel, or at least close enough to parallel with respect to the smoothing. Still, the respective plots in both graphs are similar over the ranges they share in common on the x axis.

In figure 8.9, the hazard functions are estimated over the (overall) range of observed failure times for each group, whereas the hazards in figure 9.6 are estimated over the range of observed failure times. This is one further consequence of the proportional-hazards assumption. Under a Cox model, all failure times contribute to the estimate of the baseline hazard, not just those for which `protect==0`, and the baseline hazard may in turn be transformed to the hazard for any covariate pattern using the proportionality assumption. However, when we estimate the hazard separately for each group (figure 8.9), estimates are valid only over the range of observed failure times for that particular group.

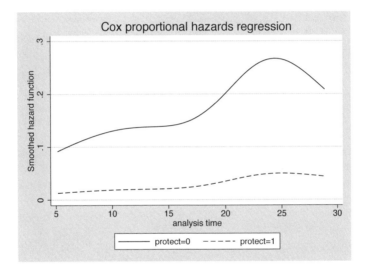

Figure 9.6. Estimated hazard functions: treatment versus control

❏ **Technical note**

The baseline hazard contributions (otherwise known as *discrete hazard components*), $\widehat{h}_{t_j}$, obtained from `stcox` are not the magnitudes of the steps of the estimated baseline cumulative hazard obtained from `stcox`. Instead, a form of the estimators derived from the estimated baseline survivor function is used, as described in Kalbfleisch and Prentice (2002, 115–116). The difference between the estimators mirrors the difference between estimating a survivor function using Kaplan–Meier and taking the Nelson–Aalen cumulative hazard and transforming it—they are asymptotically equivalent estimators of the same thing, and in practice, the difference is usually small; see section 8.3.

❏

stcurve is a wonderfully handy command for graphing estimated survivor, cumulative hazard, and hazard functions after both stcox and streg (stcurve fits parametric survival models; see chapter 12). stcurve is handy after stcox because it automates the process of taking quantities estimated at baseline by using stcox and transforming them to adhere to covariate patterns other than baseline.

stcurve can graph

1. The survivor function. Type stcurve, survival after specifying option basesurv(*newvar*) to stcox.

2. The cumulative hazard function. Type stcurve, cumhaz after specifying option basech(*newvar*) to stcox.

3. The hazard function. Type stcurve, hazard after specifying option basehc(*newvar*) to stcox.

stcurve can graph any of those functions at the values for the covariates you specify. The syntax is as follows:

stcurve, ...at(*varname*=# *varname*=# ...)

If you do not specify a variable's value, the average value is used; thus, if the at() option is omitted altogether, a graph is produced for all the covariates held at their average values. This is why we had to specify at(protect=0) when graphing the baseline hazard function; had we not, we would have obtained a graph for the average value of protect, which would not be meaningful considering protect is binary. The at() option can also be generalized to graph the function evaluated at different values of the covariates on the same graph. The syntax is

stcurve, ...at1(*varname*=# *varname*=# ...) at2(...) at3(...) ...

Earlier in this section, we graphed estimated cumulative hazard and survivor functions, and we did so manually even though we could have used stcurve. We did this not to be mysterious but to emphasize the relationship between these functions at baseline and at covariate patterns other than baseline. We could have done the same with the hazard function (i.e., manually transform the baseline hazard contributions), but then we would have had to do the smoothing ourselves. For hazard functions, we preferred to simply use stcurve.

Figures 9.1–9.4 could have been produced with stcurve without having to generate any additional variables (outside those produced by stcox). For example, figure 9.2 can be replicated by

```
. stcox protect, basechazard(H0)
. stcurve, cumhaz at1(protect=1) at2(protect=0)
```

9.1.6   The effect of units on the baseline functions

The units in which you measure covariates (kilograms or pounds, inches or centimeters) change coefficients and hazard ratios in the obvious way but do not change the baseline cumulative hazard function, survivor function, and hazard contributions.

The origin from which you measure covariates—absolute zero or the freezing point of water, absolute weight or deviation from the normal (normal being the same for everybody)—does not change coefficients $\boldsymbol{\beta}_x$ and hazard ratios (exponentiated coefficients). However, it does change the estimated baseline cumulative hazard and baseline survivor because you are changing how you define "all covariates equal to zero".

Consider fitting the model

```
. stcox protect age
```

versus

```
. gen age60 = age - 60
. stcox protect age60
```

In the first case, $h_0(t)$ corresponds to a newborn who does not wear the hip-protection device (admittedly, not an interesting hazard), and the second case refers to a 60-year-old who does not wear the hip-protection device. Not surprisingly, the baseline cumulative hazard (and the baseline survivor) functions differ.

Yet, it seems innocuous enough to type

```
. stcox protect age, basesurv(S)
. line S _t, c(J) sort
```

and wonder what is wrong with Stata because the plotted baseline survivor function varies only between 0.99375 and 0.9999546, which appears incorrect since 30% of the subjects in the data were observed to fail. (Why? You just plotted the survivor function for newborns.)

For estimating the baseline survivor function, the problem can get worse than just misunderstanding a correctly calculated result—numerical accuracy issues can arise. For example, try the following experiment:

```
. use http://www.stata-press.com/data/cggm/hip2, clear
. gen age_big = age + 300
. stcox protect age_big, basesurv(S0)
```

All we have done is change the definition of what an age of "zero" means. In this new scaling, when you are born, your value of age_big==300, and you continue to age after that, so a person who is age 60 has age_big==360. The estimates of the coefficients and their associated hazard ratios will not be affected by this (nor should they).

Look, however, at the resulting estimate of the baseline survivor function:

. list S0

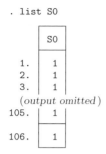

What happened? The baseline survivor function is being predicted for a subject with age_big==0, meaning age is −300. The probability of surviving is virtually 1; it is not exactly 1—it is really $1 - \epsilon$, where ϵ is a small number. Upon further investigation, you would discover that the numbers the computer listed are not exactly 1, either; they merely round to 1 when displayed in a nine-digit (%9.0g) format. In fact, you would discover that there are actually eight distinct values of S0 in the listing, corresponding to the 3 bits of precision with which the computer was left when it struggled to present these numbers so subtly different from 1 as accurately as it could.

This estimate is a poor estimate of the baseline survivor function, and it is not Stata's fault. If you push Stata the other way,

```
. gen age_small = age - 300
. stcox protect age_small, basesurv(S02)
```

you will again obtain fine estimates of β_x, but this time the baseline survivor function, corresponding to a person who is 300 years old, will be estimated to be 0 everywhere:

. list S02

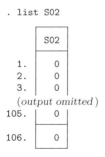

This time the numbers really are 0, even though they should not be, and even though the computer (in other circumstances) could store smaller numbers. Given the calculation formula for the baseline survivor function, this result could not be avoided.

If you intend to estimate the baseline survivor function, be sure that $\mathbf{x} = \mathbf{0}$ in your data corresponds to something reasonable. You need to be concerned about this only if you intend to estimate the baseline survivor function; the calculation of the baseline cumulative hazard (which is not bounded between 0 and 1) and the calculation of the baseline hazard contributions (upon which hazard function estimation is based) are more numerically stable.

9.2 Likelihood calculations

Cox regression results are based on forming, at each failure time, the *risk pool* or *risk set*, the collection of subjects who are at risk of failure, and then maximizing the conditional probability of failure. The times at which failures occur are not relevant in a Cox model—the ordering of the failures is. As such, when subjects are tied (fail at the same time) and the exact ordering of failure is unclear, the situation requires special treatment. We first consider, however, the case of no ties.

9.2.1 No tied failures

Consider the straightforward data

```
. list
```

	subject	t	x
1.	1	2	4
2.	2	3	1
3.	3	6	3
4.	4	12	2

```
. stset t
```
  (output omitted)

There are four failure times in these data—times 2, 3, 6, and 12—but the values of the times do not matter; only the order of the subjects matters. There are four distinct times from which we form four distinct risk pools:

1. *Time 2:*
   Risk group (those available to fail): {1,2,3,4}
   Subject #1 is observed to fail

2. *Time 3:*
   Risk group: {2,3,4}
   Subject #2 is observed to fail

3. *Time 6:*
   Risk group: {3,4}
   Subject #3 is observed to fail

4. *Time 12:*
   Risk group: {4}
   Subject #4 is observed to fail

At each of the failure times, we take as given that one of the subjects must fail, and we calculate the conditional probability of failure for the subject who actually is observed to fail. Thus we have the likelihood function

$$L(\beta) = P_1 P_2 P_3 P_4$$

where each P_i, $i = 1, \ldots, 4$ represents a conditional probability for each failure time.

The last conditional probability is the easiest to calculate. At $t = 12$, given that one failure occurs, what is the probability that it will be subject 4? The answer is $P_4 = 1$, because by that point only subject 4 is available to fail.

The calculation of P_3 (corresponding to $t = 6$) is the next easiest, and following the derivation of (3.2), we find that P_3 is the ratio

$$
\begin{aligned}
P_3 &= \frac{h(6|x_3)}{h(6|x_3) + h(6|x_4)} \\
&= \frac{\exp(x_3\beta)}{\exp(x_3\beta) + \exp(x_4\beta)}
\end{aligned}
$$

and this does not depend on the failure $t = 6$. This is fundamental to Cox regression: the ordering of the failure times is what matters, not the actual times themselves. Similar arguments lead to the expressions for P_2 and P_1:

$$
\begin{aligned}
P_2 &= \frac{\exp(x_2\beta)}{\exp(x_2\beta) + \exp(x_3\beta) + \exp(x_4\beta)} \\
P_1 &= \frac{\exp(x_1\beta)}{\exp(x_1\beta) + \exp(x_2\beta) + \exp(x_3\beta) + \exp(x_4\beta)}
\end{aligned}
$$

Thus $L(\beta) = P_1 P_2 P_3 P_4$ can be expressed as

$$L(\beta) = \prod_{j=1}^{4} \left\{ \frac{\exp(x_j\beta)}{\sum_{i \in R_j} \exp(x_i\beta)} \right\}$$

where R_j is defined to the risk set (those subjects at risk of failure) at time t_j.

Generalizing this argument to the case where we have k distinct observed failure times and multiple x variables gives the Cox likelihood function

$$L(\boldsymbol{\beta}_x) = \prod_{j=1}^{k} \left\{ \frac{\exp(\mathbf{x}_j\boldsymbol{\beta}_x)}{\sum_{i \in R_j} \exp(\mathbf{x}_i\boldsymbol{\beta}_x)} \right\} \tag{9.2}$$

Because we confine ourselves to only the individual binary analyses that occur at each failure time and make no assumption about the baseline hazard at times when failures do not occur, the Cox likelihood given in (9.2) is not strictly a likelihood—it is a *partial likelihood*. However, for all intents and purposes, (9.2) can be treated as a likelihood; that is, maximizing (9.2) results in an estimate of $\boldsymbol{\beta}_x$ that is asymptotically normal with mean $\boldsymbol{\beta}_x$ and a variance–covariance matrix equal to the inverse of the negative Hessian [matrix of second derivatives of (9.2) with respect to $\boldsymbol{\beta}_x$]. For a more technical discussion, see, for example, Kalbfleisch and Prentice (2002, 101–104, 130–133).

As with other likelihoods, estimates of $\boldsymbol{\beta}_x$ are obtained by maximizing the natural logarithm of $L(\boldsymbol{\beta}_x)$ and not $L(\boldsymbol{\beta}_x)$ itself. In figure 9.7, we plot the log likelihood and label the value of β (a scalar in our small example) that maximizes it.

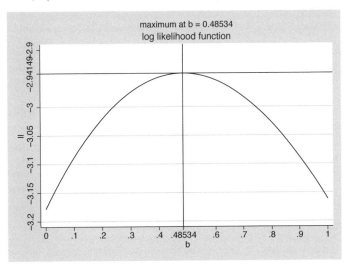

Figure 9.7. Log likelihood for the Cox model

The maximum (partial) likelihood estimate of β is $\widehat{\beta} = 0.48534$, and this number is exactly what `stcox` reports for these data:

```
. stcox x, nohr

        failure _d:  1 (meaning all fail)
   analysis time _t:  t

Iteration 0:   log likelihood = -3.1780538
Iteration 1:   log likelihood = -2.9420159
Iteration 2:   log likelihood = -2.9414857
Iteration 3:   log likelihood = -2.9414857
Refining estimates:
Iteration 0:   log likelihood = -2.9414857

Cox regression -- no ties

No. of subjects =            4                 Number of obs    =          4
No. of failures =            4
Time at risk    =           23
                                               LR chi2(1)       =       0.47
Log likelihood  =    -2.9414857                Prob > chi2      =     0.4915
```

_t _d	Coef.	Std. Err.	z	P>|z|	[95% Conf. Interval]
x	.4853405	.7326297	0.66	0.508	-.9505874 1.921268

9.2.2   Tied failures

Earlier we calculated the probability that a particular subject failed, given that one failure was to occur. To do that, we used the formula

$$\Pr(j \text{ fails}|\text{risk set } R_j) = \frac{\exp(\mathbf{x}_j\boldsymbol{\beta}_x)}{\sum_{i \in R_j} \exp(\mathbf{x}_i\boldsymbol{\beta}_x)}$$

Introducing the notation $r_j = \exp(\mathbf{x}_j\boldsymbol{\beta}_x)$ allows us to express the above formula more compactly as $\Pr(j \text{ fails}|\text{risk set } R_j) = r_j / \sum_{i \in R_j} r_i$.

In our data from the previous section, the situation never arose where two subjects failed at the same time, but we can easily imagine such a dataset:

```
. list

        subject     t    x

   1.         1     2    4
   2.         2     3    1
   3.         3     3    3
   4.         4    12    2
```

For these data, there are three risk pools:

1. *Time 2:*
   Risk group (those available to fail): {1,2,3,4}
   Subject #1 is observed to fail

2. *Time 3:*
   Risk group: {2,3,4}
   Subjects #2 and #3 are observed to fail

3. *Time 12:*
   Risk group: {4}
   Subject #4 is observed to fail

The question then becomes, "How do we calculate the probability that both subjects 2 and 3 fail, given that two subjects from the risk pool are known to fail at $t = 3$?" This is the conditional probability that we need for the second product term in (9.2).

The marginal calculation

One way we could make the calculation is to say to ourselves that subjects 2 and 3 did not really fail at the same time; we were just limited as to how precisely we could measure the failure time. In reality, one subject failed and then the other, and we just do not know the order.

Define P_{23} to be the probability that subject 2 fails and then subject 3 fails, given the current risk pool (which would no longer have subject 2 in it). Conversely, let

P_{32} be the probability that subject 3 fails and then subject 2 fails, given the risk pool reduced by the failure of subject 3. That is, if we knew that 2 failed before 3, then P_{23} is the contribution to the likelihood; if we knew that 3 failed before 2, then P_{32} is the contribution.

Following our previous logic, we find that

$$
\begin{aligned}
P_{23} &= \frac{r_2}{r_2 + r_3 + r_4} \frac{r_3}{r_3 + r_4} \\
P_{32} &= \frac{r_3}{r_2 + r_3 + r_4} \frac{r_2}{r_2 + r_4}
\end{aligned}
\tag{9.3}
$$

and in fact, if we had the exact ordering then the substitution of either P_{23} or P_{32} in (9.2) would represent the two middle failure times (now that we have separated them).

However, because we do not know the order, we can instead take the probability that we observe subjects 2 and 3 to fail in any order as $P_{23} + P_{32}$, and use this term instead. Using our data, if $\beta = 0.75$, for example, then $P_{23} + P_{32} = 0.2786$, and this is what we would use to represent $t = 3$ in the likelihood calculations.

This method of calculating the conditional probability of tied failure events is called the marginal calculation, the exact-marginal calculation, or the continuous-time calculation. The last name arises because assuming continuous times makes it mathematically impossible that failures occur at precisely the same instant. For more technical details, see Kalbfleisch and Prentice (2002, 104–105, 130–133).

Using the `exactm` option to `stcox` specifies that ties are to be treated in this manner when calculating the likelihood.

❑ **Technical note**

Actually, the name `exactm` is a bit of a misnomer. The exact marginal method as implemented in most computer software (including Stata) is only an approximation of the method we have just described. Consider the (not too unusual) case where you have 10 tied failure times. The calculation of the exact marginal would then require $10! = 3{,}628{,}800$ terms and is computationally infeasible. Instead, the sum of the probabilities of the specific orderings is approximated using Gauss–Laguerre quadrature. However, the approximation in this case is a computational issue and not one that simplifies any assumption about how we want to calculate the likelihood. When the approximation works well, we fully expect to retrieve (with negligible error) the true sum of the probability terms.

❑

The partial calculation

Another way we could proceed is to assume that the failures really did occur at the same time and treat this as a multinomial problem (Cox 1972; Kalbfleisch and Prentice 2002, 106–107). Given that two failures are to occur at the same time among subjects 2, 3, and 4, the possibilities are

- 2 and 3 fail
- 2 and 4 fail
- 3 and 4 fail

The conditional probability that 2 and 3 are observed from this set of possibilities is

$$p_{23} = \frac{r_2 r_3}{r_2 r_3 + r_2 r_4 + r_3 r_4}$$

Using $\beta = 0.75$, we would obtain $p_{23} = 0.3711$ for our data.

This is known as the partial calculation, the exact-partial calculation, the discrete-time calculation, or the conditional logistic calculation. The last name arises because this is also the calculation that conditional logistic regression uses to calculate probabilities when conditioning on more than one event taking place.

Which probability is correct: the marginal probability (0.2786) or the partial probability (0.3711)? The answer is a matter of personal taste in that you must decide how you want to think about tied failures: do they arise from imprecise measurements (marginal method), or do they arise from a discrete-time model (partial method)? In practice, the difference between the calculations is usually not that severe, and admittedly, we chose r_2, r_3, and r_4 to emphasize the difference.

Using the `exactp` option to `stcox` specifies that ties are to be treated in this manner when calculating the likelihood. The exact partial method implemented in Stata is exact; however, it can prove problematic because (1) sometimes it can take too long to calculate and (2) numerical problems in the calculation can arise, so it can produce bad results when risk pools are large and there are many ties.

The Breslow approximation

Both the exact marginal and partial calculations are so computationally intensive that it has become popular to use approximations. When no option as to how to treat tied failures is specified, Stata assumes the `breslow` option and uses the Breslow (1974) approximation. This is an approximation of the exact marginal. In this approximation, the risk pools for the second and subsequent failure events within a set of tied failures are not adjusted for previous failures. So, for example, rather than calculating P_{23} and P_{32} from (9.3), the Breslow method uses

$$P_{23} = \frac{r_2}{r_2 + r_3 + r_4} \times \frac{r_3}{r_2 + r_3 + r_4} = \frac{r_2 r_3}{(r_2 + r_3 + r_4)^2}$$

$$P_{32} = \frac{r_3}{r_2 + r_3 + r_4} \times \frac{r_2}{r_2 + r_3 + r_4} = \frac{r_2 r_3}{(r_2 + r_3 + r_4)^2}$$

and thus the contribution to the likelihood is obtained as $P_{23} + P_{32} = 2r_2 r_3/(r_2 + r_3 + r_4)^2$. Because the denominator is common to the failure events, this represents a significant reduction in the required calculations.

This approximation works well when the number of failures in the risk group is small relative to the size of the risk group itself.

The Efron approximation

Efron's method of handling ties (Efron 1977) is also an approximation to the exact marginal, except that it adjusts the subsequent risk sets using probability weights. For the two failures that occur at $t = 3$, following the first failure, the second risk set is either $\{3, 4\}$ or $\{2, 4\}$. Rather than using $r_3 + r_4$ and $r_2 + r_4$ as the denominators for the second risk set, the approximation uses the average of the two sets, $(r_3 + r_4 + r_2 + r_4)/2 = (r_2 + r_3)/2 + r_4$. Thus for our example,

$$
P_{23} = \frac{r_2}{r_2 + r_3 + r_4} \times \frac{r_3}{\frac{1}{2}(r_2 + r_3) + r_4}
$$

$$
P_{32} = \frac{r_3}{r_2 + r_3 + r_4} \times \frac{r_2}{\frac{1}{2}(r_2 + r_3) + r_4}
$$

and so

$$
P_{23} + P_{32} = \frac{2r_2 r_3}{(r_2 + r_3 + r_4)\{\frac{1}{2}(r_2 + r_3) + r_4\}}
$$

This approximation is more accurate than Breslow's approximation but takes longer to calculate. Using the `efron` option to `stcox` specifies that ties are to be treated in this manner when calculating the likelihood.

9.2.3 Summary

The hip-fracture dataset is typical for the number of tied failure times. The dataset contains 21 failure times and 31 failures, meaning an average of $31/21 = 1.48$ failures per failure time. In particular, at 12 of the times there is one failure, at 8 times there are two failures, and at 1 time there are three failures.

Table 9.1 gives the results of fitting the model `stcox protect` by using the various methods for handling tied failures:

Table 9.1. Methods for handling ties

Method	$\exp(\widehat{\beta}_x)$	95% CI		Command used
Exact marginal	0.1172	0.0481	0.2857	`stcox protect, exactm`
Exact partial	0.1145	0.0460	0.2846	`stcox protect, exactp`
Efron	0.1204	0.0503	0.2878	`stcox protect, efron`
Breslow	0.1290	0.0544	0.3059	`stcox protect`

If you do not specify otherwise, `stcox` uses the Breslow approximation. Specifying one of `efron`, `exactp`, or `exactm` will obtain results using those methods. In any case, the method used to handle ties is clearly displayed at the top of the output from `stcox`. Used with datasets with no ties, the output will state that there are no ties, in which case the specified method of handling ties becomes irrelevant.

To determine the number of ties in your data, you can do the following (*after* saving your data because this procedure will drop some observations):

```
. keep if _d                           /* Keep only the failures */
(75 observations deleted)

. sort _t
. by _t: gen number = _N               /* Count the instances of _t */
. by _t: keep if _n==1                  /* Keep one obs. representing _t */
. * What is the average number of failures per failure time?
. summarize number
```

Variable	Obs	Mean	Std. Dev.	Min	Max
number	21	1.47619	.6015852	1	3

```
. * What is the frequency of number of failures?
. tabulate number
```

number	Freq.	Percent	Cum.
1	12	57.14	57.14
2	8	38.10	95.24
3	1	4.76	100.00
Total	21	100.00	

9.3   Stratified analysis

In stratified Cox estimation, the assumption that everyone faces the same baseline hazard, multiplied by their relative hazard,

$$h(t|\mathbf{x}_j) = h_0(t)\exp(\mathbf{x}_j\boldsymbol{\beta}_x)$$

is relaxed in favor of

$$h(t|\mathbf{x}_j) = h_{01}(t)\exp(\mathbf{x}_j\boldsymbol{\beta}_x), \quad \text{if } j \text{ is in group 1,}$$
$$h(t|\mathbf{x}_j) = h_{02}(t)\exp(\mathbf{x}_j\boldsymbol{\beta}_x), \quad \text{if } j \text{ is in group 2,}$$

and so on. The baseline hazards are allowed to differ by group, but the coefficients, $\boldsymbol{\beta}_x$, are constrained to be the same.

9.3.1   Obtaining coefficient estimates

Pretend that in our hip-fracture study we had a second dataset, this one for males, which we combined with the dataset for females to produce `hip3.dta`. We could imagine using that data to estimate one effect:

```
. use http://www.stata-press.com/data/cggm/hip3, clear
. stcox protect
```

By fitting this model, however, we are asserting that males and females face the same hazard of hip fracture over time. Let's assume that, while we are willing to assume that the hip-protection device is equally effective for both males and females, we are worried that males and females might have different risks of fracturing their hips and that this difference might bias our measurement of the effectiveness of the device.

One solution would be to include the variable `male` in the model (`male==0` for females and `male==1` for males) and so fit

```
. stcox protect male
```

If we did this, we would be asserting that the hazard functions for females and for males have the same shape and that one function is merely proportional to the other. Our concern is that the hazards might be shaped differently. Moreover, in this study, we do not care how the sexes differ; we just want to ensure that we measure the effect of `protect` correctly.

In this spirit, another solution would be to fit models for males and females separately:

```
. stcox protect if male
. stcox protect if !male
```

If we did this, we would be allowing the hazards to differ (as desired), but we would also obtain separate measures of the effectiveness of our hip-protection device. We want a single, more efficient estimate, or perhaps we want to test whether the device is equally protective for both sexes (regardless of the shape of the hazard for each), and this we cannot do with separate models.

(*Continued on next page*)

The solution to our problem is stratified estimation:

```
. use http://www.stata-press.com/data/cggm/hip3
(hip fracture study)

. stcox protect, strata(male)

        failure _d:  fracture
   analysis time _t:  time1
               id:  id

Iteration 0:   log likelihood = -124.27723
Iteration 1:   log likelihood = -110.21911
Iteration 2:   log likelihood = -109.47452
Iteration 3:   log likelihood = -109.46654
Iteration 4:   log likelihood = -109.46654
Refining estimates:
Iteration 0:   log likelihood = -109.46654

Stratified Cox regr. -- Breslow method for ties

No. of subjects =          148          Number of obs   =        206
No. of failures =           37
Time at risk    =         1703
                                        LR chi2(1)      =      29.62
Log likelihood  =   -109.46654          Prob > chi2     =     0.0000
```

_t _d	Haz. Ratio	Std. Err.	z	P>|z|	[95% Conf. Interval]	
protect	.1301043	.0504375	-5.26	0.000	.0608564	.2781483

```
                                            Stratified by male
```

Using this approach, we discover that wearing the hip-protection device reduces the hazard of hip fracture to just 13% of the hazard faced when the device is not worn.

In table 9.2, we present the results of the analyses we would obtain from each approach.

Table 9.2. Various models for hip-fracture data

Command	$\exp(\widehat{\beta}_x)$	95% CI	
(1) stcox protect	0.0925	0.0442	0.1935
(2) stcox protect male	0.1315	0.0610	0.2836
(3) stcox protect if !male	0.1290	0.0544	0.3059
(4) stcox protect if male	0.1339	0.0268	0.6692
(5) stcox protect, strata(male)	0.1301	0.0609	0.2781

We obtained reasonably similar results when we fit the model separately for females and males (lines 3 and 4) but got different results when we estimated constraining the baseline hazards to be equal (line 1). Our conclusion is that the hazards functions are not the same for males and females. Regarding how the hazard functions differ by gender, it makes virtually no difference whether we constrain the hazards to be multiplicative replicas of each other (line 2) or allow them to vary freely (line 5), but we would not have known that unless we had fit the stratified model.

9.3.2 Obtaining estimates of baseline functions

We obtain estimates of the baseline functions after stratified estimation in the same way as we do after unstratified estimation—we include the `basechazard()` or `basesurv()` options when we fit the model:

```
. stcox protect, strata(male) basech(H)
(output omitted )
```

The result of adding `basech(H)` is to create the new variable `H` containing the estimated baseline cumulative hazard functions. This variable contains the baseline cumulative hazard function for females in those observations for which `male==0` and for males in those observations for which `male==1`. We can thus graph the two functions using

```
. gen H0 = H if male==0
(100 missing values generated)
. gen H1 = H if male==1
(106 missing values generated)
. line H0 H1 _t, c(J J) sort
```

which produces figure 9.8.

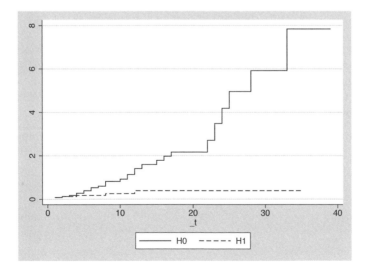

Figure 9.8. Estimated baseline cumulative hazard for males versus females

Remember, these are the baseline cumulative hazard functions and not the baseline hazard functions themselves, and smoothed hazard function estimation is not available for stratified Cox models. Thus if we think in derivatives, we can compare the baseline hazards. Concerning females (H0), there is a fairly constant slope until _t==18, and then, perhaps, a higher constant slope after _t==20, suggesting an increasing hazard rate. Concerning males (H1), there is a gentle constant slope. In any case, one could argue equally well that the hazard functions are proportional or not.

9.4   Cox models with shared frailty

The term *shared frailty* is used in survival analysis to describe regression models with random effects. A *frailty* is a latent random effect that enters multiplicatively on the hazard function. For a Cox model, the data are organized as $i = 1, \ldots, n$ groups with $j = 1, \ldots, n_i$ subjects in group i. For the jth subject in the ith group, then, the hazard is

$$h_{ij}(t) = h_0(t)\alpha_i \exp(\mathbf{x}_{ij}\boldsymbol{\beta})$$

where α_i is the group-level frailty. The frailties are unobservable positive quantities and are assumed to have mean 1 and variance θ, to be estimated from data.

For $\nu_i = \log \alpha_i$, the hazard can also be expressed as

$$h_{ij}(t) = h_0(t) \exp(\mathbf{x}_{ij}\boldsymbol{\beta} + \nu_i)$$

and thus the log frailties, ν_i, are analogous to random effects in standard linear models.

A more detailed discussion of frailty models is given in section 15.3 in the context of parametric models. Frailty is best first understood in parametric models, because these models posit parametric forms of the baseline hazard function for which the effect of frailty can be easily described and displayed graphically. Shared frailty models are also best understood if compared and contrasted with unshared frailty models, which for reasons of identifiability do not exist with Cox regression. Therefore, for a more thorough discussion of frailty, we refer you to section 15.3.

For this discussion, a Cox model with shared frailty is simply a random-effects Cox model. Shared frailty models are used to model within-group correlation; observations within a group are correlated because they share the same frailty, and the extent of the correlation is measured by θ. For example, we could have survival data on individuals within families, and we would expect (or at least be willing to allow) those subjects within each family to be correlated because some families would inherently be more frail than others. When $\theta = 0$, the Cox shared-frailty model simply reduces to the standard Cox model.

9.4.1   Parameter estimation

Consider the data from a study of kidney dialysis patients, as described in McGilchrist and Aisbett (1991) and as described in more detail in section 15.3.2 on parametric frailty models. The study is concerned with the prevalence of infection at the catheter insertion point. Two recurrence times (in days) are measured for each patient, and each recorded time is the time from initial insertion (onset of risk) to infection or censoring.

```
. use http://www.stata-press.com/data/cggm/kidney2, clear
(Kidney data, McGilchrist and Aisbett, Biometrics, 1991)
. list patient time fail age gender in 1/10, sepby(patient)
```

	patient	time	fail	age	gender
1.	1	16	1	28	0
2.	1	8	1	28	0
3.	2	13	0	48	1
4.	2	23	1	48	1
5.	3	22	1	32	0
6.	3	28	1	32	0
7.	4	318	1	31.5	1
8.	4	447	1	31.5	1
9.	5	30	1	10	0
10.	5	12	1	10	0

Each patient (`patient`) has two recurrence times (`time`) recorded, with each catheter insertion resulting in either infection (`fail==1`) or right censoring (`fail==0`). Among the covariates measured are `age` and `gender` (1 if female, 0 if male).

Note the use of the generic term "subjects": Here the subjects are taken to be the individual catheter insertions and not the patients themselves. This is a function of how the data were recorded—the onset of risk occurs at catheter insertion (of which there are two for each patient) and not, say, at the time the patient was admitted into the study. Thus we have two subjects (insertions) for each group (patient). Because each observation represents one subject, we are not required to `stset` an ID variable, although we would need to if we ever wished to `stsplit` the data later.

It is reasonable to assume independence of patients but unreasonable to assume that recurrence times within each patient are independent. One solution would be to fit a standard Cox model, adjusting the standard errors of the estimated parameters to account for the possible correlation. This is done by specifying option `vce(cluster patient)` to `stcox`. We do this below after first `stset`ting the data.

```
. stset time, failure(fail)

    failure event:  fail != 0 & fail < .
obs. time interval:  (0, time]
 exit on or before:  failure
```

```
      76  total obs.
       0  exclusions
```

```
      76  obs. remaining, representing
      58  failures in single record/single failure data
    7424  total analysis time at risk, at risk from t =          0
                            earliest observed entry t =          0
                               last observed exit t =        562
```

```
. stcox age gender, nohr vce(cluster patient)

        failure _d:  fail
   analysis time _t:  time
Iteration 0:   log pseudolikelihood = -188.44736
Iteration 1:   log pseudolikelihood = -185.36881
Iteration 2:   log pseudolikelihood = -185.11022
Iteration 3:   log pseudolikelihood = -185.10993
Refining estimates:
Iteration 0:   log pseudolikelihood = -185.10993

Cox regression -- Breslow method for ties

No. of subjects     =          76          Number of obs   =          76
No. of failures     =          58
Time at risk        =        7424
                                            Wald chi2(2)    =        2.74
Log pseudolikelihood =   -185.10993         Prob > chi2     =      0.2540

                     (Std. Err. adjusted for 38 clusters in patient)
```

_t	Coef.	Robust Std. Err.	z	P>|z|	[95% Conf. Interval]	
age	.0022426	.0078139	0.29	0.774	-.0130724	.0175575
gender	-.7986869	.487274	-1.64	0.101	-1.753726	.1563526

When you specify vce(cluster patient), you obtain the robust estimate of variance as described in the context of Cox regression by Lin and Wei (1989), with an added adjustment for clustering (see [U] **20.15 Obtaining robust variance estimates**).

If there indeed exists within-patient correlation, the standard Cox model depicted above is misspecified. However, because we specified vce(cluster patient), the standard errors of the estimated coefficients on age and gender are valid representations of the sample-to-sample variability of the obtained coefficients. We do not know exactly what the coefficients measure (for that we would need to know exactly how the correlation arises), but we can measure their variability, and often we may still be able to test the null hypothesis that a coefficient is zero. That is, in many instances, testing that a covariate effect is zero under our misspecified model is equivalent to testing that the effect is zero under several other models that allow for correlated observations.

One such model is the shared frailty model, and more specifically the model where the shared frailty is gamma distributed (with mean 1 and variance θ). We fit this model in Stata by specifying option shared(patient) in place of vce(cluster patient).

```
. stcox age gender, nohr shared(patient)

        failure _d:  fail
   analysis time _t:  time

Fitting comparison Cox model:

Estimating frailty variance:
Iteration 0:    log profile likelihood = -182.06713
Iteration 1:    log profile likelihood =  -181.9791
Iteration 2:    log profile likelihood = -181.97453
Iteration 3:    log profile likelihood = -181.97453

Fitting final Cox model:

Iteration 0:    log likelihood = -199.05599
Iteration 1:    log likelihood = -183.72296
Iteration 2:    log likelihood = -181.99509
Iteration 3:    log likelihood = -181.97455
Iteration 4:    log likelihood = -181.97453
Refining estimates:
Iteration 0:    log likelihood = -181.97453

Cox regression --
        Breslow method for ties          Number of obs      =        76
        Gamma shared frailty             Number of groups   =        38
Group variable: patient

No. of subjects =          76            Obs per group: min =         2
No. of failures =          58                          avg =         2
Time at risk    =        7424                          max =         2

                                         Wald chi2(2)       =     11.66
Log likelihood  =    -181.97453          Prob > chi2        =    0.0029
```

_t	Coef.	Std. Err.	z	P>|z|	[95% Conf. Interval]	
age	.0061825	.012022	0.51	0.607	-.0173801	.0297451
gender	-1.575675	.4626528	-3.41	0.001	-2.482458	-.6688924
theta	.4754497	.2673107				

```
Likelihood-ratio test of theta=0: chibar2(01) =     6.27 Prob>=chibar2 = 0.006
Note: standard errors of regression parameters are conditional on theta.
```

Given the estimated frailty variance, $\widehat{\theta} = 0.475$, and the significance level of the likelihood-ratio test of H_o: $\theta = 0$, we conclude that under this model there is significant within-group correlation. To interpret the coefficients, let us begin by redisplaying them as hazard ratios.

(Continued on next page)

```
. stcox

Cox regression --
           Breslow method for ties          Number of obs    =        76
           Gamma shared frailty             Number of groups =        38
Group variable: patient

No. of subjects =            76              Obs per group: min =        2
No. of failures =            58                            avg =        2
Time at risk    =          7424                            max =        2

                                            Wald chi2(2)     =     11.66
Log likelihood  =    -181.97453             Prob > chi2      =    0.0029

          _t │ Haz. Ratio   Std. Err.      z    P>|z|     [95% Conf. Interval]
─────────────┼────────────────────────────────────────────────────────────────
         age │  1.006202    .0120965     0.51   0.607     .9827701    1.030192
      gender │  .2068678    .095708     -3.41   0.001     .0835376    .5122756
─────────────┼────────────────────────────────────────────────────────────────
       theta │  .4754497    .2673108

Likelihood-ratio test of theta=0: chibar2(01) =      6.27 Prob>=chibar2 = 0.006
Note: standard errors of hazard ratios are conditional on theta.
```

The interpretation of the hazard ratios is the same as before, except that they are conditional on the frailty. For example, we interpret the hazard ratio for gender as indicating that, once we account for intragroup correlation via the shared frailty model, for a given level of frailty the hazard for females is about one-fifth that for males. Of course, a subject's frailty would have a lot to say about the hazard, and when fitting a shared frailty model, we can use stcox to also estimate the frailties (or more precisely, the log frailties ν_i) for us:

```
. stcox age gender, nohr shared(patient) effects(nu)
  (output omitted )

. sort nu

. list patient nu in 1/2

        ┌──────────────────────┐
        │ patient           nu │
        ├──────────────────────┤
    1.  │      21   -2.4487067 │
    2.  │      21   -2.4487067 │
        └──────────────────────┘

. list patient nu in 75/L

        ┌──────────────────────┐
        │ patient           nu │
        ├──────────────────────┤
   75.  │       7    .51871587 │
   76.  │       7    .51871587 │
        └──────────────────────┘
```

By specifying option effects(nu), we tell stcox to create a new variable nu containing the estimated random effects (the $\hat{\nu}_i$). After sorting, we find that the least frail (or strongest) patient is patient 21 with $\hat{\nu}_{21} = -2.45$; the most frail patient is patient 7 with $\hat{\nu}_7 = 0.52$.

❏ Technical note

Estimation for the Cox shared frailty model consists of two layers. In the outer layer, the optimization is for θ only. For fixed θ, the inner layer consists of fitting a standard Cox model via penalized likelihood, with the ν_i introduced as estimable coefficients of dummy variables identifying the groups. The penalized likelihood is simply the standard Cox likelihood with an added penalty that is a function of θ. The final estimate of θ is taken to be the one that maximizes the penalized log likelihood. Once the optimal θ is obtained, it is held fixed and a final penalized Cox model is fit. For this reason, the standard errors of the main regression parameters (or hazard ratios, if displayed as such) are treated as conditional on θ fixed at its optimal value. That is, when performing inference on these coefficients, it is with the caveat that you are treating θ as known. For more details on estimation, see chapter 9 of Therneau and Grambsch (2000).

❏

9.4.2 Obtaining estimates of baseline functions

Baseline estimates for the Cox shared frailty model are obtained in the usual way, and the definition of baseline extends to include $\nu = 0$. For example, working with our kidney data, we can obtain an estimate of the baseline survivor function via the `basesurv()` option to `stcox`, just as before. Before we do so, however, we first recenter `age` so that baseline corresponds to something more meaningful.

(Continued on next page)

```
. generate age40 = age - 40
. stcox age40 gender, nohr shared(patient) basesurv(S0)
        failure _d:  fail
   analysis time _t:  time
Fitting comparison Cox model:

Estimating frailty variance:

Iteration 0:   log profile likelihood = -182.06713
Iteration 1:   log profile likelihood = -181.9791
Iteration 2:   log profile likelihood = -181.97453
Iteration 3:   log profile likelihood = -181.97453

Fitting final Cox model:

Iteration 0:   log likelihood = -199.05599
Iteration 1:   log likelihood = -183.72296
Iteration 2:   log likelihood = -181.99509
Iteration 3:   log likelihood = -181.97455
Iteration 4:   log likelihood = -181.97453
Refining estimates:
Iteration 0:   log likelihood = -181.97453

Cox regression --
            Breslow method for ties          Number of obs    =         76
            Gamma shared frailty             Number of groups =         38
Group variable: patient

No. of subjects =          76             Obs per group: min =          2
No. of failures =          58                           avg =          2
Time at risk    =        7424                           max =          2

                                          Wald chi2(2)     =      11.66
Log likelihood  =   -181.97453            Prob > chi2      =     0.0029
```

_t	Coef.	Std. Err.	z	P>|z|	[95% Conf. Interval]	
age40	.0061825	.012022	0.51	0.607	-.0173801	.0297451
gender	-1.575675	.4626528	-3.41	0.001	-2.482458	-.6688924
theta	.4754497	.2673108				

```
Likelihood-ratio test of theta=0: chibar2(01) =    6.27 Prob>=chibar2 = 0.006
Note: standard errors of regression parameters are conditional on theta.
```

Recentering **age** has no effect on the parameter estimates, but it does produce a baseline survivor-function estimate (S0) that corresponds to a 40-year-old male patient with mean frailty, or equivalently, a log frailty of zero.

Because the estimation does not change, we know from the previous section that the estimated log frailties still range from -2.45 to 0.52. We can use this information to produce a graph comparing the survivor curves for 40-year-old males at the lowest, mean (baseline), and highest levels of frailty.

```
. gen S_low = S0^exp(-2.45)
. gen S_high = S0^exp(0.52)
. line S_low S0 S_high _t if _t<200, c(J J J) sort ytitle("Survivor function")
```

This produces figure 9.9. For these data, the least frail patients have survival experiences
that are far superior, even when compared with those with mean frailty. We have also
restricted the plot to range from time 0 to time 200, where most of the failures occur.

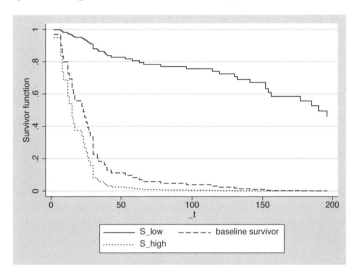

Figure 9.9. Comparison of survivor curves for various frailty values

Currently, stcurve has no facility for setting the frailty. Instead, all graphs produced
by stcurve are for $\nu = 0$. stcurve does, however, have an outfile() option for saving
the coordinates used to produce the graph, and these coordinates can be transformed
so that they correspond to other frailty values. For example, if we wanted smoothed
baseline hazard plots for the same three frailty values used above, we could perform the
following, remembering to first reestimate to obtain the baseline hazard contributions:

```
. stcox age40 gender, nohr shared(patient) basehc(h)
  (output omitted)
. stcurve, hazard kernel(gaussian) outfile(basehaz)
note: all plots evaluated at frailty equal to one
. use basehaz, clear
(Kidney data, McGilchrist and Aisbett, Biometrics, 1991)
. describe

Contains data from basehaz.dta
  obs:            83                          Kidney data, McGilchrist and
                                              Aisbett, Biometrics, 1991
  vars:            2                          6 Feb 2008 11:12
  size:          996 (99.9% of memory free)
----------------------------------------------------------------------------
              storage  display    value
variable name   type   format     label      variable label
----------------------------------------------------------------------------
haz1           float   %9.0g                 Smoothed hazard function
_t             float   %9.0g                 _t
----------------------------------------------------------------------------
Sorted by:
```

`stcurve` saved a dataset containing variable _t (time) and `haz1`, the estimated (after smoothing) hazard function. We have loaded these data into memory, and so now all we have left to do is to generate the other two hazards and then graph all three against _t.

```
. label variable haz1 "mean frailty hazard"
. gen haz_low = haz1*exp(-2.45)
. gen haz_high = haz1*exp(0.52)
. line haz_low haz1 haz_high _t if _t<200, yscale(log)
> ylabel(.005(.01).025) ylabel(.01, add) sort
> ytitle("Smoothed hazard estimate")
```

This produces figure 9.10. The comparison of hazards for the frailty extremes matches that for the survivor function; the least frail individuals are immune to infection.

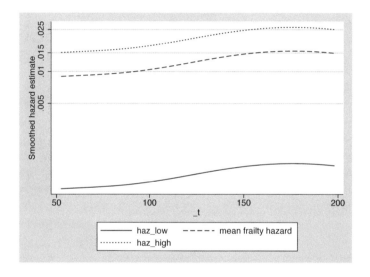

Figure 9.10. Comparison of hazards for various frailty values

9.5 Cox models with survey data

All our previous analyses relied on the assumption that survival data were obtained according to the *simple random sampling* design where each subject has the same chance of being selected into the sample. In many practical applications, the subjects may be selected into the sample based on a complex survey design with unequal sample-inclusion probabilities (*probability weights*), *clustering*, and *stratification*.

Here we will give a brief overview of some of the survey design characteristics and outline common methods used to account for them in the analysis. For more details, refer to the literature on complex survey data (e.g., Cochran 1977, Levy and Lemeshow 1999, Korn and Graubard 1999). For the purpose of this section, we strongly advise you to familiarize yourself with the *Stata Survey Data Reference Manual*.

As we mentioned above, survey data is characterized by probability weights (referred to as `pweights` in Stata), clustering, and stratification. *Probability weights* represent the chance of a subject being selected into the sample and are proportional to the inverse of the probability of being sampled. *Clustering* (or cluster sampling) arises when not the actual individuals but rather certain groups of individuals (clusters) are sampled independently with possibly further subsampling within clusters. *Stratification* arises when the population is first divided into a fixed number of groups (strata) and sampling of individuals (or clusters) is then done separately within each stratum. Strata are statistically independent.

Why take into account survey design? A failure to do so may lead to biased point estimates and underestimated standard errors, resulting in incorrect inferences about the quantities of interest. For a more detailed discussion and examples of the effect of a survey design on estimation see [SVY] **survey**, [SVY] **svy**, and the references given therein. Probability weights affect point estimates, and all three of the above characteristics affect standard errors.

Conventional estimation methods must be adjusted to produce correct estimates for survey data. Probability weights are accommodated by using weighted sample and maximum likelihood estimators. The latter give rise to so-called maximum *pseudolikelihood* estimators because in the presence of weighting the likelihood loses its usual probabilistic interpretation. To deal with clustering and stratification, various variance estimation techniques exist. The most commonly used are *balanced and repeated replication* (BRR), the *jackknife*, and Taylor *linearization*. The linearization method is the default in Stata's survey commands, but the other two methods may also be requested; see [SVY] **variance estimation** for details.

In the following sections, we demonstrate how to account for survey designs when analyzing survival data and comment on some of the issues associated with accounting for survey designs.

9.5.1 Declaring survey characteristics

Similar to the `stset` command to declare survival characteristics of the data necessary for fitting survival models, Stata has the `svyset` command to declare the survey characteristics of the data required for fitting survey models. A common syntax of `svyset` for a single-stage survey design is

`svyset` $\left[\,psu\,\right]$ $\left[\,weight\,\right]$ $\left[\,,\ \underline{\text{strata}}(varname)\ \dots\,\right]$

where *psu* is the name of the variable identifying primary sampling units (clusters at the first stage), *weight* is [pweight = *weight_varname*] and *weight_varname* is the variable containing the probability sampling weights, and strata(*varname*) specifies the variable identifying strata. For other options and a more general syntax for multistage designs see [SVY] **svyset**.

Consider the data on lung cancer, nhefs1.dta, obtained from the First National Health and Nutrition Examination Survey (NHANES I) and its 1992 Epidemiologic Follow-up Study (NHEFS); see Miller (1973), Engel et al. (1978), and Cox et al. (1997). The nhefs1.dta dataset contains variables recording sampling information, survival information, and information on such risk factors as smoking status, gender, and place of residence. The sampling design is a single-stage stratified survey design with unequal inclusion probabilities; probability weights are saved in swgt2, first-stage cluster identifiers are saved in psu2, and stratum identifiers are saved in strata2. Survival information consists of the participants' ages at the time of the interview, variable age_lung_cancer, and the disease indicator, variable lung_cancer. Smoking status is recorded in variables former_smoker and smoker identifying former and current smokers. Gender is recorded in the variable male with the base category male==0 defining females. Place-of-residence information is recorded in the variables urban1 (urban residences with less than 1 million people) and rural, with the base comparison group urban1==0 and rural==0 defining urban residences with fewer than 1 million people. For more details on data construction, see example 3 in section *Health surveys* in [SVY] **svy estimation**.

Let's now use svyset to declare the survey characteristics of our lung-cancer data:

```
. use http://www.stata-press.com/data/cggm/nhefs1, clear
. svyset psu2 [pweight=swgt2], strata(strata2)
      pweight: swgt2
          VCE: linearized
  Single unit: missing
     Strata 1: strata2
         SU 1: psu2
        FPC 1: <zero>
```

After svysetting the data, we can obtain the estimates adjusted for the sampling design by specifying the svy: prefix in front of the standard Stata estimation command; see [SVY] **svy** for details.

9.5.2   Fitting a Cox model with survey data

Let's examine the incidence of lung cancer as a function of smoking status, gender, and place of residence. The onset of risk of lung cancer is a participant's date of birth, so the age of the participant is our analysis time. The failure outcome is having lung cancer at the time of the interview.

```
. stset age_lung_cancer, failure(lung_cancer)

     failure event:  lung_cancer != 0 & lung_cancer < .
obs. time interval:  (0, age_lung_cancer]
 exit on or before:  failure
```

```
    14407  total obs.
     5126  event time missing (age_lung_cancer>=.)                PROBABLE ERROR

     9281  obs. remaining, representing
       83  failures in single record/single failure data
   599691  total analysis time at risk, at risk from t =          0
                         earliest observed entry t =              0
                               last observed exit t =             97
```

From the `stset` output, there are 5,126 observations with missing failure times. This is confirmed by the 1992 NHEFS documentation, so we can ignore the reported "probable error" message.

To study the effect of smoking, gender, and place of residence on the incidence of lung cancer, we fit a Cox regression with the risk-factor variables described in the previous subsection as predictors. To take into account the survey design defined by the above `svyset` command, we specify the prefix `svy:` with `stcox`.

```
. svy: stcox former_smoker smoker male urban1 rural
(running stcox on estimation sample)

Survey: Cox regression

Number of strata   =        35          Number of obs     =        9149
Number of PSUs     =       105          Population size   = 151327827
                                        Design df         =          70
                                        F(   5,     66)   =       14.07
                                        Prob > F          =      0.0000
```

_t	Haz. Ratio	Linearized Std. Err.	t	P>|t|	[95% Conf. Interval]	
former_smo~r	2.788113	.6205102	4.61	0.000	1.788705	4.345923
smoker	7.849483	2.593249	6.24	0.000	4.061457	15.17051
male	1.187611	.3445315	0.59	0.555	.6658757	2.118142
urban1	.8035074	.3285144	-0.54	0.594	.3555123	1.816039
rural	1.581674	.5281859	1.37	0.174	.8125799	3.078702

The first part of the output after `svy: stcox` contains generic information about the specified survey design. This output is consistent across all survey estimation commands; see [SVY] **svy estimation**. In our example, there are 35 strata, 105 clusters, and 9,149 observations representing the population of size 151,327,827, according to the specified probability sampling scheme.

The coefficient table reports results adjusted for the survey design; coefficient estimates are adjusted for the probability sampling, and standard errors are adjusted for the sampling weights, stratification, and clustering. From the output, we can see that smoking has a significant impact on the risk of having lung cancer. Interestingly, had we not accounted for the data's survey characteristics, probability weights in particular, we would have obtained different inferences about the `urban1` effect; we would have concluded that subjects living in moderately populated urban residences have lower risk of having lung cancer than those living in residences with more than one million people. We invite you to verify this by refitting the Cox model without the `svy:` prefix:

```
. stcox former_smoker smoker male urban1 rural
```

❏ **Technical note**

Notice that such survey design characteristics as probability weights and first-stage clustering can be accounted for without using the `svy` prefix. Probability weights may be specified with `stset`, and `stcox`'s option `vce(cluster varname)` may be used to adjust for clustering on *varname*. You can verify this by typing

```
. stset age_lung_cancer [pweight=swgt2], failure(lung_cancer)
. stcox former_smoker smoker male urban1 rural, vce(cluster psu2)
```

and

```
. svyset psu2 [pweight=swgt2]
. svy: stcox former_smoker smoker male urban1 rural
```

to produce the same parameter and standard-error estimates; see *Remarks* in [SVY] **svy estimation** for details.

The `svy:` prefix is indispensable, however, if you want to account for stratification (do not confuse this with `stcox`'s option `strata()`) and more complex multistage survey designs or if you want to use features specific to survey data—for example, subpopulation estimation, balanced-and-repeated-replication variance estimation, design-effects estimation, poststratification and more.

❏

9.5.3   Some caveats of analyzing survival data from complex survey designs

Here we outline some common issues you may run into when `svyset`ting your survival data.

1. We described how to fit Cox models to survey data using `svy: stcox` in this section. You can also fit parametric survival models to survey data using `svy: streg`, and extension of Stata's `streg` command for fitting these models to nonsurvey data; see chapters 12 and 13. Although not explicitly discussed, what you learn here about `svy: stcox` can also be applied to `svy: streg`.

2. Probability weights, `pweights`, may be specified in either `stset` or `svyset`. When specified in both, however, they must be in agreement; see the technical note in [SVY] **svy estimation** for details.

3. For multiple-record survival data when option `id()` is specified with `stset`, the ID variable must be nested within the final-stage sampling units.

   Consider a simple case of a stratified survey design with stratum identifiers defined by the `strataid` variable. Suppose that we have multiple records per subject identified by the `id` variable and variables `time` and `fail` recording analysis time and failure, respectively. Using the following syntax of `svyset` will produce an error when fitting the Cox (or any other survival) model.

   ```
   . svyset, strata(strataid)
   . stset time, id(id) failure(fail)
   . svy: stcox ...
   the stset ID variable is not nested within the final-stage sampling units
   r(459);
   ```

   In the above example the final-stage sampling unit is _n, i.e., the records, because

   ```
   . svyset, strata(strataid)
   ```

   is equivalent to

   ```
   . svyset _n, strata(strataid)
   ```

   The error message tells us that the `id` variable is not nested within _n; individuals are certainly not nested within records. The correct way of `svyset`ting these data is to use the `id` variable as a primary sampling unit instead of the default _n:

   ```
   . svyset id, strata(strataid)
   ```

   The above error message will also appear if, for a multistage cluster design, the ID variable is not nested within the lowest-level cluster.

4. Survey-adjusted estimates are not available with frailty models or methods other than Breslow for handling ties; see [ST] **stcox** for other restrictions on options when the `svy:` prefix is used.

5. Do not confuse `svyset`'s option `strata()` with `stcox`'s option `strata()`. The former adjusts standard errors of the estimates for the stratification within the sampling design. The latter allows separate baseline hazard estimates for each stratum.

10 Model building using stcox

stcox is extremely versatile, and you can fit many different models by controlling the covariate list that you supply to stcox. Recall the form of the Cox model,

$$h(t|\mathbf{x}) = h_0(t)\exp(\mathbf{x}\boldsymbol{\beta}_x)$$

and note that all the "modeling" that takes place in the Cox model is inherent to the linear predictor, $\mathbf{x}\boldsymbol{\beta}_x = \beta_1 x_1 + \beta_2 x_2 + \cdots + \beta_k x_k$, for k covariates.

10.1 Indicator variables

One type of variable used in modeling is the indicator variable, that is, a variable equal to 0 or 1, where a value of 1 indicates that a certain condition has been met. For example, in our hip-fracture data, we parameterize the effect of wearing the hip-protection device using the indicator variable protect. We also measure the effect of gender by using the indicator variable male.

For example, let's fit a standard Cox model with protect and male as covariates:

```
. use http://www.stata-press.com/data/cggm/hip3
(hip fracture study)
. stcox protect male

         failure _d:  fracture
   analysis time _t:  time1
                id:  id

Iteration 0:   log likelihood = -150.85015
   (output omitted )
Iteration 4:   log likelihood = -124.39469
Refining estimates:
Iteration 0:   log likelihood = -124.39469

Cox regression -- Breslow method for ties

No. of subjects =          148                   Number of obs   =        206
No. of failures =           37
Time at risk    =         1703
                                                 LR chi2(2)      =      52.91
Log likelihood  =    -124.39469                  Prob > chi2     =     0.0000
```

_t _d	Haz. Ratio	Std. Err.	z	P>|z|	[95% Conf. Interval]	
protect	.1315411	.051549	-5.18	0.000	.0610222	.2835534
male	.2523232	.1178974	-2.95	0.003	.10098	.6304912

We might wonder what would happen if we coded gender in the reverse way, that is, 1 for females and 0 for males:

```
. gen female = !male

. stcox protect female

        failure _d:  fracture
   analysis time _t:  time1
             id:  id
Iteration 0:   log likelihood = -150.85015
Iteration 1:   log likelihood = -129.05654
Iteration 2:   log likelihood = -125.53679
Iteration 3:   log likelihood =  -124.3968
Iteration 4:   log likelihood = -124.39469
Refining estimates:
Iteration 0:   log likelihood = -124.39469

Cox regression -- Breslow method for ties

No. of subjects =          148                Number of obs   =        206
No. of failures =           37
Time at risk    =         1703
                                             LR chi2(2)      =      52.91
Log likelihood  =   -124.39469               Prob > chi2     =     0.0000
```

_t	Haz. Ratio	Std. Err.	z	P>|z|	[95% Conf. Interval]	
protect	.1315411	.051549	-5.18	0.000	.0610222	.2835534
female	3.963172	1.851783	2.95	0.003	1.586065	9.902954

The estimated hazard ratio for `protect` remains unchanged (which is comforting). The hazard ratio for `male` in the first model was 0.2523, meaning that all else equal (here the same value of `protect`), males were estimated to face 0.2523 of the hazard of females. The estimated hazard ratio for `female` in the second model is 3.963, meaning that we estimate females to face 3.963 of the hazard of males. Since $1/0.2523 = 3.96$, the conclusion does not change.

If we instead look at the regression coefficients, we first obtained a coefficient for `male` of $\ln(0.2523) = -1.38$. For `female`, we obtained $\ln(3.963) = 1.38$, meaning that the magnitude of the shift in the linear predictor is the same regardless of how we code the variable.

10.2 Categorical variables

A categorical variable is a variable in which different values represent membership in different groups. Categorical variables usually take on values 1, 2, ..., but there is nothing special about those values, and a categorical variable could just as well take on 0, 1, ..., or 5, 9.2, 11.3, ... An indicator variable is just a special case of a categorical variable that has two groups, labeled 0 and 1.

The important difference between indicator and categorical variables is that you do not include categorical variables directly in your model, or at least you should not if the categorical variable contains more than two unique values. Consider a categorical variable `race` where 1 = white, 2 = black, and 3 = Hispanic. If you were to fit some model,

```
. stcox ... race ...
```

you would be constraining the effect of being Hispanic to be twice the effect of being black. There would be no justifying that because the coding 1 = white, 2 = black, 3 = Hispanic was arbitrary, and were you to use a different coding, you would be constraining some other relationship. A categorical variable that takes on m distinct values thus must be converted into $m - 1$ indicator variables:

```
. gen black = race==2
. gen hispanic = race==3
. stcox ... black hispanic ...
```

Another way of creating the variables yourself is to use the `xi:` prefix:

```
. xi: stcox ... i.race ...
```

Type `help xi` for more information on the use of `xi`.

In the example above, which uses the categorical variable `race`, we had three distinct groups (races) and included indicator variables for two of them (black and Hispanic). Which two groups we choose makes no substantive difference. Let's pretend that in the above estimation we obtained the following results:

Variable	$\exp(\widehat{\beta})$
black	1.11
hispanic	1.22

The interpretation would be that Hispanics face a 22% greater hazard than whites and that blacks face an 11% greater hazard than whites. Had we instead fit a model on `white` and `hispanic`, we would have obtained

Variable	$\exp(\widehat{\beta})$
white	0.90
hispanic	1.10

meaning that whites face 90% of the hazard that blacks (the omitted variable) face and that Hispanics face 110% of that same hazard. These results are consistent. For instance, how do Hispanics compare with whites? They face $1.10/0.90 = 1.22$ of the hazard faced by whites, and this corresponds exactly to what we estimated when we included the variables `black` and `hispanic` in the first analysis.

Which group you choose to omit makes no difference, but it does affect how results are reported. Comparisons will be made with the omitted group. Regardless of how you estimate, you can always obtain the other comparisons by following the technique we demonstrated above.

In the example we have shown, the categorical variable has no inherent order; it was arbitrary that we coded 1 = white, 2 = black, and 3 = Hispanic. Some categorical variables do have a natural order. For example, you may have a variable that is coded 1 = mild, 2 = average, and 3 = severe. Such variables are called *ordinal variables*. Even so, remember that the particular numbers assigned to the ordering are arbitrary (not in their ordinality but in their relative magnitude), and it would be inappropriate to merely include the variable in the model. Is "severe" two times "average" or four times "average"? The coefficients you obtain on the corresponding indicator variables will answer that question.

10.3 Continuous variables

Continuous variables refer to variables that are measured on a well-defined, cardinal scale, such as age in years. What distinguishes continuous variables from indicator and categorical variables is that continuous variables are measured on an interval scale. The distance between any two values of the variable is meaningful, which is not the case for indicator and categorical variables. We might have an indicator variable that assigns values 0 and 1 to each of the sexes, but the difference $1 - 0 = 1$ has no particular meaning. Or we might have a categorical variable coded 1 = mild, 2 = average, and 3 = severe, but "severe"–"mild"= $3 - 1 = 2$ has no particular meaning. On the other hand, the difference between two age values is meaningful.

Continuous variables can be directly included in your model:

```
. stcox ... age ...
```

But including continuous variables does not necessarily mean that you have correctly accounted for the effect of the variable. Including `age` by itself means that you are constraining each additional year of age to affect the hazard by a multiplicative constant that is independent of the level of age. Imagine failure processes where the effect of age increases (or decreases) with age. In such cases, it is popular to estimate the effect as quadratic in age:

```
. gen age2 = age^2
. stcox ... age age2 ...
```

This usually works well, but caution is required. Quadratic terms have a minimum or a maximum. If the extremum is in the range of the data, then you have just estimated that the effect of the variable, if it increases initially, increases and then decreases or decreases and then increases. That estimation might be reasonable for some problems, or it might simply be an artifact of using the quadratic to approximate what you believe to be a nonlinear but strictly increasing or decreasing effect.

Whether the change in direction is reasonable can be checked, and the procedure is relatively straightforward once you remember the quadratic formula. To demonstrate, we perform the following analysis and display regression coefficients instead of hazard ratios (option `nohr`):

```
. use http://www.stata-press.com/data/cggm/hip2, clear
(hip fracture study)
. gen age2 = age^2
. stcox protect age age2, nohr
        failure _d:  fracture
  analysis time _t:  time1
               id:  id
Iteration 0:   log likelihood = -98.571254
Iteration 1:   log likelihood = -82.020427
Iteration 2:   log likelihood =  -81.84397
Iteration 3:   log likelihood = -81.843652
Refining estimates:
Iteration 0:   log likelihood = -81.843652

Cox regression -- Breslow method for ties

No. of subjects =           48              Number of obs    =        106
No. of failures =           31
Time at risk    =          714
                                            LR chi2(3)       =      33.46
Log likelihood  =   -81.843652              Prob > chi2      =     0.0000
```

_t _d	Coef.	Std. Err.	z	P>|z|	[95% Conf. Interval]	
protect	-2.336337	.4640283	-5.03	0.000	-3.245816	-1.426858
age	-.9093282	.8847323	-1.03	0.304	-2.643372	.8247153
age2	.0070519	.0061416	1.15	0.251	-.0049854	.0190891

Had we forgotten to specify `nohr`, we could just replay the results in this metric (without refitting the model) by typing `stcox, nohr`.

Define b to be the coefficient for `age`, a to be the coefficient for `age2`, and $x = $ `age`. The quadratic function $ax^2 + bx$ has its maximum or minimum at $x = -b/2a$, at which point the function changes direction. From the above model we estimate $\hat{a} = 0.0070519$ and $\hat{b} = -0.9093282$, and thus we estimate $-b/2a$ to be 64.5. Up until age 64.5, the relative hazard falls, and after that, it increases at an increasing rate. What is the range of `age` in our data?

```
. summarize age
```

Variable	Obs	Mean	Std. Dev.	Min	Max
age	106	70.46226	5.467087	62	82

We estimated a turning point at 64.5 within the range of `age`. Because it is not realistic to assume that the effect of age is ever going to reduce the chances of a broken hip, our estimation results are problematic.

Here we would just drop age2 from the model, noting that the Wald test given in the output shows the effect of age2 to be insignificant given the other covariates and that age and age2 are jointly significant:

```
. test age age2
 ( 1)   age = 0
 ( 2)   age2 = 0

        chi2(  2) =    10.53
      Prob > chi2 =    0.0052
```

However, for this demonstration, let us put aside that easy solution and pretend that we really did have significant coefficients with an artifact that we do not believe.

There are many solutions to this problem, and here we explore two of them. In our first solution, it is only between ages 62 and 64.5 where we have a problem. It might be reasonable to say that there is no effect of age up until age 65, at which point age has a quadratic effect. We could try

```
. gen age65 = cond(age>65, age-65, 0)
. gen age65_2 = age65^2
. stcox protect age65 age65_2
```

and we invite you to try that solution—it produces reasonable results. We created the variable age65, which is defined as 0 for ages below 65 and as age minus 65 for ages above 65.

Our second solution would be to constrain age to have a linear effect up until age 65 and then a quadratic effect after that:

```
. gen age2gt65 = cond(age>65, age-65, 0)^2
. stcox protect age age2gt65
```

This produces even more reasonable results.

10.3.1 Fractional polynomials

In the above section, we considered simple linear and quadratic effects of a continuous variable (age) on the log hazard. In other applications, the effect of a continuous predictor may exhibit a more complex relationship with the log hazard. One approach to modeling this relationship is to use *fractional polynomials* (FP), developed by Royston and Altman (1994b).

For a continuous predictor $x > 0$, a fractional polynomial of degree m is defined as

$$\text{FP}_m\{x; (p_1, \ldots, p_m)\} = \beta_0 + \beta_1 x^{(p_1)} + \beta_2 x^{(p_2)} + \cdots + \beta_m x^{(p_m)} \tag{10.1}$$

where $p_1 \leq p_2 \leq \cdots \leq p_m$ denote integer powers, fractional powers, or both, and

$$x^{(p_j)} = \begin{cases} \ln x & \text{if } p_j = 0 \\ x^{p_j} & \text{if } p_j \neq 0 \end{cases}$$

for unique powers p_j, where $j = 1, \ldots, m$. For repeated powers, when $p_j = p_{j-1}$, the jth term $x^{(p_j)}$ in (10.1) is replaced with $x^{(p_{j-1})} \ln x$ for any $j = 2, \ldots, m$. For example, an FP of fourth degree with powers $(-1, 0, 0.5, 0.5)$, denoted as $\text{FP}_4\{x; (-1, 0, 0.5, 0.5)\}$, is $\beta_0 + \beta_1 x^{-1} + \beta_2 \ln x + \beta_3 x^{0.5} + \beta_4 x^{0.5} \ln x$.

In a Cox model, the linear predictor $\mathbf{x}\boldsymbol{\beta}_x$ is the logarithm of the relative hazard $\exp(\mathbf{x}\boldsymbol{\beta}_x)$ because the baseline hazard function is multiplicatively shifted based on this value. As such, in what follows, we use the notation LRH (log relative-hazard) to denote the linear predictor.

For simplicity, consider a single continuous predictor x. The simplest functional form of the log relative-hazard in x is linear: LRH $= x\beta$. This corresponds to an FP of first degree $(m = 1)$ with a power of one $(p_1 = 1)$, defined in (10.1). For a Cox model, we omit the intercept β_0 from the definition of an FP because it is absorbed into the estimate of the baseline hazard; see section 9.1.1. Our earlier example of LRH $= \beta_1\texttt{age} + \beta_2\texttt{age2}$ with covariates $x = \texttt{age}$ and $x^2 = \texttt{age2}$ is an FP of second degree with powers $(1,2)$. One can extend this further and consider a more general form of LRH expressed as an FP from (10.1):

$$\text{LRH} = \text{FP}\{m; (p_1, \ldots, p_m)\} = \sum_{j=1}^{m} \beta_j x^{(p_j)}$$

The objective of the method of fractional polynomials is to obtain the best values of the degree m and of powers $p_1, \ldots, p_m$ by comparing the deviances of different models. For a given m, the best-fitting powers $p_1, \ldots, p_m$ corresponding to the model with the smallest deviance are obtained from the restricted set $\{-2, -1, -0.5, 0, 0.5, 1, 2, 3, \ldots, \max(3, m)\}$. For technical details, see Royston and Altman (1994b).

In Stata, you can use the `fracpoly` command to fit fractional polynomials in a covariate of interest adjusted for other covariates in the model. For our hip-fracture dataset, let's investigate the nonlinearity of the effect of age:

(Continued on next page)

```
. fracpoly stcox age protect, nohr compare
........
-> gen double Iage__1 = X^3-349.8402563 if e(sample)
-> gen double Iage__2 = X^3*ln(X)-683.0603763 if e(sample)
   (where: X = age/10)

          failure _d:  fracture
    analysis time _t:  time1
                  id:  id

Iteration 0:   log likelihood = -98.571254
Iteration 1:   log likelihood = -82.024371
Iteration 2:   log likelihood = -81.826922
Iteration 3:   log likelihood = -81.826562
Refining estimates:
Iteration 0:   log likelihood = -81.826562

Cox regression -- Breslow method for ties

No. of subjects =          48                Number of obs   =        106
No. of failures =          31
Time at risk    =         714
                                             LR chi2(3)      =      33.49
Log likelihood  =   -81.826562               Prob > chi2     =     0.0000
```

_t	Coef.	Std. Err.	z	P>|z|	[95% Conf. Interval]
Iage__1	-.1119713	.1288941	-0.87	0.385	-.3645991 .1406566
Iage__2	.0513411	.0556334	0.92	0.356	-.0576984 .1603806
protect	-2.33403	.4634473	-5.04	0.000	-3.24237 -1.42569

```
Deviance:   163.65. Best powers of age among 44 models fit: 3 3.
Fractional polynomial model comparisons:
```

age	df	Deviance	Dev. dif.	P (*)	Powers
Not in model	0	172.738	9.085	0.059	
Linear	1	164.941	1.287	0.732	1
m = 1	2	164.472	0.819	0.664	3
m = 2	4	163.653	—	—	3 3

```
(*) P-value from deviance difference comparing reported model with m = 2 model
```

The syntax of `fracpoly` requires specifying the estimation command to be used for fitting, the dependent variable (when required), the covariate of interest to be transformed to an FP (the first listed covariate), and other covariates to be used in the model. In our example, the estimation command is `stcox`, the dependent variables are omitted in the syntax of survival estimation commands, the covariate of interest is `age`, and the other covariate included in the Cox model is `protect`.

By default, `fracpoly` finds the best-fitting powers for a second-degree fractional polynomial. In our example these powers are (3,3). The new variables `Iage__1` and `Iage__2` contain the best-fitting powers of `age`, additionally centered and scaled by `fracpoly`; see [R] **fracpoly** for details. The transformations of `age` used in the model are reported in the first part of the output. The second part of the output consists of the iteration log and the results from fitting the final model. The final part of the output is the table comparing the best-fitting FP models for each degree $k < m$ (including models without the covariate of interest x and linear in x) with the best-fitting FP model of degree m. This comparison is what we requested by specifying fracpoly's option compare. From this table we see that, although the $\text{FP}_2\{\texttt{age};(3,3)\}$ model has the smallest deviance of 163.65, the improvement in model fit compared with the best-fitting FP model of first degree, $\text{FP}_1\{\texttt{age};(3)\}$, with deviance 164.47, is not significant at the 5% level. Refitting the model as an FP of first degree by specifying option degree(1),

```
. fracpoly stcox age protect, nohr compare degree(1)
-> gen double Iage__1 = X^3-349.8402563 if e(sample)
   (where: X = age/10)
   (output omitted)
Fractional polynomial model comparisons:
```

age	df	Deviance	Dev. dif.	P (*)	Powers
Not in model	0	172.738	8.266	0.016	
Linear	1	164.941	0.468	0.494	1
m = 1	2	164.472	—	—	3

```
(*) P-value from deviance difference comparing reported model with m = 1 model
```

and evaluating the results from the FP model comparisons table, we conclude that the cubic-in-age model, which is the best-fitting FP model of degree 1, does not provide a better fit compared with a simple linear-in-age model at the 5% level; the corresponding p-value of the asymptotic deviance-difference test is 0.49.

The method of fractional polynomials can also be applied to nonpositive covariates after the appropriate transformation; see [R] **fracpoly** for details.

For more details and examples of using fractional polynomials, see Becketti (1995), Royston and Altman (1994a; 1994b), Hosmer, Lemeshow, and May (2008, 136f), Ambler and Royston (2001), and Royston and Sauerbrei (2007a). Fractional polynomials have also been used to model time-varying effects of covariates, which will be discussed later in this chapter (for examples, see Sauerbrei, Royston, and Look 2007; Berger, Schäfer, and Ulm 2003; and Lehr and Schemper 2007).

Royston and Altman (1994b) also describe the application of the method of fractional polynomials to more than one continuous covariate in a model. You can read more about *multivariable fractional polynomial* models, for example, in Sauerbrei and Royston (1999; 2002), Royston and Sauerbrei (2005), and Royston, Reitz, and Atzpodien (2006). In Stata, these models can be fit by using the mfp command; see [R] **mfp** for details on syntax and examples.

An alternative to using fractional polynomials to model the functional form of the log relative-hazard is to use regression splines; see, for example, Royston and Sauerbrei (2007b), Therneau and Grambsch (2000), and Andersen and Keiding (2006).

10.4   Interactions

The job of the analyst using Cox regression is to choose a parameterization of $\mathbf{x}\boldsymbol{\beta}_x$ that fairly represents the process at work. It is not sufficient to say that the outcome is a function of variables A, B, and C, and then parameterize $\mathbf{x}\boldsymbol{\beta}_x = \beta_A A + \beta_B B + \beta_C C$.

For instance, the effect of A may be constant (as we have it), or it may increase or diminish with the level of A. If the effect increases or diminishes, we must allow for that in our parameterization, perhaps by approximating the effect by including A^2 in the model. Similarly, the effect of A may increase or diminish with the level of another factor, say, B; then it is common to approximate that effect by inclusion of the terms such as AB, i.e., the *interaction* of A and B.

Returning to our example, if we wanted to entertain the possibility that the effect of wearing a hip-protection device might increase or decrease with age, we could include the term `protect*age` in the model:

```
. gen pXage = protect*age
. stcox protect age pXage
```

The sign of the coefficient for the variable `pXage` will tell whether the effect of `protect` increases or decreases with `age` because the Cox model fit by the above commands corresponds to the LRH:

$$\text{LRH} = \beta_1 \texttt{protect} + \beta_2 \texttt{age} + \beta_3 \texttt{protect} * \texttt{age}$$

This equation implies that the discrete change in LRH when going from `protect==0` to `protect==1` is

$$\text{LRH}(\texttt{protect==1}) - \text{LRH}(\texttt{protect==0}) = \beta_1 + \beta_3 \texttt{age}$$

If $\beta_3 > 0$, the effect of `protect` increases with `age`; if $\beta_3 < 0$, it decreases; and if $\beta_3 = 0$, the effect of `protect` is constant. If you fit this model using `hip2.dta`, we would estimate $\widehat{\beta}_3 = 0.036$ with a 95% confidence interval $(-0.11, 0.18)$. Because this confidence interval includes zero, we cannot reject the hypothesis (at the 5% level) that there is no interaction effect.

Continuous variables can interact with other continuous variables and with indicator variables, and indicator variables can interact with other indicator variables. Consider an indicator variable `one_if_female` and an indicator variable `one_if_over65`. The interaction `one_if_female*one_if_over65` takes on two values, 0 and 1, with 1 meaning that the subject is both female and over 65.

It is common to read that interaction effects should be included in the model only when the corresponding main effects are also included, but there is nothing wrong with including interaction effects by themselves. In some particular problem, being female *and* over 65 may be important, whereas either fact alone could be of little or no importance. The goal of the researcher is to parameterize what is reasonably likely to be true for the data considering the problem at hand and not merely following a prescription.

In developing parameterizations and in interpreting parameterizations developed by others, we find it easiest to think in terms of derivatives—in terms of changes. Consider the hypothetical parameterization

$$\text{LRH} = \beta_1\texttt{weight} + \beta_2\texttt{age} + \beta_3\texttt{age}^2 + \beta_4\texttt{protect} + \beta_5\texttt{female} * \texttt{protect}$$

The easiest way to interpret this parameterization is to consider each of its component derivatives or, for indicator variables, the discrete change from zero to one. What is the effect of `weight` in this model? The derivative of the above with respect to `weight` is

$$\frac{\partial \text{LRH}}{\partial \texttt{weight}} = \beta_1$$

so the effect of `weight` is constant.

What is the effect of `age` in this model? It is

$$\frac{\partial \text{LRH}}{\partial \texttt{age}} = \beta_2 + 2\beta_3\texttt{age}$$

meaning that the effect is not constant but changes with `age`, either decreasing ($\beta_3 < 0$) or increasing ($\beta_3 > 0$).

The effect of `protect` in this model is

$$\text{LRH}(\texttt{protect==1}) - \text{LRH}(\texttt{protect==0}) = \beta_4 + \beta_5\texttt{female}$$

meaning that `protect` has the effect β_4 for males and $\beta_4 + \beta_5$ for females. If $\beta_5 = 0$, the effect for males and females is the same.

The effect of `female` is

$$\text{LRH}(\texttt{female==1}) - \text{LRH}(\texttt{female==0}) = \beta_5\texttt{protect}$$

meaning that there is an effect only if `protect==1`; males and females are otherwise considered to be identical. That is odd but perhaps not unreasonable for some problems.

One can build models by reversing this process. For instance, we start by thinking how the variables `weight`, `age`, `protect`, and `female` affect the outcome. From that, we can build a specification of LRH by

(1) thinking carefully about each variable separately and writing down how we think a change in the variable affects the outcome,

(2) multiplying each of the formulas developed in (1) by its corresponding variable, and

(3) adding the resulting terms together and simplifying.

Let us demonstrate this process. We begin by stepping through each of the variables (`weight`, `age`, `protect`, `female`) one at a time and considering how changes in each affect the outcome:

(1) What is the effect of `weight`? We think it is constant and call it α_1.

(2) What is the effect of `age`? We think the effect increases with `age` and express it as $\alpha_2 + \alpha_3$`age`.

(3) What is the effect of `protect`? We think `protect` has one constant effect for males and another for females, so the effect is $\alpha_4 + \alpha_5$`female`.

(4) What is the effect of `female`? We do not think there is one, other than what we have already considered. If we wanted to be consistent about it, an implication of (3) is that the effect must be α_5`female`, but we need not worry about consistency here. Each variable may be considered in isolation, and later, when we assemble the results, inconsistencies will work themselves out. The effect of `female`, therefore, is zero.

Now you must think carefully about each variable in isolation, but you can be sloppy in your thinking between variables. Do not worry about consistency—the next step will resolve that. You can write down that there is one effect at step 3 and then forget that you did that in step 4, as we did, or you can include the same effect more than once; it will not matter.

So, that is the first step. Write down all the derivatives/discrete changes without worrying about consistency. Now we have

> `weight`: α_1
>
> `age`: $\alpha_2 + \alpha_3$`age`
>
> `protect`: $\alpha_4 + \alpha_5$`female`
>
> `female`: 0

Importantly, in addition to these derivatives, we have a story to go along with each because we have thought carefully about them. All the above are marginal effects, or said differently, they are derivatives, or they are the predicted effects of a change in the variable; so the next step is to integrate each with respect to its variable. Actually, multiplication by the variable will be good enough here. That is not equivalent to integration, yet the difference is consummate to the inclusion of the correct constant term, which the Cox model just subsumes into the baseline hazard anyway.

Therefore, multiplying each of our derivatives by the variable of interest and summing, we get

$$\text{LRH} = \alpha_1\texttt{weight} + (\alpha_2 + \alpha_3\texttt{age})\texttt{age} + (\alpha_4 + \alpha_5\texttt{female})\texttt{protect} + 0$$

which simplifies to

$$\text{LRH} = \alpha_1\texttt{weight} + \alpha_2\texttt{age} + \alpha_3\texttt{age}^2 + \alpha_4\texttt{protect} + \alpha_5\texttt{protect} * \texttt{female}$$

There is our model. If you go through this process with your own model, you might end up with a term such as $(\alpha_2 + \alpha_3)$`weight`, but if so, that does not matter. Just treat

$(\alpha_2 + \alpha_3)$ as one separate coefficient when you fit the model. When you interpret the results, this one coefficient represents the cumulative effect of the two terms that arose because you inconsistently wrote the same effect twice.

Models specified by LRH are not a random assortment of terms manufactured according to some predetermined set of rules, such as "include this if you include that" or "do not include this unless you include that". Models are developed by carefully and substantively thinking about the problem, and thinking of derivatives with respect to the component variables in your data is a useful way to proceed.

10.5   Time-varying variables

Time-varying variables (covariates) are handled automatically by Stata's st family of commands, including stcox. If you have data of the form

```
id   _t0   _t   age   calcium   protect   fracture
...
12    0     5    64    11.77       0          0
12    5     8    64    11.34       0          1
...
```

and you fit the model

```
. stcox protect age calcium
```

the variable calcium will take on the value 11.77 for subject 12 during the period $(0, 5]$ and the value 11.34 during $(5, 8]$.

Thus the only trick to using time-varying covariates is understanding what they mean and concocting the time-varying covariates that you want.

For meaning, just understand that the hazard of the risk changes the instant the variable changes values; there is no anticipation of the change, nor is there any delay in the change taking effect. For instance, if you are an economist analyzing time to employment for the unemployed, you might theorize that whether the person is receiving unemployment payments would affect the chances of accepting employment. Then you might include a time-varying variable receiving_benefits, and that variable would change from 1 to 0 when the benefits ran out. Let us further assume that the estimated coefficient on receiving_benefits is negative. The instant benefits stop being paid, the hazard of employment increases. There would be nothing in the model that would say that the hazard of employment increases before the benefits ran out—in anticipation of that event—unless you included some other variable in your model.

Alternatively, imagine you are testing generators in overload situations. You run a generator under its approved load for a while and then increase the load. You have a time-varying covariate recording load in kilovolt-amperes. Then you would be asserting that the instant the load increased, the hazard of failure increased, too. There would be no delay even though failure might be caused by overheated bearings and heat takes time to accumulate.

Usually, researchers ignore anticipation or delay effects, assuming they do not amount to much. That is fine as long as care is taken to ensure that any possible delay in the effect is insignificant.

Relatedly, it is useful conceptually to distinguish between time-varying covariates and time-varying coefficients, even though the latter can be estimated using the former. With time-varying covariates (also known as time-dependent covariates), the marginal effect remains the same but the variable changes, so the aggregate effect changes. With time-varying coefficients (also known as time-dependent coefficients), the variable remains the same but the marginal effect changes, so the aggregate effect still changes.

Suppose you hypothesize that the effect of some variable x is β_1 up to $t = 5$ and $\beta_1 + \beta_2$ thereafter. You can fit such a model by specifying

$$\text{LRH} = \beta_1 x + \beta_2 I(t > 5)x$$

where $I()$ is the indicator function equal to 1 if the condition is true ($t > 5$ in this case) and 0 otherwise. If your data looked like

id	_t0	_t	x	xtgt5	_d
1	0	3	22	0	1
2	0	4	17	0	0
3	0	5	23	0	0
3	5	12	23	23	1
4	0	5	11	0	0
4	5	21	11	11	0
...					

you could type

```
. stcox x xtgt5
```

and be done with it. That is, each subject who is known to have survived beyond $t = 5$ has a record for $t \le 5$ and a record thereafter, so the values of the variable `xtgt5` can be recorded. Probably, however, your data look like

id	_t0	_t	x	_d
1	0	3	22	1
2	0	4	17	0
3	0	12	23	1
4	0	21	11	0
...				

in which case, fitting a Cox model with time-varying variables would require a little work, which we discuss next.

10.5.1   Using stcox, tvc() texp()

Consider the dataset presented above in its more common form. It is fairly easy to fit the Cox model with the time-varying covariate

$$\text{LRH} = \beta_1 \mathbf{x} + \beta_2 I(t > 5)\mathbf{x}$$

without having to concoct a new (expanded) dataset with the generated interaction variable xtgt5. The interaction term $I(t > 5)\mathbf{x}$ is of the form

$$variable\_in\_my\_data \times some\_function\_of\_analysis\_time$$

where *variable_in_my_data* is x and *some_function_of_analysis_time* is $I(t > 5)$. Because our interaction takes this specific form, we can just type

```
. stcox x, tvc(x) texp(_t>5)
```

The first x in the variable list for stcox represents the main effect term $\beta_1 \mathbf{x}$. The option tvc(x) specifies that we want x to be interacted with some function of analysis time, and the option texp(_t>5) specifies the function of time to be used (_t represents analysis time here because this is how stset defines it).

That the function _t > 5 is an indicator with only one change point masks a lot of what can be done using tvc() and texp(). Let us consider a more complex example:

Consider a more complete version of our hip-fracture data with another variable, init_drug_level, which gives the initial dosage level for a new experimental bone-fortifying drug:

```
. use http://www.stata-press.com/data/cggm/hip4, clear
(hip fracture study)
. list id _t0 _t _d init_drug_level in 1/10
```

	id	_t0	_t	_d	init_d~l
1.	1	0	1	1	50
2.	2	0	1	1	50
3.	3	0	2	1	50
4.	4	0	3	1	50
5.	5	0	4	1	100
6.	6	0	4	1	50
7.	7	0	5	1	100
8.	8	0	5	1	50
9.	9	0	5	0	50
10.	9	5	8	1	50

The initial dosage of the drug comes in one of two levels, either 50 mg or 100 mg. To analyze the effect of this drug, we could easily fit the Cox model with

$$\text{LRH} = \beta_1 \mathbf{protect} + \beta_2 \mathbf{init\_drug\_level}$$

using

```
. stcox protect init_drug_level
```

but we would be assuming that the relative hazard between those with an initial dosage of 50 mg and those with an initial dosage of 100 mg would be constant over time. This would be fine if we were willing to assume that the hazard function for the duration of the study was completely determined by the initial dosage, but it is perhaps more reasonable to think that the hazard function changes with the *current* level of the drug in the patient's bloodstream, which we know to decay (nonlinearly) as time passes.

Let's instead pretend that the drug is administered only once upon entry to the study and that its level in the bloodstream decays at an exponential rate such that the current level of the drug is equal to $\texttt{init\_drug\_level} \times \exp(-0.35t)$. For example, at $t = 2$, the current drug level would be $\exp(-0.7) = 0.5$ times its original level (2 days is the half-life of the drug). We wish to fit the Cox model with $\texttt{protect}$ and the current (time-varying) drug level as covariates, or equivalently,

$$\text{LRH} = \beta_1 \texttt{protect} + \beta_2 \texttt{init\_drug\_level} \exp(-0.35t)$$

This is accomplished by typing

```
. stcox protect, tvc(init_drug_level) texp(exp(-0.35*_t))
         failure _d:  fracture
   analysis time _t:  time1
                id:   id
Iteration 0:   log likelihood = -98.571254
Iteration 1:   log likelihood = -83.895138
Iteration 2:   log likelihood = -83.241951
Iteration 3:   log likelihood = -83.214617
Iteration 4:   log likelihood = -83.214437
Refining estimates:
Iteration 0:   log likelihood = -83.214437

Cox regression -- Breslow method for ties

No. of subjects =          48                  Number of obs   =        106
No. of failures =          31
Time at risk    =         714
                                               LR chi2(2)      =      30.71
Log likelihood  =   -83.214437                 Prob > chi2     =     0.0000
```

_t	Haz. Ratio	Std. Err.	z	P>|z|	[95% Conf. Interval]	
rh						
protect	.1183196	.0518521	-4.87	0.000	.050122	.2793091
t						
init_drug_~l	.8848298	.0601786	-1.80	0.072	.7744052	1.011

```
Note: second equation contains variables that continuously vary with respect
      to time; variables are interacted with current values of exp(-0.35*_t).
```

Now our estimated hazard ratios are split into two categories: those for constant-with-time variables ($\texttt{rh}$) and those for time-varying variables ($\texttt{t}$), and we are reminded

at the bottom of the output which function of analysis time was used to create the
time-varying covariate. The hazard ratio 0.8848 is now interpreted to mean that those
with higher drug levels *in their bloodstreams* have a lower risk of having a hip fracture.

10.5.2 Using stsplit

Recall our simpler example from the beginning of section 10.5, where our data looked
like

```
id   _t0   _t    x    _d
 1    0     3    22    1
 2    0     4    17    0
 3    0    12    23    1
 4    0    21    11    0
...
```

and we were fitting a Cox model with

$$\text{LRH} = \beta_1 x + \beta_2 I(t > 5) x$$

Earlier we mentioned that using `stcox, tvc(x) texp(_t>5)` avoided having to concoct
the expanded dataset

```
id   _t0   _t    x    xtgt5    _d
 1    0     3    22     0       1
 2    0     4    17     0       0
 3    0     5    23     0       0
 3    5    12    23    23       1
 4    0     5    11     0       0
 4    5    21    11    11       0
...
```

and fit the same model with `stcox x xtgt5`.

Although not having to expand the data proves convenient, it is worth learning how
to expand a dataset so that time-varying covariates may be introduced. Suppose that
we instead wished to fit a Cox model using

$$\text{LRH} = \beta_1 x + \beta_2 I(t > 5) x + \beta_3 I(t > 10) x$$

so that the effect of x is β_1 over $(0, 5]$, $\beta_1 + \beta_2$ over $(5, 10]$, and $\beta_1 + \beta_2 + \beta_3$ over $(10, \infty)$.
One limitation of `stcox, tvc() texp()` is that only one function of analysis time may
be specified, and here we have two: $I(t > 5)$ and $I(t > 10)$.

Another limitation of `stcox, tvc() texp()` is that it is unavailable if the exact
marginal (`exactm`) or exact partial method (`exactp`) is used for treating ties. For these
reasons, it is important to learn how to manually replicate the calculations performed
by the convenience command `stcox, tvc() texp()` so that they may be generalized to
cases where the convenience command is not available.

Toward this goal, Stata has the powerful `stsplit` command. For a simple illustration
of `stsplit`, let's return to our model,

$$\text{LRH} = \beta_1 \mathbf{x} + \beta_2 I(t > 5)\mathbf{x}$$

and pretend that estimation was not possible via `stcox, tvc() texp()`. `stsplit` takes
data such as

```
id   _t0    _t     x    _d
 1    0      3     22     1
 2    0      4     17     0
 3    0     12     23     1
 4    0     21     11     0
...
```

and if you type

```
. stsplit new, at(5)
```

the data will be divided at analysis time 5, and a new variable `new` will be created so
that `new==0` for $t <= 5$ and `new==5` for $t > 5$:

```
. list id _t0 _t x _d
```

	id	_t0	_t	x	_d
1.	1	0	3	22	1
2.	2	0	4	17	0
3.	3	0	12	23	1
4.	4	0	21	11	0

```
. stsplit new, at(5)
(2 observations (episodes) created)
. list id _t0 _t x _d new, sep(0)
```

	id	_t0	_t	x	_d	new
1.	1	0	3	22	1	0
2.	2	0	4	17	0	0
3.	3	0	5	23	0	0
4.	3	5	12	23	1	5
5.	4	0	5	11	0	0
6.	4	5	21	11	0	5

`stsplit` automatically fills in the time and failure variables appropriately on the new
records. Once the data have been split on $t = 5$, creating the `xtgt5` variable and fitting
the model is easy:

```
. gen xtgt5 = x*(new==5)
. stcox x xtgt5
```

Or we could type

```
. gen xtgt5 = x*(_t>5)
. stcox x xtgt5
```

Whether we use the **new** variable `stsplit` created for us or the original `_t` makes no difference.

Although we illustrate `stsplit` on a simple dataset, the original data with which we start could be inordinately complicated. The original might already have multiple records per subject, and some subjects might already have records split at $t = 5$ and some not. None of this will cause `stsplit` any difficulty.

`stsplit` can split at multiple times simultaneously. You could type

```
. stsplit cat, at(5 10 25 100)
```

and `stsplit` would split the appropriate records at the indicated times. The split points do not have to be equally spaced. For each subject, the new variable `cat` will contain 0 over the interval $(0, 5]$, 5 over $(5, 10]$, 10 over $(10, 25]$, 25 over $(25, 100]$, and 100 for records over the interval $(100, \infty)$. Any valid numlist may be used with `stsplit`'s `at()` option; see `help numlist`.

❏ **Technical note**

`stjoin` is the inverse of `stsplit`, and it will safely rejoin split records; "safely" means that `stjoin` will verify that, in the rejoining, no information will be lost. `stjoin` does this by making sure no variables differ (other than those specifying the time span) for any records it considers joining. For instance, looking through the data, `stjoin` would join the two records

id	_t0	_t	_d	x1	x2
57	5	7	0	3	8
57	7	9	0	3	8

to produce

id	_t0	_t	_d	x1	x2
57	5	9	0	3	8

but it would not have joined the records if either `x1` or `x2` varied over the two records.

Thus to use `stjoin` after `stsplit`, you must drop the new variable that `stsplit` created because it will differ over the now split records for a particular subject.

You can use `stjoin` anytime, not only after `stsplit`. It is always safe to type "`stjoin`".

❏

When the time-varying covariates vary continuously with time, `stsplit`, in an alternative syntax, proves even more useful. Returning to our hip-fracture example with time-varying covariates model,

$$\text{LRH} = \beta_1\text{protect} + \beta_2\text{init\_drug\_level}\exp(-0.35t)$$

we can use `stsplit` to manually replicate what was done by typing

```
. stcox protect, tvc(init_drug_level) texp(exp(-0.35*_t))
```

Cox regression operates only on times when failures actually occur. `stsplit` has an option that will split your records at all observed failure times (for this particular use of `stsplit`, no new variable is specified):

```
. use http://www.stata-press.com/data/cggm/hip4, clear
(hip fracture study)
. stsplit, at(failures)
(21 failure times)
(452 observations (episodes) created)
. gen current_drug_level = init_drug_level*exp(-0.35*_t)
. stcox protect current_drug_level

        failure _d:  fracture
   analysis time _t:  time1
               id:  id
Iteration 0:   log likelihood = -98.571254
Iteration 1:   log likelihood = -83.895138
Iteration 2:   log likelihood = -83.241951
Iteration 3:   log likelihood = -83.214617
Iteration 4:   log likelihood = -83.214437
Refining estimates:
Iteration 0:   log likelihood = -83.214437

Cox regression -- Breslow method for ties

No. of subjects =          48              Number of obs   =        558
No. of failures =          31
Time at risk    =         714
                                           LR chi2(2)      =      30.71
Log likelihood  =   -83.214437             Prob > chi2     =     0.0000
```

_t	Haz. Ratio	Std. Err.	z	P>|z|	[95% Conf. Interval]	
protect	.1183196	.0518521	-4.87	0.000	.050122	.2793091
current_dr~l	.8848298	.0601786	-1.80	0.072	.7744052	1.011

Except for some stylistic changes in the output, we have the same results as previously reported. The disadvantages to performing the analysis this way are that (1) it took more effort on our part, and (2) we had to expand our dataset from 106 to 558 observations. Such expansion is not always feasible (think of very large datasets with lots of failure times). The advantages of expanding the data are that (1) we are not limited to how many functions of analysis time we can use in our Cox models, and (2) we can estimate using any method to handle ties and are not limited to the `breslow` or `efron` methods.

`stsplit, at(failures)` is indeed a powerful tool because it uses the minimum amount of record splitting possible while still enabling your Cox model to fully capture the continuously changing nature of your time-varying covariate. When you have large datasets, this minimalistic approach to record splitting may prove invaluable.

In any case, just remember that when `stcox, tvc() texp()` does not apply, use `stsplit, at(failures)`.

❏ **Technical note**

`stsplit` has another neat feature. Typing

    . stsplit, at(failures) riskset(*newvar*)

will not only split the records at every observed failure time but will also create the new variable *newvar*, which uniquely identifies the risk sets. For every failure time, there is a group of subjects who were at risk of failure, and *newvar* describes this grouping.

Once we have the risk sets organized into groups, we can (for example) use this to verify something we hinted at in section 1.2, namely, that Cox regression is really just a collection of conditional logistic regressions, one taking place at each failure time.

Try the following:

    . use http://www.stata-press.com/data/cggm/hip2, clear
    . stsplit, at(failures) riskset(riskid)
    . stcox age protect, exactp nohr
    . clogit _d age protect, group(riskid)

and verify that you get equivalent regression results. Because we do have ties, we had to tell `stcox` to handle them the discrete way—the same way a conditional logistic model would—using option `exactp`.   ❏

10.6 Modeling group effects: fixed-effects, random-effects, stratification, and clustering

Suppose we collected data measuring time (variable `time`) from the onset of risk at time zero until occurrence of an event of interest (variable `fail`) on patients from different hospitals (variable `hospital`). We want to study patients' survival as a function of some risk factors, say age and gender (variables `age` and `gender`). We `stset` our hypothetical survival data and estimate the effect of predictors on survival by fitting a Cox model:

    . stset time, failure(fail)
    . stcox age gender

In the above commands, we ignored the fact that patients come from different hospitals and therefore assumed that hospitals have no effect on the results. However, if we believe that there might be a group effect, the effect of a hospital in our example, we should take it into account in our analysis.

There are various ways of adjusting for group effects. Each depends on the nature
of the grouping of subjects and on the assumptions we are willing to make about the
effect of grouping on subjects' survival. For example, a grouping can emerge from a
sampling design or it can simply be recorded along with other information about the
subjects. In the former case, groups carry information about how the data were sampled,
and we already know from the previous chapter that this could affect the estimates of
standard errors. In the latter case, we may want to account for possible intragroup
correlation. In both cases, groups may also have a direct effect on the hazard function.
Below we demonstrate several ways of modeling group effects. We intentionally choose
to use a fictional example rather than a real dataset in the following discussions to
emphasize the "modeling philosophy" and not be distracted by the interpretation of the
obtained results. For the real-data applications of the models considered here, refer to
the respective chapters discussed previously.

1. *Sampling design.* Let's start with the sampling stage of a study design and investigate how the data were collected. Suppose that we identified a fixed number of
   hospitals and then sampled our patients within each hospital, that is, we stratified
   on hospitals in our sampling design. Then we can adjust for the homogeneity of
   patients within a stratum (a hospital) as follows:

   ```
   . svyset, strata(hospital)
   . svy: stcox age gender
   ```

   That is, by first typing `svyset, strata(hospital)`, we account for the stratified
   sampling design.

   If, instead, we randomly selected these hospitals from a pool of all hospitals, we
   can account for the increased sample-to-sample variability as follows:

   ```
   . svyset hospital
   . svy: stcox age gender
   ```

   That is, by first typing `svyset hospital`, we account for the clustered sampling
   at the hospital level.

   Of course, in practice the sampling design is likely to be more complex with hospitals being nested within randomly sampled cities, which are themselves sampled
   within counties, and so on, and possibly with subjects selected into the sample
   according to the assigned probability weights. In this complex case, the effect
   of a hospital should be considered in conjunction with the higher-level grouping
   such as cities and counties via a multistage sampling design; see [SVY] **svyset** for
   details.

2. *Adjusting for possible correlation.* Suppose that we sampled patients but believe
   that there is possible dependence among patients within a hospital. We can adjust
   for this by clustering on hospitals:

   ```
   . stcox age gender, vce(cluster hospital)
   ```

   Although the sampling is done differently in this example, the results will be
   identical to those obtained from the previous example of a cluster sampling design

(as we also mentioned in sec. 9.5.2). All we know is that subjects might be correlated, either because of how we sampled our data or because of some other reasons specific to the nature of the grouping. In the above examples, we simply adjust the precision of the point estimates for the possible correlation without making any parametric assumptions about the correlation process. In the context of survival data, by saying that subjects are correlated we mean that subjects' failure times are correlated.

3. *Random effects.* Alternatively, we can model correlation by assuming that it is induced by an unobservable hospital-level random effect, or frailty, and by specifying the distribution of this random effect. The effect of a hospital is assumed to be random and have a multiplicative effect on the hazard function.

   Until now, the group effect had an impact only on the standard errors of the point estimates of predictors `age` and `gender` but not on the point estimates themselves. Here the effect of a hospital is directly incorporated into the hazard function, resulting in a different model specification for the survival data: a shared-frailty model. As such, both point estimates and their standard errors will change.

   Shared-frailty models were discussed in detail in section 9.4. Here we only provide a syntax for fitting the shared-frailty model for our example:

   ```
   . stcox age gender, shared(hospital)
   ```

   In the above command, the effect of a hospital is governed by the gamma distribution with mean 1 and variance θ, which is estimated from the data. If the estimated $\widehat{\theta}$ is not significantly different from zero, the correlation because of the hospital in our example can be ignored. We can also obtain the effect specific to each hospital by specifying option `effects()` in the above `stcox` command.

   To obtain reliable estimates of the parameters, point estimates, standard errors, or any combination of these, in all the above examples, one needs to have a sufficient number of groups (clusters).

4. *Fixed effects.* Suppose now that we are only interested in the effect of our observed hospitals rather than in making inferences about the effect of all hospitals based on the observed random sample of hospitals. In this case, the effects of all hospitals are treated as fixed, and we can account for them by including indicator variables identifying hospitals in our model:

   ```
   . xi: stcox age gender i.hospital
   ```

   In the above command, we assume that the hospitals have a direct multiplicative effect on the hazard function. That is, patients in all hospitals share the same baseline hazard function, and the effect of a hospital multiplies this baseline hazard function up or down depending on the sign of the estimated coefficients for the hospital indicator variables.

   Following the discussions in section 10.4, we may also include various interactions of hospitals with other predictors, such as `i.hospital*age` or `i.hospital*gender` in the above Cox model depending on whether we believe that the effect of a hospital modifies the effect of the other predictors `age` and `gender`.

5. *Stratified model.* Alternatively, we can account for the effect of a hospital on the hazard function by stratifying on `hospital` in the Cox model:

   ```
   . stcox age gender, strata(hospital)
   ```

   In this case, as also discussed in section 9.3, we allow baseline hazards to be different (to have a different shape) for each hospital rather than constraining them to be multiplicative versions of each other as in the previous example. This approach allows more flexibility in modeling the shape of the hazard function for each hospital. Of course, the flexibility comes with a price—the loss of the simplicity in the interpretation of the effect of the hospital, described by the hazard ratios, on the survival of subjects. If, however, your main focus is on the effect of other predictors, `age` and `gender` in our example, you may benefit from accounting for the group-specific effects in a more general way—especially if these effects vary with time—by stratifying on the group.

6. *Other models.* Now you may ask, What if we combine the above two approaches?

   ```
   . xi: stcox age gender i.hospital, strata(hospital)
   ```

   The hazard ratios for hospitals will all be estimated to 1. Specifying `i.hospital` in the above command is redundant because the effect of a hospital, including the shift, has already been absorbed by the hospital-specific hazard function. You may, however, instead include an interaction term `i.hospital*age`, which will result in a different model: the effect of a hospital is absorbed in the baseline hazard but the effect of `age` is allowed to vary with hospitals. Similarly, you can include an interaction term with another predictor `i.hospital*gender`, or you can include both interaction terms.

   Recall the note about options `strata()` of `svyset` and `stcox` from section 9.5.3. We can even consider the model

   ```
   . svyset, strata(hospital)
   . svy: stcox age gender, strata(hospital)
   ```

   provided we stratified on hospitals at a sampling stage of the design. This model not only takes into account the reduced variability because of the stratified design but also allows each stratum (hospital) to have a more flexible shape of the baseline hazard functions.

   Similarly, you can construct many more models by combining the various approaches discussed above.

In summary, there is no definitive recommendation on how to account for the group effect and on which model is the most appropriate when analyzing data. Certainly, if the grouping information comes from a survey design, this must be accounted for by employing the survey-specific methods provided by the `svy:` prefix. If you believe that there is a dependence among subjects within a group, you can account for the increased variability of the point estimates by adjusting standard errors (option `vce(cluster varname)`). You can directly model the dependence by using shared-frailty models (option `shared()`). A direct fixed-group effect may be modeled by including the group-specific indicator variables in the model or by stratifying on the groups (`stcox`'s option `strata()`). Finally, the combination of some of the above approaches may result in a model more suitable for the analysis of your failure data. The decision is yours, and you should base it on your science, on your judgment guided by experience, and on anything else that gives you an insight about the process generating failures. You can confirm the reasonableness of the selected models and choose the "best" model among the viable alternatives by performing the various postestimation diagnostics discussed in the next chapter.

11   The Cox model: Diagnostics

Despite its semiparametric nature, the Cox model is no different from an ordinary least-squares model in that there are many diagnostics that will check for model misspecification, outliers, influential points, etc. All the usual notions of what constitutes a well-fitting model are also true for Cox models; the difference is in the details.

11.1   Testing the proportional-hazards assumption

Despite the suggestive name "testing the proportional-hazards assumption", these tests are really just model specification tests that verify you have adequately parameterized the model and you have chosen a good specification for $\mathbf{x}\boldsymbol{\beta}_x$. It just so happens that, with Cox models, specification tests often go under the name "tests of the proportional-hazards assumption".

As with specification tests for other models, there are many ways to test the proportional-hazards assumption, and passing one test does not necessarily mean that others would be passed also. There are many ways that you can misspecify a model.

11.1.1   Tests based on reestimation

The way specification tests generally work is that one searches for variables to add to the model. Under the assumption that the model is correctly specified, these added variables will add little or no explanatory power, so one tests that these variables are "insignificant". Even tests of the proportional-hazards assumption follow that scheme.

The first test that we strongly recommend—for all models, not just Cox models or survival models—is called a link test. This test is easy to do and is remarkably powerful. Type

```
. stcox ...              /* fit your Cox model */
. linktest
```

which is really just shorthand for

```
. stcox ...
. predict s, xb          /* obtain linear predictor */
. gen s2 = s^2
. stcox s s2
```

The link test verifies that the coefficient on the squared linear predictor (s2) is insignificant. The basis for this test is that one first estimates $\boldsymbol{\beta}_x$ from the standard Cox model and then estimates β_1 and β_2 from a second-round model

$$\text{LRH} = \beta_1(\mathbf{x}\widehat{\boldsymbol{\beta}}_x) + \beta_2(\mathbf{x}\widehat{\boldsymbol{\beta}}_x)^2$$

Under the assumption that $\mathbf{x}\boldsymbol{\beta}_x$ is the correct specification, $\beta_1 = 1$ and $\beta_2 = 0$. Thus one tests that $\beta_2 = 0$.

This test is reasonably powerful for errors in specifying $\mathbf{x}\boldsymbol{\beta}_x$ under the assumption that $\mathbf{x}$ at least has the right variables in it, or more correctly, that the test is weak in terms of detecting the presence of omitted variables. If you fail this test, you need to go back and reconsider the specifications of the effects (derivatives) for each of the variables.

A popular way to directly go after the proportional-hazards assumption is to interact analysis time with the covariates and verify that the effects of these interacted variables are not different from zero because the proportional-hazards assumption states that the effects do not change with time except in ways that you have already parameterized. Or, more generally, if you have the correct specification the effect will not change with time, and moreover, any variable you can think to add will have no explanatory power.

There are many ways to proceed. You could, as with the link test, start with the predicted $\mathbf{x}\boldsymbol{\beta}_x$ from a first-round model and estimate

$$\text{LRH} = \beta_1(\mathbf{x}\widehat{\boldsymbol{\beta}}_x) + \beta_2(\mathbf{x}\widehat{\boldsymbol{\beta}}_x)t$$

and test that $\beta_2 = 0$, or you could refit the entire model in one swoop and include many interactions,

$$\text{LRH} = \mathbf{x}\boldsymbol{\beta}_{x1} + \mathbf{x}\boldsymbol{\beta}_{x2}t$$

and test $\boldsymbol{\beta}_{x2} = \mathbf{0}$. However, the most popular method is to fit separately one model per covariate and then perform separate tests of each,

- Fit the model $\text{LRH} = \mathbf{x}\boldsymbol{\beta}_x + \beta_1(x_1 t)$ and test $\beta_1 = 0$
- Fit the model $\text{LRH} = \mathbf{x}\boldsymbol{\beta}_x + \beta_2(x_2 t)$ and test $\beta_2 = 0$

and so on. If you are going to follow this approach, and if you have many covariates, we recommend that you worry about the effect of performing so many tests as if each were the only test you were performing. Some sort of adjustment, such as a Bonferroni adjustment, is called for. In the Bonferroni adjustment, if you are testing at the 0.05 level and performing k tests, you use $0.05/k$ as the critical level for each.

You could instead simply include all the covariates, interacted with _t, at once. We have been fitting a model on **age** and **protect**, and so we could type

```
. use http://www.stata-press.com/data/cggm/hip2
. stcox protect age, tvc(age protect) texp(_t)
. test [t]
```

If you instead want to follow the individual-regression approach, you would type

```
. stcox protect age, tvc(age) texp(_t)
. stcox protect age, tvc(protect) texp(_t)
```

Doing that, you will discover that neither term significantly interacts with time (even without a Bonferroni adjustment).

Note the following: (1) We could have omitted the `texp(_t)` option in the above because this is the default function of analysis time used by `stcox`, `tvc()` and (2) we handled tied failures using the default Breslow approximation. Had we instead wanted to use the marginal or partial methods for ties, we would need to `stsplit` our records at all observed failure times and generate the interactions ourselves; see section 10.5.2.

Earlier we stated that a popular test of the proportional-hazards assumption is to interact analysis time with the covariates (individually) and to verify that the interacted variables have no significant effect. Actually, any function of analysis time may be interacted with the covariates, and not all variables need to be interacted with the same function of time. It is all a matter of personal taste, and researchers often use $\ln(t)$ instead of t in the interactions. Doing this in Stata is easy enough—just change the specification of `texp()`. For example,

```
. use http://www.stata-press.com/data/cggm/hip2
. stcox age protect, tvc(age) texp(ln(_t))
```

In any case, such tests have been found to be sensitive to the choice of function, and there are no clear guidelines for choosing functions. A more complete approach would be to try t, $\ln(t)$, and then any other functions of time that occur to you.

A more flexible method, however, is to let the function of time be a step function (of a few steps) and let the magnitudes of the steps be determined by the data:

```
. use http://www.stata-press.com/data/cggm/hip2
(hip fracture study)
. stsplit cat, at(5 8 15)
(30 observations (episodes) created)
. gen age2 = age*(cat==5)
. gen age3 = age*(cat==8)
. gen age4 = age*(cat==15)
. stcox protect age age2-age4, noshow
Iteration 0:   log likelihood = -98.571254
Iteration 1:   log likelihood = -81.803706
Iteration 2:   log likelihood = -81.511169
Iteration 3:   log likelihood =   -81.5101
Refining estimates:
Iteration 0:   log likelihood =   -81.5101
```

```
Cox regression -- Breslow method for ties

No. of subjects =           48                 Number of obs   =         136
No. of failures =           31
Time at risk    =          714
                                               LR chi2(5)      =       34.12
Log likelihood  =     -81.5101                 Prob > chi2     =      0.0000
```

_t	Haz. Ratio	Std. Err.	z	P>|z|	[95% Conf. Interval]	
protect	.1096836	.0496909	-4.88	0.000	.0451351	.266544
age	1.17704	.0789269	2.43	0.015	1.032081	1.34236
age2	.937107	.1024538	-0.59	0.552	.7563579	1.16105
age3	.8677277	.0911832	-1.35	0.177	.7062149	1.066179
age4	.9477089	.0948454	-0.54	0.592	.7789111	1.153087

```
. testparm age2-age4

 ( 1)   age2 = 0
 ( 2)   age3 = 0
 ( 3)   age4 = 0

         chi2(  3) =      1.83
       Prob > chi2 =    0.6079
```

This, we argue, is a better test. Time is allowed to have whatever effect it wants although the cost is that the time function is only crudely defined over broad ranges of the data, meaning we misfit those broad ranges. For what we are testing, however, that cost should be minimal. We do not need to find the best model; we just need to know if there is evidence as to the existence of another model that is better than our current one.

To summarize, the basis of specification tests is the consideration of models of the form

$$\text{LRH} = \mathbf{x}_j \boldsymbol{\beta}_x + \beta_2 q_j$$

where q_j is something else. Under the assumption that $\mathbf{x}\boldsymbol{\beta}_x$ is the correct specification, β_2 will be zero, and you can test for that. Because the choice of q_j is entirely up to the user, this forms a large class of tests. You can also use fractional polynomial methods, discussed in section 10.3.1, to directly choose the form of q_j.

11.1.2 Test based on Schoenfeld residuals

Another way of checking the proportional-hazards assumption (specification) is based on analysis of residuals. This is like reestimation. The idea is to retrieve the residuals, fit a smooth function of time to them, and then test whether there is a relationship. Stata's `estat phtest` command does this and is based on the generalization by Grambsch and Therneau (1994). They showed that many of the popular tests for proportional hazards are in fact a test of nonzero slope in a generalized linear regression of the scaled Schoenfeld (1982) residuals on functions of time.

In its simplest form, when there are no tied failure times, the Schoenfeld residual for covariate x_u, $u = 1, \ldots, p$, and for observation j observed to fail is

$$r_{uj} = x_{uj} - \frac{\sum_{i \in R_j} x_{ui} \exp(\mathbf{x}_i \widehat{\boldsymbol{\beta}}_x)}{\sum_{i \in R_j} \exp(\mathbf{x}_i \widehat{\boldsymbol{\beta}}_x)}$$

That is, r_{uj} is the difference between the covariate value for the failed observation and the weighted average of the covariate values (weighted according to the estimated relative hazard from a Cox model) over all those subjects at risk of failure when subject j failed.

Let us momentarily allow that the coefficient on x_u does vary with time (in contradiction of the proportional-hazards assumption) so that β_u is actually

$$\beta_u(t) = \beta_u + q_j g(t)$$

where q_j is some coefficient and $g(t)$ is some function of time. Under the proportional-hazards assumption, $q_j = 0$. Grambsch and Therneau (1994) provide a method of scaling the Schoenfeld residual to form r_{uj}^*, and the scaling provides that

$$E(r_{uj}^* + \beta_u) = \beta_u(t)$$

Consequently, a graph of r_{uj}^* versus t_j (or some function of t_j) provides an assessment of the proportional-hazards assumption. Under the null hypothesis of proportional hazards, we expect the curve to have zero slope. In fact, r_{uj}^* versus t_j can be fairly rough, so it is common to add to the graph some sort of smooth curve through the points and to base one's judgment on that. You can base your judgment on the graph or on a formal statistical test of $H_o : q_j = 0$, the common test being based on a linear regression estimate of r_{uj}^* on t_j or $g(t_j)$. In the testing framework, the choice of $g()$ is important. When you look at a graph, checking for a nonzero slope by inspection is sufficient. If you plan on performing the test after looking at a graph, choose a form of $g()$ that transforms the relationship to a linear one, and then the test based on linear regression can detect a nonzero slope, given that the relationship is linear.

Stata's `estat phtest` automates this process, graphing and testing for individual covariates and, globally, the null hypothesis of zero slope. This test is equivalent to testing that the log hazard-ratio function is constant over time. Thus rejection of the null hypothesis indicates a deviation from the proportional-hazards assumption. By default, `estat phtest` sets $g(t) = t$, but you can change this by specifying the `time()` option; see `help estat phtest` for details.

`estat phtest` works after you have fit your model using `stcox`, but to use `estat phtest`, you must specify both the `schoenfeld()` and `scaledsch()` options at the time you estimate using `stcox`. These options save the Schoenfeld and scaled Schoenfeld residuals (respectively), which are quantities required by `estat phtest`. Actually, you need to specify `schoenfeld()` to perform the global test and `scaledsch()` if you want the variable-by-variable tests.

```
. use http://www.stata-press.com/data/cggm/hip2, clear
(hip fracture study)
. stcox protect age, schoenfeld(sch*) scaledsch(sca*)
  (output omitted)
```

The effect of the above commands, in addition to fitting the model, is to create four new variables in the dataset: sch1 and sch2 are created by schoenfeld(sch*) and store the Schoenfeld residuals for variables protect and age, respectively; sca1 and sca2 are created by scaledsch(sca*) and store the scaled Schoenfeld residuals. You could have given any new variable names you wanted, but it is convenient to provide these options with a stub (sch*, for example) and to let Stata number the variables for you.

Having estimated the Schoenfeld and scaled Schoenfeld residuals, we may now use estat phtest. The detail option reports the variable-by-variable tests along with the overall test:

```
. estat phtest, detail
    Test of proportional-hazards assumption
    Time:   Time
```

	rho	chi2	df	Prob>chi2
protect	0.00889	0.00	1	0.9627
age	-0.11519	0.43	1	0.5140
global test		0.44	2	0.8043

We find no evidence that our specification violates the proportional-hazards assumption.

Had we typed just estat phtest, only the global (combined) test would have been reported:

```
. estat phtest
    Test of proportional-hazards assumption
    Time:   Time
```

	chi2	df	Prob>chi2
global test	0.44	2	0.8043

For this test we would need only the schoenfeld() option when we fit the model.

Also we can see each of the graphs by specifying estat phtest's plot() option,

```
. estat phtest, plot(age)
```

which produces figure 11.1. We could also see the graph for the other covariate in our model using estat phtest, plot(protect). Examining these graphs will help you to choose a transform of t that makes the curve roughly linear. This transform may then be used by estat phtest to perform a formal statistical test.

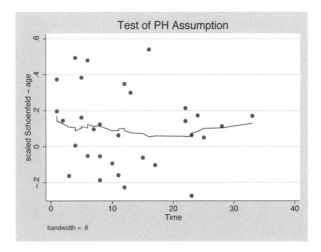

Figure 11.1. Test of the proportional-hazards assumption for `age`

❑ Technical note

The Grambsch and Therneau tests of the proportional-hazards assumption as implemented in Stata assume homogeneity of variance across risk sets. This allows the use of the estimated overall (pooled) variance–covariance matrix in the equations. Although these tests have been shown to be fairly robust to departures from this assumption, care must be exercised where this assumption may not hold, particularly when used with a stratified Cox model. Then we recommend that the proportional-hazards assumption be checked separately for each stratum.

❑

11.1.3   Graphical methods

Of the many graphical methods proposed throughout the literature for assessing the proportionality of hazards, two are available in Stata: the commands `stphplot` and `stcoxkm`. Both are intended for use with discrete covariates. For a good introduction and review of graphical methods, see Hess (1995) and Garrett (1997).

`stphplot` plots an estimate of $-\ln[-\ln\{\widehat{S}(t)\}]$ versus $\ln(t)$ for each level of the covariate in question, where $\widehat{S}(t)$ is the Kaplan–Meier estimate of the survivor function. For proportional hazards, $h(t|\mathbf{x}) = h_0(t)\exp(\mathbf{x}\boldsymbol{\beta}_x)$, and thus

$$S(t|\mathbf{x}) = S_0(t)^{\exp(\mathbf{x}\boldsymbol{\beta}_x)}$$

which implies that

$$-\ln[-\ln\{S(t|\mathbf{x})\}] = -\ln[-\ln\{S_0(t)\}] - \mathbf{x}\boldsymbol{\beta}_x$$

Under the proportional-hazards assumption, the plotted curves should thus be parallel. For example,

```
. use http://www.stata-press.com/data/cggm/hip2, clear
(hip fracture study)
. stphplot, by(protect)
        failure _d:  fracture
   analysis time _t:  time1
              id:  id
```

which produces figure 11.2. We see that the curves are roughly parallel, providing evidence in favor of the proportional-hazards assumption for the effect of wearing the hip-protection device.

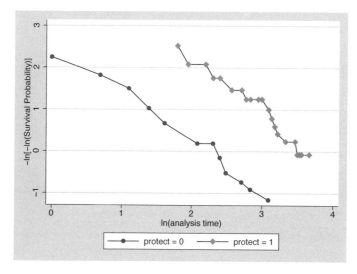

Figure 11.2. Test of the proportional-hazards assumption for `protect`

There is nothing magical being done by `stphplot` that we could not do ourselves. Basically, `stphplot` calculates a Kaplan–Meier curve $\widehat{S}(t)$ for each level of the variable specified in `by()`, calculates for each curve the transformation $-\ln[-\ln\{\widehat{S}(t)\}]$, and plots these curves with $\ln(t)$ on the x axis.

The model for the hip-protection device we have been considering is

$$\text{LRH} = \beta_1 \texttt{protect} + \beta_2 \texttt{age}$$

Figure 11.2 does not address the question of whether the effect of `protect` is constant *conditional* on `age` being in the model. `stphplot` works differently from `estat phtest` in that it does not make use of the `stcox` model that we might have previously fit. We can obtain the graph for `protect` conditional on `age` being in the model by

```
. stphplot, strata(protect) adjust(age)
   (output omitted)
```

which produces figure 11.3.

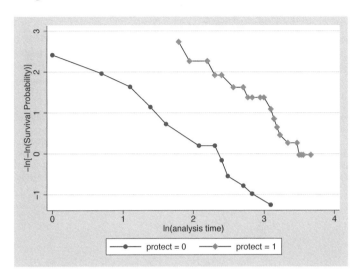

Figure 11.3. Test of the proportional-hazards assumption for `protect`, controlling for `age`

In general, you should specify whatever other variables are in your model in the `adjust()` option. Specifying `adjust()` tells Stata not to estimate the survivor function using Kaplan–Meier but instead to estimate it using `stcox` so that we can introduce other variables into the model. In any case, the procedure followed by `stphplot, adjust() strata()` is as follows:

1. Fit a Cox model on the `adjust()` variables as covariates and on the `strata()` variable as a stratification variable. Here the command `stcox age, strata(protect)` is executed. Thus the baseline hazard is allowed to be different for different values of `protect`, and it is left to the user to decide whether the hazards are really proportional.

2. Obtain the baseline survivor functions for each of the strata directly from `stcox` and then calculate $-\ln[-\ln\{\widehat{S}(t)\}]$ for each.

3. Plot the transformed (stratified) survivor curves on the same graph, with $\ln(t)$ on the x axis.

Another method of graphically assessing proportional hazards is to compare separately estimated Kaplan–Meier curves (which are model agnostic) with estimates of $S_0(t)^{\exp(\mathbf{x}\boldsymbol{\beta}_x)}$ from a Cox model, which does impose the model assumption of proportional hazards. `stcoxkm` does this,

```
. stcoxkm, by(protect)
        failure _d:  fracture
  analysis time _t:  time1
              id:  id
```

which produces figure 11.4.

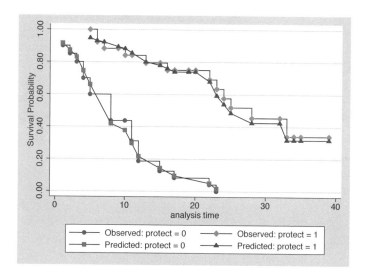

Figure 11.4. Comparison of Kaplan–Meier and Cox survivor functions

The way this graph is implemented in Stata, however, does not allow its use with other covariates in the model. Actually, the way this assessment has been developed in the literature does not allow its use when other covariates are included in the model, but generalizing it to that would be easy enough. Rather than using the Kaplan–Meier estimates to compare against, you could use the baseline survivor curve from a stratified Cox model.

11.2 Residuals

Although the uses of residuals vary and depend on the data and user preferences, some suggested traditional uses are as follows:

1. Cox–Snell residuals are useful in assessing overall model fit.

2. Martingale residuals are useful in determining the functional form of the covariates to be included in the model.

3. Schoenfeld residuals (both scaled and unscaled), score residuals, and efficient score residuals are useful for checking and testing the proportional-hazards assumption, examining leverage points, and identifying outliers.

4. Deviance residuals are useful in examining model accuracy and identifying outliers.

Efficient score residuals, martingale residuals, Schoenfeld residuals, and scaled Schoenfeld residuals are obtained when we fit the Cox proportional hazards model via options to `stcox`. Cox–Snell and deviance residuals are obtained after estimation using `predict`. When there are multiple observations per subject, cumulative martingale and cumulative Cox–Snell residuals are obtained after estimation using `predict`, and accumulation takes place over each subject's set of records.

Reye's syndrome data

We illustrate some of the possible applications of residuals using a study of 150 children diagnosed with Reye's syndrome.

Reye's syndrome is a rare disease, usually affecting children under the age of 15 who are recovering from an upper respiratory illness, chicken pox, or flu. The condition causes severe brain swelling and inflammation of the liver. This acute illness requires immediate and aggressive medical attention. The earlier the disease is diagnosed, the better the chances of a successful recovery. Treatment protocols include drugs to control the brain swelling and intravenous fluids to restore normal blood chemistry.

For this study of a new medication to control the brain swelling, and thus to prevent death, 150 Reye's syndrome patients were randomly allocated at the time of hospital admission to either the standard high-dose barbiturate treatment protocol or to a treatment protocol that included the new experimental drug. The time from treatment allocation to death or the end of follow-up was recorded in days. Here are a few of the records.

```
. use http://www.stata-press.com/data/cggm/reyes
(Reye´s syndrome data)

. list id days dead treat age sex ftliver ammonia sgot in 1/6, noobs sep(0)
```

id	days	dead	treat	age	sex	ftliver	ammonia	sgot
1	8	1	1	13	1	1	14.3	300
2	82	1	0	13	1	0	.9	287
3	19	1	0	16	1	1	1.2	298
6	46	0	1	15	1	0	.6	286
7	33	0	1	13	1	0	.8	270
9	44	1	0	10	1	0	3	280

The variable `treat` equals 0 for patients on the standard protocol and 1 for patients on the experimental protocol. The variable `days` indicates the number of follow-up days from treatment allocation to death (`death==1`) or to censoring (`death==0`). In addition to `age` and `sex`, several laboratory measurements including blood `ammonia` level and serum `sgot` were performed at the time of initial hospitalization. All patients involved also underwent a liver biopsy to assess the presence of fatty liver disease (`ftliver`) within 24 hours of treatment allocation. All covariates in this study are fixed.

We are interested in determining whether the new experimental treatment is more effective in prolonging the life expectancy of these children. We begin addressing this question by `stsetting` our dataset and fitting a Cox proportional hazards model with the treatment variable only.

```
. stset days, id(id) failure(dead)
  (output omitted)
. stcox treat, efron
         failure _d:  dead
   analysis time _t:  days
                 id:  id
Iteration 0:    log likelihood = -253.90166
Iteration 1:    log likelihood = -252.86308
Iteration 2:    log likelihood = -252.86284
Refining estimates:
Iteration 0:    log likelihood = -252.86284
Cox regression -- Efron method for ties
No. of subjects =            150            Number of obs    =          150
No. of failures =             58
Time at risk    =           4971
                                            LR chi2(1)       =         2.08
Log likelihood  =    -252.86284            Prob > chi2      =       0.1495
```

_t _d	Haz. Ratio	Std. Err.	z	P>|z|	[95% Conf. Interval]	
treat	.6814905	.1831707	-1.43	0.154	.4024151	1.154105

We observe that although there is some indication that the new treatment reduces the risk of dying when compared with the standard treatment (hazard ratio = 0.68), there is not strong evidence suggesting that the new therapy is superior to the standard treatment. We could have arrived at a similar conclusion had we performed a log-rank test by typing `sts test treat`.

11.2.1   Determining functional form

Previous studies on children with Reye's syndrome have identified elevated `ammonia` and `sgot` levels along with the presence of fatty liver disease (`ftliver`) as important predictors of final outcome. Thus, in our analysis, we would like to compare the two treatment protocols while adjusting for the effects of these covariates.

What is the best functional form to use for each covariate? We can use martingale residuals to help answer this question. Martingale residuals are obtained by specifying the `mgale()` option when fitting the Cox model. These residuals can be interpreted simply as the difference between the observed number of failures in the data and the number of failures predicted by the model.

Martingale residuals are also helpful in determining the proper functional form of the covariates to be included in the model. That is, should we include in the model the variable `age` directly, or should we perhaps include ln(`age`) or some other function of `age`?

Let M_i be the martingale residual of the ith observation obtained when no covariates are included in the model. Assume a true relative hazard model of the form

$$h(t|\mathbf{x}_i) = h_0(t) \exp\{f(\mathbf{x}_i)\}$$

where $\mathbf{x}_i$ is the covariate vector for the ith subject and $f()$ is some function. It may then be shown that M_i is approximately $kf(\mathbf{x}_i)$, where k is a constant that depends on the number of censored observations. The above equation implies the existence of a linear relationship between $f(\mathbf{x}_i)$ and M_i. When used with a single predictor, x, a smooth plot of martingale residuals versus x consequently may provide a visual indication of the transformation needed, if any, to obtain a proper functional form for x. The goal is to determine a transformation that will result in an approximately straight curve.

Returning to our data, we apply this method to determine the functional form for the `ammonia` variable. We begin by fitting a Cox model without covariates and obtaining the martingale residuals using the `mgale(`*newvar*`)` option. We handle ties using Efron's method (for no real reason other than variety). It is necessary to use the `estimate` option when fitting null models.

```
. stcox, mgale(mg) efron estimate

        failure _d:  dead
   analysis time _t:  days
              id:  id

Iteration 0:   log likelihood = -253.90166
Refining estimates:
Iteration 0:   log likelihood = -253.90166

Cox regression -- Efron method for ties

No. of subjects =         150              Number of obs   =         150
No. of failures =          58
Time at risk    =        4971
                                           LR chi2(0)      =        0.00
Log likelihood  =    -253.90166            Prob > chi2     =           .

         _t
         _d │   Haz. Ratio    Std. Err.      z    P>|z|     [95% Conf. Interval]
────────────┼─────────────────────────────────────────────────────────────────
```

The newly created variable, `mg`, contains the values of the martingale residuals from this null model. We can now plot each variable whose functional form we are interested in determining against `mg`. We use the `lowess` command to obtain the graph along with a running-mean smoother to ease interpretation. We could also use a local polynomial (see `help lpoly`) or other smoother. We type

```
. lowess mg ammonia
```

which produces figure 11.5.

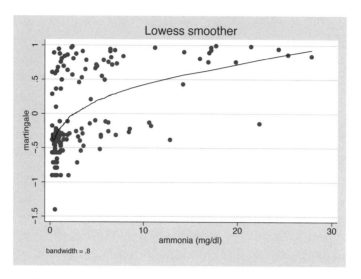

Figure 11.5. Finding the functional form for `ammonia`

We see that the resulting smooth plot is not linear at lower values of `ammonia`. Here a simple log transformation of the variable will yield a more linear result. We type

```
. gen lamm=ln(ammonia)
. lowess mg lamm
```

which produces figure 11.6.

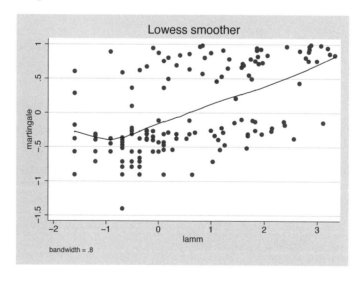

Figure 11.6. Using the log transformation

This yields an approximately linear smooth plot and leads us to believe that the log transformed `ammonia` variable is better for inclusion in the model.

We now fit the Cox model with the treatment and `lamm` covariates.

```
. stcox treat lamm, efron noshow

Iteration 0:   log likelihood = -253.90166
Iteration 1:   log likelihood = -222.27086
Iteration 2:   log likelihood = -221.77893
Iteration 3:   log likelihood = -221.77885
Refining estimates:
Iteration 0:   log likelihood = -221.77885

Cox regression -- Efron method for ties

No. of subjects =          150              Number of obs   =          150
No. of failures =           58
Time at risk    =         4971
                                            LR chi2(2)      =        64.25
Log likelihood  =   -221.77885              Prob > chi2     =       0.0000
```

_t _d	Haz. Ratio	Std. Err.	z	P>|z|	[95% Conf. Interval]	
treat	.5943764	.1656545	-1.87	0.062	.3442144	1.026347
lamm	2.546743	.3201795	7.44	0.000	1.99054	3.25836

We see that (as shown by other studies) increased levels of blood ammonia are strongly associated with increases in the hazard of dying. More importantly, however, we see that if we adjust by this important covariate, the treatment effect becomes significant at the 10% level. This finding indicates a lack of homogeneity of patients assigned to the treatment modalities. We can observe this heterogeneity by tabulating the mean ammonia level by treatment level and performing a t test.

```
. ttest ammonia, by(treat) uneq

Two-sample t test with unequal variances
```

Group	Obs	Mean	Std. Err.	Std. Dev.	[95% Conf. Interval]	
0	71	2.683099	.392177	3.304542	1.900926	3.465271
1	79	5.143038	.7920279	7.039698	3.566231	6.719844
combined	150	3.978667	.4661303	5.708907	3.057587	4.899746
diff		-2.459939	.8838049		-4.210858	-.7090201

```
        diff = mean(0) - mean(1)                             t =  -2.7834
Ho: diff = 0                      Satterthwaite's degrees of freedom =  113.345

     Ha: diff < 0                  Ha: diff != 0                  Ha: diff > 0
  Pr(T < t) = 0.0032         Pr(|T| > |t|) = 0.0063         Pr(T > t) = 0.9968
```

We find that, on average, patients with lower ammonia levels ended up on the standard treatment, whereas those with higher ammonia levels (with a worse prognosis) were assigned to the experimental group. The fact that this imbalance exists can be significant in assessing the validity of this study and of the treatment allocation protocol.

We used this same method to examine the functional forms of the other covariates and found it unnecessary to transform the other covariates. We chose, however, to center `sgot` at its mean so that the baseline hazard corresponds to `treat==0`, `lamm==0`, `ftliver==0`, and `sgot` equal to its mean value. This last part we did by typing

```
. quietly summarize sgot
. replace sgot = sgot - r(mean)
(150 real changes made)
```

Finally, we fit a Cox model using `treat`, `lamm`, `sgot` (now with 0 corresponding to its mean level), and `ftliver`.

```
. stcox treat lamm sgot ftliver, efron

         failure _d:  dead
   analysis time _t:  days
                 id:  id

Iteration 0:    log likelihood = -253.90166
Iteration 1:    log likelihood = -219.18093
Iteration 2:    log likelihood =  -215.1553
Iteration 3:    log likelihood = -215.14935
Refining estimates:
Iteration 0:    log likelihood = -215.14935

Cox regression -- Efron method for ties

No. of subjects =          150              Number of obs   =          150
No. of failures =           58
Time at risk    =         4971
                                            LR chi2(4)      =        77.50
Log likelihood  =   -215.14935             Prob > chi2     =       0.0000
```

_t _d	Haz. Ratio	Std. Err.	z	P>|z|	[95% Conf. Interval]	
treat	.4899586	.1439631	-2.43	0.015	.2754567	.8714962
lamm	2.194971	.2831686	6.09	0.000	1.704578	2.826444
sgot	1.049024	.0183381	2.74	0.006	1.01369	1.085589
ftliver	2.14807	.6931377	2.37	0.018	1.141257	4.043091

We observe that after adjusting for ammonia, serum concentration, and the presence of fatty liver disease, the new therapy reduces the risk of death by 51%.

❏ **Technical note**

Alternatively, you can use multivariable fractional polynomials (MFP)—discussed in section 10.3.1—to determine the functional form of the covariates in the model. For the above model we use `mfp`,

```
. mfp stcox treat ammonia sgot ftliver, efron adjust(ammonia:no, sgot:no)

Deviance for model with all terms untransformed =   441.673, 150 observations
  (output omitted)

Transformations of covariates:

-> gen double Iammo__1 = ln(X) if e(sample)
   (where: X = ammonia/10)

Final multivariable fractional polynomial model for _t
```

Variable	——Initial——			——Final——		
	df	Select	Alpha	Status	df	Powers
treat	1	1.0000	0.0500	in	1	1
ammonia	4	1.0000	0.0500	in	2	0
sgot	4	1.0000	0.0500	in	1	1
ftliver	1	1.0000	0.0500	in	1	1

```
Cox regression -- Efron method for ties
Entry time _t0                              Number of obs   =        150
                                            LR chi2(4)      =      77.50
                                            Prob > chi2     =     0.0000
Log likelihood = -215.14935                 Pseudo R2       =     0.1526
```

_t	Coef.	Std. Err.	z	P>|z|	[95% Conf. Interval]	
treat	-.7134343	.2938271	-2.43	0.015	-1.289325	-.1375437
Iammo__1	.7861686	.1290079	6.09	0.000	.5333177	1.03902
sgot	.0478598	.0174811	2.74	0.006	.0135974	.0821222
ftliver	.76457	.3226792	2.37	0.018	.1321303	1.39701

```
Deviance:   430.299.
```

and arrive at the same functional form as was chosen based on the martingale residuals plots. We specify no within the adjust option to request that no adjustment be made for variables ammonia and sgot. The MFP approach selects the log transformation of ammonia to be the best-fitting functional form for this variable among a fixed set of power transformations. It also confirms the linearity of the effect of sgot. The results from the selected final model are the same as from the above stcox but displayed in a log-hazard (coefficient) metric.

❑

11.2.2   Goodness of fit

We now turn our attention to the evaluation of the overall model fit using Cox–Snell (Cox and Snell 1968) residuals. It has been shown that if the Cox regression model fits the data, then the true cumulative hazard function conditional on the covariate vector has an exponential distribution with a hazard rate of 1.

The Cox–Snell residual for the jth observation is defined as

$$\text{CSr}_j \;=\; \widehat{H}_0(t_j)\exp(\mathbf{x}_j\widehat{\boldsymbol{\beta}}_x) \qquad (11.1)$$

where both $\widehat{H}_0()$ and $\widehat{\beta}_x$ are obtained from the Cox model fit. If the model fits the data well, then the Cox–Snell residuals should have a standard exponential distribution with hazard function equal to 1 for all t, and thus the cumulative hazard of the Cox–Snell residuals should be a straight 45° line. We can verify the model fit by estimating the empirical Nelson–Aalen cumulative hazard function with the Cox–Snell residuals as the time variable along with the data's original censoring variable.

We will fit two models. In the first model, we include the covariates `treat` and `ammonia`, and in the second model, we will substitute `lamm` for `ammonia`.

In Stata, Cox–Snell residuals are computed from the martingale residuals. Thus when we fit the model using `stcox`, we save the martingale residuals and then use `predict` to obtain the Cox–Snell residuals. If we forget to save the martingale residuals at estimation, `predict` will refuse to calculate Cox–Snell residuals.

```
. use http://www.stata-press.com/data/cggm/reyes, clear
(Reye´s syndrome data)

. stcox treat ammonia, efron mgale(mg)

         failure _d:  dead
   analysis time _t:  days
              id:  id
Iteration 0:   log likelihood = -253.90166
Iteration 1:   log likelihood = -230.10661
Iteration 2:   log likelihood = -227.58788
Iteration 3:   log likelihood = -227.58783
Refining estimates:
Iteration 0:   log likelihood = -227.58783

Cox regression -- Efron method for ties

No. of subjects =          150                 Number of obs   =         150
No. of failures =           58
Time at risk    =         4971
                                               LR chi2(2)      =       52.63
Log likelihood  =   -227.58783                 Prob > chi2     =      0.0000
```

_t _d	Haz. Ratio	Std. Err.	z	P>|z|	[95% Conf. Interval]	
treat	.4295976	.1300496	-2.79	0.005	.2373463	.777573
ammonia	1.171981	.0233525	7.96	0.000	1.127094	1.218657

Now we use `predict` to compute the Cox–Snell residuals, and then we `stset` the data using the Cox–Snell residuals as the time variable and our original failure variable, `dead`, as the failure indicator.

```
. predict cs, csnell

. stset cs, fail(dead)

     failure event:  dead != 0 & dead < .
obs. time interval:  (0, cs]
 exit on or before:  failure
─────────────────────────────────────────────────────────────────────────
        150  total obs.
          0  exclusions
─────────────────────────────────────────────────────────────────────────
        150  obs. remaining, representing
         58  failures in single record/single failure data
         58  total analysis time at risk, at risk from t =         0
                              earliest observed entry t =         0
                                 last observed exit t =   2.902519
```

We can now use `sts generate` to generate the Nelson–Aalen cumulative hazard function and plot it against the Cox–Snell residuals:

```
. sts gen H = na

. line H cs cs, sort ytitle("") legend(cols(1))
```

This produces figure 11.7.

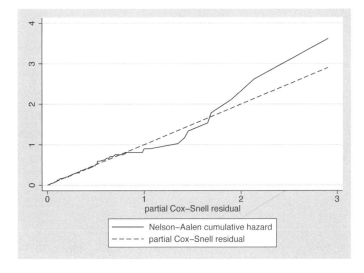

Figure 11.7. Cumulative hazard of Cox–Snell residuals (`ammonia`)

In the `line` command, we specified `cs` twice so as to get a 45° reference line. Comparing the jagged line to the reference line, we observe what could be considered a concerning lack of fit.

We repeat the above steps for the second model,

```
. use http://www.stata-press.com/data/cggm/reyes, clear
. gen lamm = ln(ammonia)
. stcox treat lamm, efron mgale(mg)
. predict cs, csnell
. stset cs, fail(dead)
. sts gen H = na
. line H cs cs, sort ytitle("") legend(cols(1))
```

which produces figure 11.8.

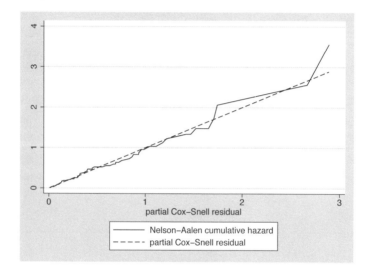

Figure 11.8. Cumulative hazard of Cox–Snell residuals (lamm)

When we use lamm=ln(ammonia) instead of ammonia in our model, the model does not fit the data too badly.

❏ **Technical note**

When plotting the Nelson–Aalen cumulative hazard estimator for Cox–Snell residuals, even if we have a well-fitting Cox model, some variability about the 45° line is still expected, particularly in the right-hand tail. This is because of the reduced effective sample caused by prior failures and censoring.

❏

We can also evaluate the predictive power of the above Cox model by computing the Harrell's C concordance statistic (Harrell et al. 1982; Harrell, Lee, and Mark 1996). This statistic measures the agreement of predictions with observed failure order. It is defined as the proportion of all usable subject pairs in which the predictions and outcomes are concordant; for more technical details see [ST] **stcox postestimation**.

Stata's estat concordance command can be used to obtain the concordance measures after stcox.

```
. use http://www.stata-press.com/data/cggm/reyes, clear
(Reye's syndrome data)

. gen lamm = ln(ammonia)

. stcox treat lamm, efron

  (output omitted)

. estat concordance

  Harrell's C concordance statistic

          failure _d:  dead
    analysis time _t:  days
                  id:  id

  Number of subjects (N)                =      150
  Number of comparison pairs (P)        =     5545
  Number of orderings as expected (E) =      4399
  Number of tied predictions (T)        =       28

          Harrell's C = (E + T/2) / P =    .7959
                             Somers' D =    .5917
```

The values of C range between 0 and 1. In our example it is estimated to be 0.796, indicating that by using predictors `treat` and `lamm` in the model we correctly identify the order of the survival times for pairs of patients roughly 80% of the time.

`estat concordance` also reports the value of Somers' D rank correlation, if you prefer to think of a correlation coefficient ranging from -1 to 1. The two measures are closely related: $D = 2(C - 0.5)$. In general, a value of 0.5 of Harrell's C and 0 of Somers' D indicate no predictive ability of the model. For a discussion of other ways of determining the predictive accuracy of survival models, see Korn and Simon (1990).

11.2.3 Outliers and influential points

In evaluating the adequacy of the fit model, it is important to determine if any one or any group of observations has a disproportionate influence on the estimated parameters. This is known as influence or leverage analysis. The preferred method of performing leverage analysis is to compare the estimated parameter $\widehat{\beta}_x$ obtained from the full data with the estimated parameters $\widehat{\beta}_x^{(i)}$ obtained by fitting the model to the $n-1$ observations remaining after the ith observation is removed. If $\widehat{\beta}_x - \widehat{\beta}_x^{(i)}$ is close to zero, then the ith observation has little influence on the estimate. The process is repeated for all observations included in the original model.

To compute these differences for a dataset with n observations, we would have to execute `stcox` $n + 1$ times, which could be impractical for large datasets. Instead, an approximation to the difference $\widehat{\beta}_x - \widehat{\beta}_x^{(i)}$ based on the efficient score residuals can be calculated as

$$\mathbf{D} \times \mathbf{V}(\widehat{\beta}_x)$$

where $\mathbf{V}(\widehat{\boldsymbol{\beta}}_x)$ is the variance–covariance matrix of $\widehat{\boldsymbol{\beta}}_x$ and $\mathbf{D}$ is the matrix of efficient score residuals. The difference $\widehat{\boldsymbol{\beta}}_x - \widehat{\boldsymbol{\beta}}_x^{(i)}$ is commonly referred to as "dfbeta" in the literature.

Stata saves efficient score residuals in variables generated at the time of estimation when you specify option esr(*newvars*). One efficient score residual variable is created for each regressor in the model; the first new variable corresponds to the first regressor, the second to the second, and so on. Therefore, if you want all the efficient score residuals, then you must either explicitly specify one new variable name for each regressor in your model or specify esr(*stub**), where *stub* is a name of your choosing. Stata then creates the variables *stub*1, *stub*2, etc.

❑ **Technical note**

Be careful when asking for efficient score residuals. stcox may drop variables from the model because of collinearity. This is a desirable feature, yet a side effect is that the score residual variables (and the Schoenfeld residual variables and the scaled Schoenfeld residual variables) may not align with the regressors in the way you expect. Say that you fit a model by typing

```
. stcox x1 x2 x3, esr(r1 r2 r3)
```

Usually r1 will contain the residual associated with x1, r2 the residual associated with x2, and r3 the residual associated with x3. But assume that x2 is dropped because of collinearity. Then r1 will correspond to x1, as before, but r2 will correspond to x3 (and r3 will contain 0). This happens because, after omitting the collinear variables, there are only two variables in the model: x1 and x3.

❑

Returning to the Reye's syndrome data, we can obtain the efficient score residuals by typing

```
. use http://www.stata-press.com/data/cggm/reyes, clear
. gen lamm = ln(ammonia)
. quietly summarize sgot
. replace sgot = sgot - r(mean)
. stcox treat lamm sgot ftliver, efron esr(esr*)
```

This command creates four new variables: esr1, esr2, esr3, and esr4, corresponding to the four covariates in the model. We next use the matrix command mkmat to create a matrix of the efficient score residuals, then we obtain the variance–covariance matrix from the fit model and multiply the two matrices:

```
. mkmat esr1 esr2 esr3 esr4, matrix(esr)
. mat V = e(V)
. mat Inf = esr*V
. svmat Inf, names(s)
```

The last command saves the estimates of $\widehat{\boldsymbol{\beta}}_x - \widehat{\boldsymbol{\beta}}_x^{(i)}$ in the variables s1, s2, s3, and s4. We then label these new variables just so that we do not forget their contents:

```
. label var s1 "dfbeta - treat"
. label var s2 "dfbeta - lamm"
. label var s3 "dfbeta - sgot"
. label var s4 "dfbeta - ftliver"
```

In any case, a graph of any of these variables versus time or observation numbers can be used to identify observations with disproportionate influence, for example,

```
. scatter s3 _t, yline(0) mlabel(id) msymbol(i)
```

which produces figure 11.9.

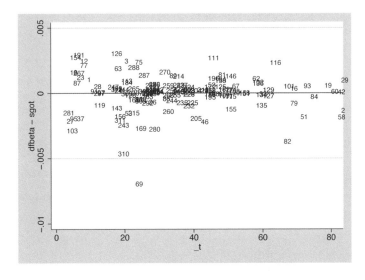

Figure 11.9. Dfbeta(`sgot`) for Reye's syndrome data

12 Parametric models

12.1 Motivation

Compared with the Cox model, parametric models are different in the way they exploit the information contained in the data. As an extreme case, consider the following data:

```
. use http://www.stata-press.com/data/cggm/odd
. list id _t0 _t female _d
```

	id	_t0	_t	female	_d
1.	1	0	2	0	1
2.	2	3	5	1	0
3.	3	6	8	0	1
4.	4	9	10	1	1

In this data, there are no overlaps in our observations of the subjects:

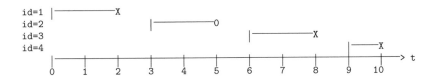

Here t is analysis time, not calendar time. We observed subject 1 from $(0, 2]$, at which point the subject fails; we observe subject 2 from $(3, 5]$, at which point the subject is censored; and so on. What makes these data so odd is that at any particular analysis time we observe at most one subject because most of our subjects arrived late to our study.

In any case, given these odd data, the risk groups so important to semiparametric analysis are

1. Failure at $t = 2$: risk group contains only subject id==1.
   $\Pr(\mathtt{id} == 1 \text{ fails}|\text{one failure}) = 1$.

2. Failure at $t = 8$: risk group contains only subject id==3.
   $\Pr(\mathtt{id} == 3 \text{ fails}|\text{one failure}) = 1$.

3. Failure at $t = 10$: risk group contains only subject id==4.
   $\Pr(\mathtt{id} == 4 \text{ fails}|\text{one failure}) = 1$.

and the conditional probability of observing what we did is 1, regardless of any parameter we might want to estimate, because the probability that "our" subject fails given one subject fails is 1 if "our" subject is the only one at risk at that time. We could not, for instance, fit a Cox model on `female`:

```
. stcox female
          failure _d:  dead
    analysis time _t:  time1
Iteration 0:   log likelihood =              0
Refining estimates:
Iteration 0:   log likelihood =              0
Cox regression -- no ties
No. of subjects =            4                    Number of obs   =          4
No. of failures =            3
Time at risk    =            7
                                                 LR chi2(1)      =       0.00
Log likelihood  =            0                    Prob > chi2     =     1.0000
```

_t _d	Haz. Ratio	Std. Err.	z	P>|z|	[95% Conf. Interval]
female	1	.	.	.	. .

Moreover, we could not even obtain a Kaplan–Meier estimate of the overall survivor function:

```
. sts list
          failure _d:  dead
    analysis time _t:  time1
```

Time	Beg. Total	Fail	Net Lost	Survivor Function	Std. Error	[95% Conf. Int.]
2	1	1	0	0.0000	.	. .
3	0	0	-1	0.0000	.	. .
5	1	0	1	0.0000	.	. .
6	0	0	-1	0.0000	.	. .
8	1	1	0	0.0000	.	. .
9	0	0	-1	0.0000	.	. .
10	1	1	0	0.0000	.	. .

This extreme dataset has no information that nonparametric and semiparametric methods can exploit; namely, there is never more than one subject at risk at any particular time, which makes comparing subjects within risk sets difficult. Yet parametric methods have no difficulty with these data:

```
. streg female, dist(exponential)

        failure _d:  dead
   analysis time _t:  time1

Iteration 0:    log likelihood = -.46671977
Iteration 1:    log likelihood = -.41031745
Iteration 2:    log likelihood = -.40973293
Iteration 3:    log likelihood = -.40973283

Exponential regression -- log relative-hazard form

No. of subjects =            4              Number of obs    =         4
No. of failures =            3
Time at risk    =            7
                                            LR chi2(1)       =      0.11
Log likelihood  =   -.40973283             Prob > chi2      =    0.7357
```

_t	Haz. Ratio	Std. Err.	z	P>|z|	[95% Conf. Interval]
female	.6666667	.8164966	-0.33	0.741	.0604511 7.352135

We began with this odd example not because it demonstrates the superiority of parametric methods, which have their disadvantages. We used this example only to emphasize the different way parametric methods exploit the information in the data to obtain estimates of the parameters.

Nonparametric and semiparametric methods compare subjects at the times when failures happen to occur. To emphasize that, we have carefully concocted an example where no such comparison is possible.

Parametric methods, on the other hand, do not base their results on such comparisons. Rather, for each record in the data spanning $(t_{0j}, t_j]$, parametric estimation schemes use probabilities that depict what occurs over the whole interval given what is known about the subject during this time $(\mathbf{x}_j)$. For example, if it is observed that the subject was censored at time t_j, then the contribution to the likelihood for this record is

$$L_j = S(t_j | t_{0j}, \mathbf{x}_j) = \frac{S(t_j | \mathbf{x}_j)}{S(t_{0j} | \mathbf{x}_j)}$$

The implications of this difference in procedure are dramatic, but they are only matters of efficiency. For example, consider a survivor model where failure depends on a (possibly time-varying) covariate x. Pretend that one of the subjects in the dataset has the following x profile:

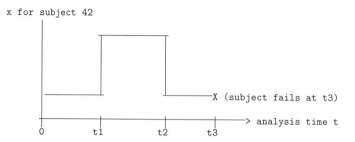

In semiparametric analysis (Cox regression), if no other subject fails between `t1` and `t2`, it simply does not matter that x blipped up for this subject because no comparisons will be made in that interval using the temporarily higher value of x. We would obtain the same Cox regression results if the blip in the time profile for this subject did not exist, i.e., if x remained at its initial value throughout.

The blip in x, however, would be of importance in a parametric model, regardless of whether other failures occurred in the interval, because the parametric model would exploit all the information.

Cox regression is not making an error by ignoring the blip—it is merely being inefficient. Suppose that higher values of x increase failure rates. Conditional on having survived beyond `t2`, the blip's occurrence becomes irrelevant for subsequent survival. The information in the blip is that it indeed occurred and the subject managed to survive it, which means that this subject provides evidence that higher values of x really are not so bad. Cox regression would ignore that unless other failures occurred in the interval, in which case some amount of the information contained in the interval would be exploited in improving the estimate of the effect of x. Parametric methods would ignore no part of that information.

12.2   Classes of parametric models

Parametric models are written in a variety of ways. As we stated in chapter 1, linear regression is an example of a survival model:

$$t_j = \mathbf{x}_j \boldsymbol{\beta}_x + \epsilon_j \qquad \epsilon_j \sim \mathrm{N}(0, \sigma^2)$$

Other models are written in the log-time metric (also known as the accelerated failure-time metric),

$$\ln(t_j) = \mathbf{x}_j \boldsymbol{\beta}_x + \epsilon_j \qquad \epsilon_j \sim \text{oddly, but not odd given the context}$$

and others in the hazard metric,

$$h(t|\mathbf{x}_j) = h_0(t) \exp(\mathbf{x}_j \boldsymbol{\beta}_x), \qquad h_0(t) = \text{some functional form}$$

and, in fact, you can imagine parametric models written in other ways. Parametric models are written in the way natural to the parameterization. Some models may be written in more than one way because they fit more than one style of thinking. It is still the same model and is just being expressed differently. Stata can fit the following parametric models:

1. Time parameterization:

   a. linear regression
      `regress, cnreg, intreg, tobit`

2. Log-time parameterization:

   a. exponential
      `streg, dist(exponential) time`
   b. Weibull
      `streg, dist(weibull) time`
   c. lognormal
      `streg, dist(lognormal)`
   d. loglogistic
      `streg, dist(loglogistic)`
   e. gamma
      `streg, dist(gamma)`

3. `hazard` parameterization:

   a. exponential
      `streg, dist(exponential)`
   b. Weibull
      `streg, dist(weibull)`
   c. Gompertz
      `streg, dist(gompertz)`

Except for linear-regression time-parameterization models (which we do not discuss further), parametric models in Stata are fit using the `streg` command, and options control the particular parameterization used. As with `stcox`, you must `stset` your data prior to using `streg`. Although some models (e.g., exponential and Weibull) fit into more than one parameterization, they are still the same model.

12.2.1 Parametric proportional hazards models

Proportional hazards models are written

$$h(t|\mathbf{x}_j) = h_0(t) \exp(\mathbf{x}_j \boldsymbol{\beta}_x)$$

In the Cox model, $h_0(t)$ was simply left unparameterized, and through conditioning on failure times, estimates of $\boldsymbol{\beta}_x$ were obtained anyway. In the parametric approach, a functional form for $h_0(t)$ is specified. For example, if we assume $h_0(t) = \exp(a)$ for some a, then we have the exponential model. The baseline hazard is assumed constant over time, and there is an extra parameter, a, to estimate. When we fit this model, we are just estimating $(a, \boldsymbol{\beta}'_x)$.

If we assume

$$h_0(t) = pt^{p-1}\exp(a)$$

then we have the Weibull model. This model has two ancillary parameters, a and p. When we fit this model, we are estimating $(a, p, \boldsymbol{\beta}'_x)$.

If we assume

$$h_0(t) = \exp(a)\exp(\gamma t)$$

then we obtain the Gompertz model, another model with two ancillary parameters, this time a and γ. Estimated from that data is $(a, \gamma, \boldsymbol{\beta}'_x)$.

There are other possibilities, and, of course, you could make up your own. Whatever function is chosen for $h_0(t)$ should parameterize adequately what the baseline hazard really is.

In any case, all these models produce results that are directly comparable to those produced by Cox regression. In all these models, $\mathbf{x}\boldsymbol{\beta}_x$ is the log relative-hazard and the elements of $\boldsymbol{\beta}_x$ have the standard interpretation, meaning that $\exp(\beta_i)$ is the hazard ratio for the ith coefficient. Also parametric models produce estimates of the ancillary parameters, and from that you can obtain the predicted baseline hazard function $h_0(t)$ and any of the other related functions, such as the cumulative hazard and the survivor function.

The (direct) comparability to Cox regression is probably the most appealing feature of the parametric proportional hazards model. When you engage in this kind of parametric estimation, it is prudent to compare the estimated coefficients, $\widehat{\boldsymbol{\beta}}_x$, with those from a Cox model fit to verify that they are roughly similar. If they prove not to be similar, then this is evidence of a misparameterized underlying baseline hazard.

Stata makes this comparison easy because, by default, `streg` reports hazard ratios (exponentiated coefficients) when fitting models in this metric. For example, in our discussion of Cox regression, we repeatedly examined the following model:

```
. use http://www.stata-press.com/data/cggm/hip2
(hip fracture study)
```

```
. stcox protect age
           failure _d:  fracture
     analysis time _t:  time1
                   id:  id
Iteration 0:   log likelihood = -98.571254
Iteration 1:   log likelihood = -82.735029
Iteration 2:   log likelihood = -82.471037
Iteration 3:   log likelihood = -82.470259
Refining estimates:
Iteration 0:   log likelihood = -82.470259

Cox regression -- Breslow method for ties

No. of subjects =           48            Number of obs    =        106
No. of failures =           31
Time at risk    =          714
                                          LR chi2(2)       =      32.20
Log likelihood  =   -82.470259            Prob > chi2      =     0.0000
```

_t _d	Haz. Ratio	Std. Err.	z	P>|z|	[95% Conf. Interval]	
protect	.1046812	.0475109	-4.97	0.000	.043007	.2547989
age	1.110972	.0420079	2.78	0.005	1.031615	1.196434

If we now fit this same model using exponential regression (meaning that we assume the baseline hazard $h_0(t)$ is constant), we would obtain

```
. streg protect age, dist(exponential)
           failure _d:  fracture
     analysis time _t:  time1
                   id:  id
Iteration 0:   log likelihood = -60.067085
Iteration 1:   log likelihood = -54.034598
Iteration 2:   log likelihood = -47.553588
Iteration 3:   log likelihood = -47.534671
Iteration 4:   log likelihood = -47.534656
Iteration 5:   log likelihood = -47.534656

Exponential regression -- log relative-hazard form

No. of subjects =           48            Number of obs    =        106
No. of failures =           31
Time at risk    =          714
                                          LR chi2(2)       =      25.06
Log likelihood  =   -47.534656            Prob > chi2      =     0.0000
```

_t	Haz. Ratio	Std. Err.	z	P>|z|	[95% Conf. Interval]	
protect	.1847118	.0684054	-4.56	0.000	.0893849	.3817025
age	1.084334	.0371696	2.36	0.018	1.013877	1.159688

Note how different is the estimated hazard ratio for protect in the two models: stcox reports 0.105, streg reports 0.185. This inconsistency points out the inadequacy of a model that assumes a constant baseline hazard, and in fact, when we previously considered this Cox model, we noted that the estimated cumulative hazard seemed to be increasing at an increasing rate. Thus the assumption of constant hazard is probably incorrect here, and we would obtain better estimates by choosing a parameterization for $h_0(t)$ that would allow it to grow.

The Weibull model will do that. Using that distribution, we would obtain

```
. streg protect age, dist(weibull)

        failure _d:  fracture
   analysis time _t:  time1
                id:  id

Fitting constant-only model:

Iteration 0:   log likelihood = -60.067085
Iteration 1:   log likelihood =  -59.30148
Iteration 2:   log likelihood = -59.298481
Iteration 3:   log likelihood = -59.298481

Fitting full model:

Iteration 0:   log likelihood = -59.298481
Iteration 1:   log likelihood = -54.887563
Iteration 2:   log likelihood = -42.123875
Iteration 3:   log likelihood = -41.993012
Iteration 4:   log likelihood = -41.992704
Iteration 5:   log likelihood = -41.992704

Weibull regression -- log relative-hazard form
```

No. of subjects =	48			Number of obs	=	106
No. of failures =	31					
Time at risk =	714					
				LR chi2(2)	=	34.61
Log likelihood =	-41.992704			Prob > chi2	=	0.0000

_t	Haz. Ratio	Std. Err.	z	P>|z|	[95% Conf. Interval]	
protect	.1099611	.0448214	-5.42	0.000	.0494629	.2444548
age	1.117186	.0423116	2.93	0.003	1.03726	1.203271
/ln_p	.5188694	.1376486	3.77	0.000	.2490831	.7886556
p	1.680127	.2312671			1.282849	2.200436
1/p	.5951931	.0819275			.4544553	.7795152

and we now obtain an estimate of the hazard ratio for `protect` of 0.110, which is more in line with the Cox regression result.

If you are looking for a parameterization of $h_0(t)$ that has considerable flexibility and has no restrictions on the shape of the hazard that you want to impose upon the model, we suggest you stay with Cox regression. That is what Cox regression does, and Cox regression does it well. Parametric estimation is appropriate when you do have an idea of what the baseline hazard looks like and you want to impose that idea to (1) obtain the most efficient estimates of $\boldsymbol{\beta}_x$ possible and (2) obtain an estimate of $h_0(t)$ subject to that constraint.

We mentioned above that Stata makes comparing Cox with parametric estimation easy because `streg` reports hazard ratios when fitting models that have a natural proportional hazards parameterization (exponential, Weibull, Gompertz). Parametric models, as we have suggested, sometimes have more than one parameterization. The Weibull model is an example of this, and a popular parameterization of the Weibull is that which carries the accelerated failure-time interpretation. Stata can also report results in this alternate metric (obtained by specifying the `time` option):

```
. streg protect age, dist(weibull) time

         failure _d:  fracture
   analysis time _t:  time1
                 id:  id

Fitting constant-only model:

Iteration 0:   log likelihood = -60.067085
Iteration 1:   log likelihood =  -59.30148
Iteration 2:   log likelihood = -59.298481
Iteration 3:   log likelihood = -59.298481

Fitting full model:

Iteration 0:   log likelihood = -59.298481
Iteration 1:   log likelihood = -54.887563
Iteration 2:   log likelihood = -42.123875
Iteration 3:   log likelihood = -41.993012
Iteration 4:   log likelihood = -41.992704
Iteration 5:   log likelihood = -41.992704

Weibull regression -- accelerated failure-time form

No. of subjects =          48            Number of obs    =        106
No. of failures =          31
Time at risk    =         714
                                         LR chi2(2)       =      34.61
Log likelihood  =    -41.992704          Prob > chi2      =     0.0000
```

_t	Coef.	Std. Err.	z	P>|z|	[95% Conf. Interval]	
protect	1.313965	.2366229	5.55	0.000	.8501928	1.777737
age	-.0659554	.0221171	-2.98	0.003	-.1093041	-.0226067
_cons	6.946524	1.575708	4.41	0.000	3.858192	10.03486
/ln_p	.5188694	.1376486	3.77	0.000	.2490831	.7886556
p	1.680127	.2312671			1.282849	2.200436
1/p	.5951931	.0819275			.4544553	.7795152

These results look nothing like those reported by `stcox`, nor should they. These results are reported in a different metric, but it is merely a different way of reporting the same information. In fact, the above coefficients may be transformed back into the proportional hazards metric using the relationship $\mathrm{HR} = \exp(-\widehat{p}\widehat{\beta}_{\mathrm{AFT}})$, where HR is a single hazard ratio reported in the proportional hazards metric and $\widehat{\beta}_{\mathrm{AFT}}$ is the corresponding regression coefficient from an accelerated failure-time model.

However, we do not expect you to remember that relationship, which is unique to the Weibull. If a model can be cast in the proportional hazards metric, then `streg` can report results in that metric, and those results are directly comparable with those produced by `stcox`. Even if you find it otherwise desirable to view and think about these results in the accelerated failure-time metric, you can still compare the results reported in the hazard metric with those of `stcox`, and then, after convincing yourself that they are reasonable, return to the accelerated failure-time metric. The models are the same—they just look different.

❑ **Technical note**

When fitting models that have both a proportional hazards and an accelerated failure-time parameterization, switching from one parameterization to the other requires refitting the entire model. Using `streg` in replay mode will not suffice. Although both models are equivalent and transformation from one form to the other is possible without reestimation, there are many aspects of Stata's postestimation commands (such as `test` and `predict`) that are tailored to one particular parameterization. As such in practice, reestimation is required. View this as a shortcoming of Stata, not that the models are different in any substantive way.

❑

Returning to our discussion of the proportional hazards metric, recall that `stcox` by default reported hazard ratios (exponentiated coefficients) but that, if you specified the `nohr` option, it would report the coefficients themselves. `streg` works the same way, and you can specify `nohr` when fitting the model or when you redisplay results. For example, when we fit our exponential model in the proportional hazards metric, by default we displayed hazard ratios. If we specify the `nohr` upon replay, we get the coefficients themselves:

```
. streg protect age, dist(exponential)
  (output omitted)
. streg, nohr

Exponential regression -- log relative-hazard form
No. of subjects =          48                Number of obs   =       106
No. of failures =          31
Time at risk    =         714
                                             LR chi2(2)      =     25.06
Log likelihood  =   -47.534656               Prob > chi2     =    0.0000
```

_t	Coef.	Std. Err.	z	P>|z|	[95% Conf.	Interval]
protect	-1.688958	.3703357	-4.56	0.000	-2.414803	-.9631137
age	.0809663	.0342787	2.36	0.018	.0137813	.1481514
_cons	-7.892737	2.458841	-3.21	0.001	-12.71198	-3.073498

When comparing this output with the previous one from exponential regression, $\exp(-1.688958) = 0.1847$, the reported hazard ratio for `protect`.

We see one thing, however, that is new: `streg` reported a coefficient for `_cons`, something to which `stcox` has no counterpart. The coefficient on `_cons` has to do with the estimation of $h_0(t)$. In particular, the exponential model is

$$h(t|\mathbf{x}_j) = h_0(t)\exp(\mathbf{x}_j\boldsymbol{\beta}_x)$$

where $h_0(t) = \exp(a)$, meaning that

$$\begin{aligned} h(t|\mathbf{x}_j) &= \exp(a + \mathbf{x}_j\boldsymbol{\beta}_x) \\ &\equiv \exp(\beta_0 + \mathbf{x}_j\boldsymbol{\beta}_x) \end{aligned}$$

from which we get our intercept term `_cons`. The intercept in the exponential model has to do with the level of the baseline hazard function, and here we estimate $h_0(t) = \exp(-7.892737)$. This is a very small hazard, but remember, just as with `stcox`, the baseline hazard is a reflection of the hazard when all covariates equal zero. In particular, for our model `age==0` is an absurd value. Just as with `stcox`, if you want to normalize the baseline hazard to reflect the hazard for reasonable values of the covariates, adjust the covariates so that the zeros are themselves reasonable:

```
. gen age60 = age-60

. streg protect age60, dist(exponential) nohr noshow

Iteration 0:    log likelihood = -60.067085
Iteration 1:    log likelihood = -54.034598
Iteration 2:    log likelihood = -47.553588
Iteration 3:    log likelihood = -47.534671
Iteration 4:    log likelihood = -47.534656
Iteration 5:    log likelihood = -47.534656

Exponential regression -- log relative-hazard form

No. of subjects =          48              Number of obs   =        106
No. of failures =          31
Time at risk    =         714
                                           LR chi2(2)      =      25.06
Log likelihood  =    -47.534656            Prob > chi2     =     0.0000
```

_t	Coef.	Std. Err.	z	P>|z|	[95% Conf. Interval]
protect	-1.688958	.3703357	-4.56	0.000	-2.414803 -.9631137
age60	.0809663	.0342787	2.36	0.018	.0137813 .1481514
_cons	-3.034758	.4536734	-6.69	0.000	-3.923942 -2.145575

The baseline hazard is now a more reasonable $\exp(-3.034758) = 0.048$ failures per month.

12.2.2   Accelerated failure-time models

Accelerated failure-time models, also known as accelerated-time models or ln(time) models, follow the parameterization

$$\ln(t_j) = \mathbf{x}_j\boldsymbol{\beta}_x + \epsilon_j \qquad \epsilon_j \sim \text{oddly, but not odd given the context}$$

The word "accelerated" is used in describing these models because—rather than assuming that failure time t_j is exponential, Weibull, or some other form—a distribution is instead assumed for

$$\tau_j = \exp(-\mathbf{x}_j\boldsymbol{\beta}_x)t_j$$

and $\exp(-\mathbf{x}_j\boldsymbol{\beta}_x)$ is called the acceleration parameter.

- If $\exp(-\mathbf{x}_j\boldsymbol{\beta}_x) = 1$, then $\tau_j = t_j$, and time passes at its "normal" rate.
- If $\exp(-\mathbf{x}_j\boldsymbol{\beta}_x) > 1$, then time passes more quickly for the subject (time is accelerated), and so failure would be expected to occur sooner.
- If $\exp(-\mathbf{x}_j\boldsymbol{\beta}_x) < 1$, then time passes more slowly for the subject (time is decelerated), and so failure would be expected to occur later.

The derivation of these models is straightforward. If $\tau_j = \exp(-\mathbf{x}_j\boldsymbol{\beta}_x)t_j$, then $t_j = \exp(\mathbf{x}_j\boldsymbol{\beta}_x)\tau_j$, and

$$\ln(t_j) = \mathbf{x}_j\boldsymbol{\beta}_x + \ln(\tau_j) \tag{12.1}$$

The random quantity $\ln(\tau_j)$ has a distribution determined by what is assumed about the distribution of τ_j, and in the usual nomenclature of these models, it is the distribution of τ_j that is specified. For example, in a lognormal model, τ_j follows a lognormal distribution, which implies that $\ln(\tau_j)$ follows a normal distribution, which makes (12.1) analogous to linear regression.

❏ Technical note

In comparing our derivation of this model with that found elsewhere, you may find that other sources report the result

$$\ln(t_j) = -\mathbf{x}_j\boldsymbol{\beta}_x + \ln(\tau_j)$$

rather than (12.1), but that is only because they started with the assumption that $\tau_j = \exp(\mathbf{x}_j\boldsymbol{\beta}_x)t_j$ rather than $\tau_j = \exp(-\mathbf{x}_j\boldsymbol{\beta}_x)t_j$. The sign change is not important. If one really thought in the accelerated failure-time metric, it would indeed be more natural to follow this alternate specification because, in our developments, it is $-\boldsymbol{\beta}_x$ that corresponds to how time is accelerated.

However, the use of (12.1) is justified on the grounds of predicted time. A positive coefficient in $\boldsymbol{\beta}_x$ serves to increase the expected value of $\ln$(time to failure). This is the view we chose to accept at the outset, and thus we use $\tau_j = \exp(-\mathbf{x}_j\boldsymbol{\beta}_x)t_j$.

❏

There is a second sense in which these models are "accelerated". The effect of a change in one of the $\mathbf{x}$ variables, measured in time units, increases with t. For example, pretend that $\beta_1 = 0.5$, meaning that a 1-unit increase in x_1 increases the expected value of $\ln(t)$ by 0.5. For a subject predicted to fail at $t = 1$, this 1-unit increase would delay the predicted time of failure to $\exp\{\ln(1) + 0.5\} = 1.65$. For a subject predicted to fail at $t = 5$, this 1-unit increase would delay the predicted time of failure to 8.24. That is, the marginal effect of x_1 accelerates. For larger t, we expect a longer delay in failure due to a 1-unit increase in x_1.

In the accelerated failure-time metric, exponentiated coefficients have the interpretation of time ratios for a 1-unit change in the corresponding covariate. For a subject with covariate values $\mathbf{x} = (x_1, x_2, \ldots, x_k)$,

$$t_j = \exp(\beta_1 x_1 + \beta_2 x_2 + \cdots + \beta_k x_k)\tau_j$$

If the subject had x_1 increased by 1, then

$$t_j^* = \exp\{\beta_1(x_1 + 1) + \beta_2 x_2 + \cdots + \beta_k x_k\}\tau_j$$

and the ratio of t_j^* to t_j [and, more importantly, the ratio of $E(t_j^*)$ and $E(t_j)$] is $\exp(\beta_1)$. As such, the time ratio, $\exp(\beta_1)$, can be interpreted as the factor by which the expected time-to-failure is multiplied as a result of increasing x_1 to $x_1 + 1$. If $\exp(\beta_1) < 1$ ($\beta_1 < 0$), the expected time decreases; if $\exp(\beta_1) > 1$ ($\beta_1 > 0$), it increases.

By default, `streg`, when used to fit accelerated failure-time (AFT) models, reports coefficients and not exponentiated coefficients, but you can specify the `tr` option either when you estimate or when you redisplay results to see results in time ratios. The default is surprising given that, by default, `streg`, when used to fit proportional hazards (PH) models, reports exponentiated coefficients, but that is because most authors do not even mention the time-ratio interpretation of exponentiated coefficients. However, because of the ease of interpretation, specifying option `tr` and reporting time ratios is recommended.

12.2.3 Comparing the two parameterizations

The PH metric is used mainly as an analog to Cox regression when the researcher wishes to gain insight into the actual risk process (the hazard function) that causes failure and to gain insight into how the risk changes with the values of covariates in the model. As with Cox regression, little attention is paid to the actual failure times, and predictions of these failure times are seldom desired.

The AFT metric, however, gives a more prominent role to analysis time. The typical AFT model is of the form

$$\ln(t_j) = \mathbf{x}_j \boldsymbol{\beta}_x + \ln(\tau_j)$$

and by specifying a model in this form, one basically is asserting an interest in what happens to $E\{\ln(t_j)|\mathbf{x}_j\}$ for different values of $\mathbf{x}_j$. With such an interest usually comes a desire to predict either failure time or the logarithm of failure time, and there are instances when such predictions can prove problematic.

One difficulty with predicting time to failure has to do with time-varying covariates. Time-varying covariates cause no theoretical difficulty in accelerated failure-time models, nor are they difficult to fit (in fact, Stata does all the work). The problem occurs when one goes back to construct predictions.

In our hip-fracture data, we have a time-varying covariate, calcium. It turned out not to be significant in a Cox model, and the same is true if we estimate using a Weibull model. However, if we omit age from the model, we can get calcium to be significant to allow us to make our point about prediction with time-varying covariates. We begin by fitting the model:

```
. use http://www.stata-press.com/data/cggm/hip2, clear
(hip fracture study)
. streg protect calcium, time dist(weibull)
        failure _d:  fracture
   analysis time _t:  time1
              id:  id

Fitting constant-only model:

Iteration 0:   log likelihood = -60.067085
Iteration 1:   log likelihood =  -59.30148
Iteration 2:   log likelihood = -59.298481
Iteration 3:   log likelihood = -59.298481

Fitting full model:

Iteration 0:   log likelihood = -59.298481
Iteration 1:   log likelihood = -54.764667
Iteration 2:   log likelihood = -42.840111
Iteration 3:   log likelihood = -42.728013
Iteration 4:   log likelihood = -42.727796
Iteration 5:   log likelihood = -42.727796

Weibull regression -- accelerated failure-time form

No. of subjects =           48              Number of obs    =         106
No. of failures =           31
Time at risk    =          714
                                            LR chi2(2)       =       33.14
Log likelihood  =   -42.727796             Prob > chi2      =      0.0000
```

_t	Coef.	Std. Err.	z	P>|z|	[95% Conf. Interval]	
protect	1.300919	.241717	5.38	0.000	.8271626	1.774676
calcium	.238073	.0904338	2.63	0.008	.060826	.4153199
_cons	-.0700465	.9016456	-0.08	0.938	-1.837239	1.697146
/ln_p	.4849596	.1363349	3.56	0.000	.2177481	.752171
p	1.624109	.2214228			1.243274	2.121601
1/p	.6157221	.0839444			.4713421	.804328

This model has been fit in the AFT metric. Thus the model states that

$$\ln(t_j) = -0.07 + 1.30\text{protect}_j + 0.24\text{calcium}_j + \ln(\tau_j)$$

The estimate $\hat{p} = 1.624$ has to do with the distribution of $\ln(\tau)$ and need not concern us right now.

We now obtain the predicted times to failure (which we choose to be the conditional mean given the covariates), something easy enough to do in Stata:

```
. predict t_hat, mean time
```

Let's look at the predictions for subject id==10, one of our subjects who has time-varying covariate calcium.

```
. list id _t0 _t protect calcium _d t_hat if id==10
```

	id	_t0	_t	protect	calcium	_d	t_hat
11.	10	0	5	0	9.69	0	8.384427
12.	10	5	8	0	9.47	0	7.956587

Subject 10 was observed over the period $(0, 8]$ and then was censored; at time 0 and 5, we had measurements of calcium on her. The variable t_hat contains the predicted (mean) failure times.

The interpretation of t_hat in the first observation is that if we had a subject who had protect==0 and calcium==9.69, and those covariates were fixed, the predicted time of failure would be 8.38. In the second observation, the prediction is that if we had a subject who was under continual observation from time 0 forward, and if the subject had fixed covariates protect==0 and calcium==9.47, then the predicted time of failure would be 7.96.

Neither of those predictions really has to do with subject 10. To obtain the predicted time to failure for a subject under continual observation whose value of calcium starts at 9.69 and then changes to 9.47 at $t = 5$ is a truly miserable calculation, and Stata has no automated way to calculate this for you. If you really wanted it, here is what you would have to do:

1. Start by translating results back to the hazard metric.
2. Write down the hazard function $h(t)$ for subject 10. It will be a function of the estimated parameters, and it will have a discontinuity at time 5.
3. Integrate $h(t)$ to obtain the cumulative hazard $H(t)$.
4. Obtain the density function of t, $f(t) = h(t) \exp\{-H(t)\}$.
5. Having $f(t)$, now obtain its expected value.

It would be a lot of work. Even after you had that, you would have to think carefully if that is what you want. Do you really want to assume that calcium changes from 9.69 to 9.47 at $t = 5$, or would you also like to add that it continues to decline, perhaps declining to 9.25 at time 10, and so on? Changing the future profile of the variable will change the expected value.

Thus time can be a difficult metric in which to interpret regression results. The above difficulties are not the fault of the AFT metric—we would have the same problem with a PH model—it is just that AFT models tend to place the emphasis on time, and the above predictions thus become more desirable.

For this reason, it is often preferable to think of AFT models not as linear models on ln(time) but instead to use something analogous to the PH interpretation. Namely, for an AFT model we can think in terms of survivor functions:

$$S(t|\mathbf{x}) = S_0\{\exp(-\mathbf{x}\boldsymbol{\beta}_x)t\}$$

The probability of survival past time t for an individual with covariates $\mathbf{x}$ is equivalent to the probability of survival past time $\exp(-\mathbf{x}\boldsymbol{\beta}_x)t$ for an individual at baseline (one with covariate values all equal to zero). For example, take $t = 2$ and $\exp(-\mathbf{x}\boldsymbol{\beta}_x) = 0.7$. This would mean that an individual with covariates $\mathbf{x}$ has probability $S_0(1.4)$ of surviving past time $t = 2$. An individual at baseline has probability $S_0(2)$ of surviving past time $t = 2$. Because $S_0()$ is increasing, $S_0(1.4) > S_0(2)$, and our individual with covariates $\mathbf{x}$ is better off.

Some AFT models, namely the exponential and Weibull, have both a hazard interpretation and an accelerated failure-time interpretation. The other AFT models that Stata can fit—lognormal, loglogistic, and gamma—have no natural PH interpretation. This is not to deny that you could work out the hazard function corresponding to these models, but that function would not be a simple function of the regression coefficients, and in no sense would it be easy to interpret.

13 A survey of parametric regression models in Stata

The parametric regression models in Stata work like the Cox model in that they can handle all the same problems—time-varying covariates, delayed entry, gaps, and right censoring. If your dataset is appropriate for use with `stcox`, it is ready for use with `streg`.

Datawise, there is one important difference that you will need to consider. In Cox regression, the definition of `origin()`—the definition of when analysis time $t = 0$—plays no real role, whereas in parametric models, it can be (and usually is) vital.

The only role `origin()` plays in `stcox` is that Stata applies the rule that subjects are not and cannot be at risk prior to $t = 0$. Aside from that, the definition of the origin really does not matter. If you took a dataset, ran `stcox`, and then added 50 to all time variables and reran the analysis, nothing would change. The only role played by time in the Cox model is to determine who is to be compared with whom; the magnitude of the time variable does not matter.

Try the same experiment with a parametric model—add 50 to the time variables—and results will change, unless you are fitting an exponential regression model. In parametric models, time plays a real role, and the point when $t = 0$ determines when risk begins accumulating. Adding 50 to all the time variables changes that accumulated risk nonproportionally. You need to think about when the onset of risk really is.

Both the Cox model and parametric models are invariant to multiplicative transforms of time: it makes no difference whether you measure time in minutes, hours, days, or years. It actually is not a theoretical constraint on parametric models that they be scale invariant, but all the standard models are because, were they not, they would hardly be reasonable or useful.

The likelihood functions of the parametric models—regardless of the particular one under consideration—all follow the same general form:

$$L_j(\boldsymbol{\beta}_x, \boldsymbol{\Theta}) = \frac{\{S(t_j|\mathbf{x}_j\boldsymbol{\beta}_x, \boldsymbol{\Theta})\}^{1-d_j}\{f(t_j|\mathbf{x}_j\boldsymbol{\beta}_x, \boldsymbol{\Theta})\}^{d_j}}{S(t_{0j}|\mathbf{x}_j\boldsymbol{\beta}_x, \boldsymbol{\Theta})} \tag{13.1}$$

where $f()$ is the density function of the assumed distribution, $S()$ is the corresponding survivor function, and $(t_{0j}, t_j, d_j, \mathbf{x}_j)$ is the information on the jth observation. The parameters $\boldsymbol{\beta}_x$ and $\boldsymbol{\Theta}$ are estimated from the data: $\boldsymbol{\beta}_x$ are the coefficients on $\mathbf{x}$, and $\boldsymbol{\Theta}$ are ancillary parameters, if any, required by the assumed distribution. For instance, for the Weibull distribution $\boldsymbol{\Theta} = (\beta_0, p)$, where β_0 is the scale parameter and p is the shape parameter, and we choose β_0 to denote the scale because, as we show later, for this model the scale can also be thought of as an intercept term for the linear predictor $\mathbf{x}\boldsymbol{\beta}_x$.

The triple (t_{0j}, t_j, d_j) summarizes the survival experience for the observation: the subject was observed and known not to fail during the period $t_{0j} < t < t_j$, and then at $t = t_j$, the subject either failed $(d_j = 1)$ or was censored $(d_j = 0)$. Thus the powers $(1 - d_j)$ and d_j in (13.1) serve to select either $S()$ or $f()$ as the numerator of the ratio. If censored, $S()$ is chosen, and that is the probability that the subject survives from 0 to t_j without failure. If $d_j = 1$—if the subject fails—$f()$ is chosen, and that is the "probability" of failure at time t_j. Either way, the numerator is divided by $S(t_{0j}|\mathbf{x}_j\boldsymbol{\beta}_x, \boldsymbol{\Theta})$, which is the probability of surviving up to time t_{0j}; thus, whichever is the numerator, it is converted to a conditional probability or probability density for the time span under consideration. When $t_{0j} = 0$, $S(t_{0j}|\mathbf{x}_j\boldsymbol{\beta}_x, \boldsymbol{\Theta}) = 1$.

The terms of the likelihood function are stated for observations and not subjects: there may be more than one observation on a subject. In simple survival data, there is a one-to-one correspondence between observations and subjects, and $t_{0j} = 0$. But in more complex cases, a subject may have multiple observations, as we have seen previously:

```
id    _t0   _t    x    _d
101   0     5     4    0
101   5     7     3    0
101   7     8     4    1
```

Thus parametric models are generalized to allow time-varying covariates.

Equation (13.1) may be equivalently written as

$$L_j(\boldsymbol{\beta}_x, \boldsymbol{\Theta}) = \frac{S(t_j|\mathbf{x}_j\boldsymbol{\beta}_x, \boldsymbol{\Theta})}{S(t_{0j}|\mathbf{x}_j\boldsymbol{\beta}_x, \boldsymbol{\Theta})} \{h(t_j|\mathbf{x}_j\boldsymbol{\beta}_x, \boldsymbol{\Theta})\}^{d_j}$$

which you can obtain by substitution using the formulas given in chapter 2. This variation is as easily thought about. The first part, $S(t_j|\ldots)/S(t_{0j}|\ldots)$, is the probability of survival from t_{0j} until t_j. The last part, $h(t|\ldots)^{d_j}$, becomes $h(t_j|\ldots)$ if the span ends in failure (which is the corresponding risk of that event at time t_j), or 1 if the span ends in censoring.

All parametric likelihoods are of the above form, and the only difference among the models is how $S()$ [and therefore $f()$ and $h()$] is chosen.

streg can fit any of six parametric models: exponential, Weibull, Gompertz, lognormal, loglogistic, and generalized gamma.

13.1   The exponential model

13.1.1   Exponential regression in the PH metric

The exponential model is the simplest of the parametric survival models because it assumes that the baseline hazard is constant,

$$
\begin{aligned}
h(t|\mathbf{x}_j) &= h_0(t)\exp(\mathbf{x}_j\boldsymbol{\beta}_x) \\
&= \exp(\beta_0)\exp(\mathbf{x}_j\boldsymbol{\beta}_x) \\
&= \exp(\beta_0 + \mathbf{x}_j\boldsymbol{\beta}_x)
\end{aligned}
$$

for some constant β_0. We use the notation β_0 to emphasize that the constant may also be thought of as an intercept term from the linear predictor. Using the well-known relationships for the exponential model,

$$
\begin{aligned}
H(t|\mathbf{x}_j) &= \exp(\beta_0 + \mathbf{x}_j\boldsymbol{\beta}_x)t \\
S(t|\mathbf{x}_j) &= \exp\{-\exp(\beta_0 + \mathbf{x}_j\boldsymbol{\beta}_x)t\}
\end{aligned}
\tag{13.2}
$$

If you fit a model in this metric and display the regression coefficients instead of hazard ratios, the intercept term _b[_cons] is the estimate of β_0. To do this, specify the nohr option either during estimation or upon replay:

```
. use http://www.stata-press.com/data/cggm/hip2
(hip fracture study)

. streg age protect, dist(exp) nohr

        failure _d:  fracture
   analysis time _t:  time1
             id:  id

Iteration 0:   log likelihood = -60.067085
Iteration 1:   log likelihood = -54.034598
Iteration 2:   log likelihood = -47.553588
Iteration 3:   log likelihood = -47.534671
Iteration 4:   log likelihood = -47.534656
Iteration 5:   log likelihood = -47.534656

Exponential regression -- log relative-hazard form

No. of subjects =           48             Number of obs    =        106
No. of failures =           31
Time at risk    =          714
                                           LR chi2(2)       =      25.06
Log likelihood  =    -47.534656            Prob > chi2      =     0.0000
```

_t	Coef.	Std. Err.	z	P>|z|	[95% Conf. Interval]	
age	.0809663	.0342787	2.36	0.018	.0137813	.1481514
protect	-1.688958	.3703357	-4.56	0.000	-2.414803	-.9631137
_cons	-7.892737	2.458841	-3.21	0.001	-12.71198	-3.073498

Translating the above to our mathematical notation,

$$
\mathbf{x}_j\widehat{\boldsymbol{\beta}}_x = -1.69\text{protect}_j + 0.08\text{age}_j
$$

and $\widehat{\beta}_0 = -7.89$. Thus our estimate of the baseline hazard is $\widehat{h}_0(t) = \exp(-7.89) = 0.00037$, and our estimate of the overall hazard is

$$h(t|\mathbf{x}_j) = 0.00037 \exp(-1.69\texttt{protect}_j + 0.08\texttt{age}_j)$$

In the exponential model, $h_0(t)$ being constant means that the failure rate is independent of time; thus, the failure process is said to lack memory. As such, you may be tempted to view exponential regression as suitable for use only in the simplest of cases.

That would be unfair. There is another sense in which the exponential model is the basis of all other models. The baseline hazard $h_0(t)$ is constant, and given the proportional hazards model,

$$h(t|\mathbf{x}_j) = h_0(t) \exp(\mathbf{x}_j \boldsymbol{\beta}_x)$$

the way in which the overall hazard varies with time is purely a function of how $\mathbf{x}\boldsymbol{\beta}_x$ varies with time, if at all. The overall hazard need not be constant with time; it is just that every little bit of how the hazard varies must be explained by $\mathbf{x}\boldsymbol{\beta}_x$. If you fully understand a process, you should be able to do that. When you do not fully understand a process, you are forced to assign a role to time (the hazard increases with time, the hazard decreases with time, etc.), and in that way, you hope, you put your ignorance aside and still describe the part of the process you do understand.

It is a rare process that is fully understood, but your goal should be to understand the process well enough that, some day, you could fit an exponential model. Once you do fully understand a process, time plays no role. Rather, you have a model in which it is the accumulation of this toxin or of a particular kind of information or of something else that accounts for the apparent role of time, and you can describe exactly how the toxin, knowledge, or whatever accumulates with time.

Also exponential models can be used to model the overall hazard as a function of time if they include t or functions of t as covariates.

For example, for our hip-fracture data we could fit a model in which we simply claim that the hazard is constant with time,

$$h(t|\mathbf{x}_j) = h_0 \exp(\beta_1 \texttt{protect}_j + \beta_2 \texttt{age}_j)$$

for some constant h_0, or we could directly include t in the model,

$$h(t|\mathbf{x}_j) = h_0 \exp(\beta_1 \texttt{protect}_j + \beta_2 \texttt{age}_j + \beta_3 t) \tag{13.3}$$

As a computer issue, we will have to construct the time-varying covariate for t, but that is easy (although it may require considerable memory depending on your dataset. `streg` has no `tvc()` option like `stcox`, and so you must `stsplit` the data):

```
. summarize _t

    Variable |      Obs        Mean   Std. Dev.        Min        Max
-------------+-----------------------------------------------------
          _t |      106     11.5283    8.481024          1         39

. stsplit myt, at(1(1)39)
(608 observations (episodes) created)

. streg protect age myt, dist(exp) nohr

         failure _d:  fracture
   analysis time _t:  time1
                 id:  id

Iteration 0:   log likelihood = -60.067085
Iteration 1:   log likelihood = -51.705495
Iteration 2:   log likelihood = -43.731761
Iteration 3:   log likelihood = -43.704655
Iteration 4:   log likelihood = -43.704635
Iteration 5:   log likelihood = -43.704635

Exponential regression -- log relative-hazard form

No. of subjects =           48                 Number of obs   =        714
No. of failures =           31
Time at risk    =          714
                                               LR chi2(3)      =      32.72
Log likelihood  =    -43.704635                Prob > chi2     =     0.0000
```

_t	Coef.	Std. Err.	z	P>|z|	[95% Conf. Interval]	
protect	-2.225445	.4381355	-5.08	0.000	-3.084175	-1.366715
age	.1023765	.036895	2.77	0.006	.0300636	.1746895
myt	.066876	.0234889	2.85	0.004	.0208387	.1129133
_cons	-9.862588	2.694014	-3.66	0.000	-15.14276	-4.582417

We used `stsplit` to split the records at each time in $\{1, 2, 3, \dots, 39\}$ and create the time-varying covariate $\mathtt{myt} = t$.

Given the above estimation results, we can rewrite the model in (13.3) as

$$h(t|\mathbf{x}_j) = \{h_0 \exp(\beta_3 t)\} \exp(\beta_1 \mathtt{protect}_j + \beta_2 \mathtt{age}_j)$$

and, in a logical sense, $h_0 \exp(\beta_3 t)$ is our baseline hazard. Because $h_0 = \exp(\beta_0)$ in an exponential model, for our estimates we can calculate and graph the baseline hazard,

```
. gen hazard = exp(_b[_cons]) * exp(_b[myt] * myt)
. line hazard myt, c(J) sort l1title("baseline hazard")
```

which produces figure 13.1.

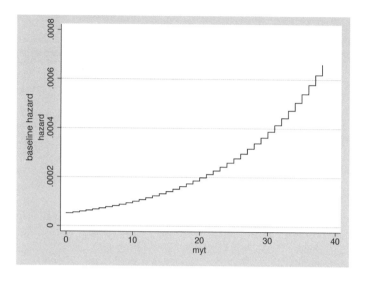

Figure 13.1. Estimated baseline hazard function

We did not really fit a smooth function of t for the baseline hazard in streg because that would require stsplitting our records infinitely finely. However, the "grid" size we chose (every integer) seems to be adequate for our purposes.

Here including t in the model resulted in a hazard function that not only increases with t but increases at an increasing rate. That effect, however, is due to the functional form we chose and may not really reflect what is going on in the data. In our parameterization, $h_0 \exp(\beta_3 t)$, the only way the hazard function can increase is at an increasing rate.

Continuing this development, let's try $\ln(\text{myt} + 1)$ instead of myt as a covariate so that the model becomes

$$
\begin{aligned}
h(t|\mathbf{x}_j) &= h_0 \exp\{\beta_1 \text{protect}_j + \beta_2 \text{age}_j + \beta_3 \ln(\text{myt} + 1)\} \\
&= [h_0 \exp\{\beta_3 \ln(\text{myt} + 1)\}] \exp(\beta_1 \text{protect}_j + \beta_2 \text{age}_j)
\end{aligned}
$$

where we now treat $h_0 \exp\{\beta_3 \ln(\text{myt} + 1)\}$ as the de facto baseline hazard,

```
. gen lmyt = ln(myt+1)
. streg protect age lmyt, dist(exp)
(output omitted)
. gen hazard2 = exp(_b[_cons] + _b[lmyt]*lmyt)
. line hazard2 myt, c(J) sort l1title("baseline hazard")
```

which produces figure 13.2.

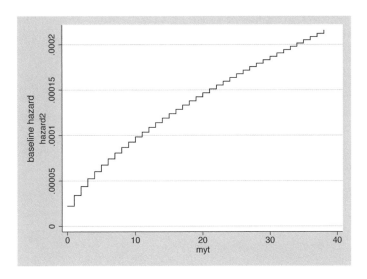

Figure 13.2. Estimated baseline hazard function using $\ln(\mathtt{myt} + 1)$

Here we get a hazard that increases at a (gently) decreasing rate, but once again, this could be an artifact of our parameterization.

Rather than specifying a particular functional form of myt to include in our model, we can get a better idea of the hazard function preferred by the data by including a crude step function of myt instead, in which the height of each step is its own model parameter:

```
. gen in_t1 =  0<=myt & myt<10
. gen in_t2 = 10<=myt & myt<20
. gen in_t3 = 20<=myt & myt<30
. gen in_t4 = 30<=myt
. streg protect age in_t2 in_t3 in_t4, dist(exp)
  (output omitted)
. gen hazard3 = exp(_b[_cons] + _b[in_t2]*in_t2 + _b[in_t3]*in_t3 +
> _b[in_t4]*in_t4)
. line hazard3 myt, c(J) sort l1title("baseline hazard")
```

This produces figure 13.3.

(Continued on next page)

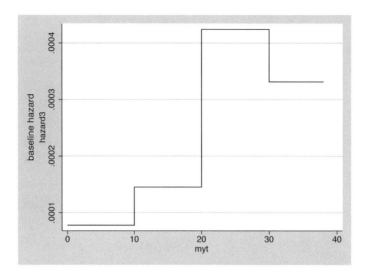

Figure 13.3. Estimated baseline hazard function using a step function

For reasons of collinearity, we do not include the first step indicator variable in_t1 in our model—the intercept term takes care of that step for us.

From figure 13.3, it appears that the hazard is increasing at an increasing rate but then jumps down at the end. This graph, however, can be misleading because there are different numbers of observations in each of the steps. A better graph would define the steps such that approximately 25% of the data are in each step:

```
. summarize myt, detail

                            myt

        Percentiles      Smallest
  1%           0              0
  5%           0              0
 10%           1              0        Obs              714
 25%           3              0        Sum of Wgt.      714

 50%           8                       Mean         10.61765
                           Largest     Std. Dev.    8.678827
 75%          16             35
 90%          23             36        Variance     75.32204
 95%          28             37        Skewness     .7923223
 99%          33             38        Kurtosis      2.79787

. gen in_t5 =   0<=myt & myt<3

. gen in_t6 =   3<=myt & myt<8

. gen in_t7 =   8<=myt & myt<16

. gen in_t8 = 16<=myt

. streg protect age in_t6 in_t7 in_t8, dist(exp)

  (output omitted )
```

```
. gen hazard4 = exp(_b[_cons] + _b[in_t6]*in_t6 + _b[in_t7]*in_t7 +
> _b[in_t8]*in_t8)
. line hazard4 myt, c(J) sort l1title("baseline hazard")
```

This produces figure 13.4.

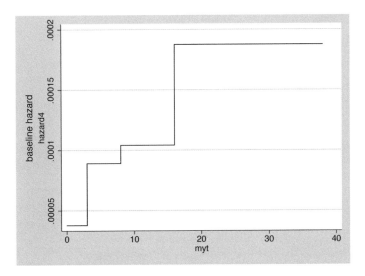

Figure 13.4. Estimated baseline hazard function using a better step function

Looking at this graph, we now know what we believe: the hazard increases at a decreasing rate. Actually, a better statement would be that up until $t = 16$, we believe it certainly is not increasing faster than linearly (probably increasing at a decreasing rate), and after that, we really do not know. In any case, we compare this hazard with the result we obtained when we used $\ln(\texttt{myt} + 1)$,

```
. line hazard2 hazard4 myt, c(J J) sort l1title("baseline hazard")
```

which produces figure 13.5.

(Continued on next page)

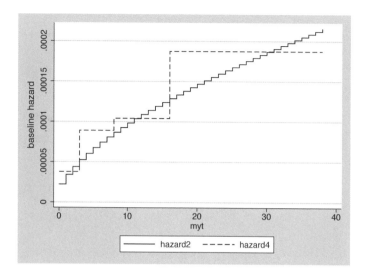

Figure 13.5. Comparison of estimated baseline hazards

It is a fair fit, except perhaps in the second interval.

The point of the above exercise is not that you should use exponential regression when you have no idea of what the baseline hazard looks like—use Cox regression instead. Exponential regression can be used to fit models in which the hazard varies with time, and that may be a reasonable thing to do, especially if you want to verify the fit of another parametric model. For instance, pretend that you have strong reason to believe that the formulation ought to be Weibull. Even after fitting a Weibull model, you could use the exponential model with dummy variables for intervals to verify that the Weibull fit was reasonable.

The method we have used—splitting the data on each integer time point—uses considerable memory. Depending on the size of your dataset, that may not matter, but the piecewise-constant hazard with steps at the 25th, 50th, and 75th percentiles of time would have required splitting the data on only three time points.

13.1.2   Exponential regression in the AFT metric

In the accelerated failure-time formulation of section 12.2.2, we have $\tau_j = \exp(-\mathbf{x}_j\boldsymbol{\beta}_x)t_j$, and in the exponential model, it is assumed that

$$\tau_j \sim \text{Exponential}\{\exp(\beta_0)\}$$

τ_j is distributed as exponential with mean $\exp(\beta_0)$. This implies that

$$
\begin{aligned}
\ln(t_j) &= \mathbf{x}_j\boldsymbol{\beta}_x + \ln(\tau_j) \\
&= \mathbf{x}_j\boldsymbol{\beta}_x + \beta_0 + u_j
\end{aligned}
$$

where u_j follows the extreme value (Gumbel) distribution, the result of which is that

$$E\{\ln(t_j)|\mathbf{x}_j\} = \beta_0 + \mathbf{x}_j\boldsymbol{\beta}_x + \Gamma'(1)$$

where $\Gamma'(1)$ is the negative of Euler's constant, obtained in Stata via the function call digamma(1), e.g.,

```
. display digamma(1)
-.57721566
```

❏ Technical note

The Gumbel distribution gets the name "extreme value distribution" because it can be shown to be the limiting distribution (as $n \to \infty$) of the minimum value from a sample of n identically distributed random deviates with a continuous probability distribution supported on the real line.

❏

We can also derive the accelerated failure-time (AFT) formulation by accelerating the effect of time on survival experience. At baseline values of $\mathbf{x}$, $\tau_j = t_j$ because all covariates are equal to zero. Thus the baseline survivor function of t_j is that from an exponential distribution with mean $\exp(\beta_0)$. That is,

$$S_0(t_j) = \exp\{-\exp(-\beta_0)t_j\}$$

In an AFT model, the effect of the covariates is to accelerate time by a factor of $\exp(-\mathbf{x}_j\boldsymbol{\beta}_x)$. Thus for the AFT model,

$$
\begin{aligned}
S(t_j|\mathbf{x}_j) &= S_0\{\exp(-\mathbf{x}_j\boldsymbol{\beta}_x)t_j\} \\
&= \exp\{-\exp(-\beta_0)\exp(-\mathbf{x}_j\boldsymbol{\beta}_x)t_j\} \\
&= \exp\{-\exp(-\beta_0 - \mathbf{x}_j\boldsymbol{\beta}_x)t_j\}
\end{aligned}
$$

and when we compare with (13.2), we find that the transformation from the proportional hazards (PH) model to the AFT metric (for the exponential model) is simply one of changing the signs of the regression coefficients.

Using our hip-fracture data, we can obtain exponential regression estimates in the AFT metric:

```
. use http://www.stata-press.com/data/cggm/hip2, clear
(hip fracture study)

. streg protect age, dist(exp) time

        failure _d:  fracture
   analysis time _t:  time1
              id:  id

Iteration 0:   log likelihood = -60.067085
Iteration 1:   log likelihood = -54.034598
Iteration 2:   log likelihood = -47.553588
Iteration 3:   log likelihood = -47.534671
Iteration 4:   log likelihood = -47.534656
Iteration 5:   log likelihood = -47.534656
```

```
Exponential regression -- accelerated failure-time form
No. of subjects =          48              Number of obs   =        106
No. of failures =          31
Time at risk    =         714
                                           LR chi2(2)      =      25.06
Log likelihood  =   -47.534656             Prob > chi2     =     0.0000
```

_t	Coef.	Std. Err.	z	P>|z|	[95% Conf. Interval]	
protect	1.688958	.3703357	4.56	0.000	.9631137	2.414803
age	-.0809663	.0342787	-2.36	0.018	-.1481514	-.0137813
_cons	7.892737	2.458841	3.21	0.001	3.073498	12.71198

From this, it follows that

$$\ln(t_j) = 7.89 + 1.69\text{protect}_j - 0.08\text{age}_j + u_j$$

or, if you prefer,

$$\tau_j = \exp(-1.69\text{protect}_j + 0.08\text{age}_j)t_j$$

The effect of `protect` is to slow down time, and the effect of `age` is to accelerate it; equivalently, the effect of `protect` is to delay failure, and that of `age` is to hasten it.

As earlier in section 13.1.1, we do not believe the assumption of constant hazard is appropriate for these data, and our previous concerns apply as much to the exponential model estimated in the AFT metric as they do to the model fit in the PH metric.

The approach we suggested for relaxing the assumption of constant hazard in the previous section, although applicable in this metric, is nearly impossible to interpret, and so we do not recommend it. You could, however, `stsplit` the data and introduce t as a variable in the model. If you do that, you will then discover that time speeds up with time, which you will have to think about carefully. When you go to evaluate the fit—to find out whether time should accelerate with time or, say, ln(time)—you find that nearly impossible to do because `predict` will not help you, or at least it will not help you if you stick with the time metric and predict failure times. In time-varying data, `predict, time` calculates predictions for each observation as if the covariates are constant, and that is not what you need. If you want to use the exponential model to explore the hazard, you will find that much easier to do in the PH metric.

13.2   Weibull regression

13.2.1   Weibull regression in the PH metric

The Weibull model assumes a baseline hazard of the form $h_0(t) = pt^{p-1}\exp(\beta_0)$, where p is some ancillary shape parameter estimated from the data and the scale parameter is parameterized as $\exp(\beta_0)$. Given a set of covariates, $\mathbf{x}_j$, under the PH model,

$$\begin{aligned} h(t|\mathbf{x}_j) &= h_0(t)\exp(\mathbf{x}_j\boldsymbol{\beta}_x) \\ &= pt^{p-1}\exp(\beta_0 + \mathbf{x}_j\boldsymbol{\beta}_x) \end{aligned}$$

and this yields

$$\begin{aligned} H(t|\mathbf{x}_j) &= \exp(\beta_0 + \mathbf{x}_j\boldsymbol{\beta}_x)t^p \\ S(t|\mathbf{x}_j) &= \exp\{-\exp(\beta_0 + \mathbf{x}_j\boldsymbol{\beta}_x)t^p\} \end{aligned} \qquad (13.4)$$

The estimated scale parameter thus is obtained by exponentiating the estimated intercept coefficient.

The Weibull distribution can provide a variety of monotonically increasing or decreasing shapes of the hazard function, and their shape is determined by the estimated parameter p. Figure 13.6 gives a few examples.

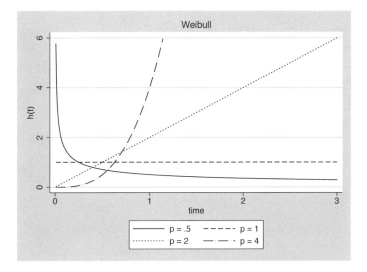

Figure 13.6. Weibull hazard function for various p

When $p = 1$, the hazard is constant, so the Weibull model reduces to the exponential model. For other values of p, the Weibull hazard is not constant; it is monotone decreasing when $p < 1$ and monotone increasing when $p > 1$. The Weibull is suitable for modeling data that exhibit monotone hazard rates.

Using `streg, dist(weibull)` with our hip-fracture data, type

```
. use http://www.stata-press.com/data/cggm/hip2
(hip fracture study)
. streg protect age, dist(weibull)
        failure _d:  fracture
   analysis time _t:  time1
              id:  id

Fitting constant-only model:

Iteration 0:   log likelihood = -60.067085
Iteration 1:   log likelihood =  -59.30148
Iteration 2:   log likelihood = -59.298481
Iteration 3:   log likelihood = -59.298481

Fitting full model:

Iteration 0:   log likelihood = -59.298481
Iteration 1:   log likelihood = -54.887563
Iteration 2:   log likelihood = -42.123875
Iteration 3:   log likelihood = -41.993012
Iteration 4:   log likelihood = -41.992704
Iteration 5:   log likelihood = -41.992704

Weibull regression -- log relative-hazard form

No. of subjects =              48             Number of obs    =         106
No. of failures =              31
Time at risk    =             714
                                             LR chi2(2)       =       34.61
Log likelihood  =    -41.992704             Prob > chi2      =      0.0000
```

_t	Haz. Ratio	Std. Err.	z	P>|z|	[95% Conf. Interval]	
protect	.1099611	.0448214	-5.42	0.000	.0494629	.2444548
age	1.117186	.0423116	2.93	0.003	1.03726	1.203271
/ln_p	.5188694	.1376486	3.77	0.000	.2490831	.7886556
p	1.680127	.2312671			1.282849	2.200436
1/p	.5951931	.0819275			.4544553	.7795152

In the output above, Stata reports a Wald test for H_o: $\ln(p) = 0$ for which the test statistic is 3.77 and reports that we can reject the null hypothesis. This is equivalent to testing H_o: $p = 1$, and thus we can reject that the hazard is constant.

❑ Technical note

We now see that the results for three parameterizations of p are given: $\ln(p)$, p itself, and $1/p$. The first parameterization, $\ln(p)$, represents the metric in which the model is actually fit. By estimating in this metric, we are assured of obtaining an estimate of p that is positive, and the estimate of p is obtained by transforming $\ln(p)$ postestimation. The third parameterization, $1/p$, is given so that one may compare these results with those of other researchers who commonly choose to parameterize the shape in this manner.

❑

By default, `streg` reports hazard ratios (exponentiated coefficients) when estimating in the PH metric. We see that wearing the hip-protection device reduces the hazard of hip fracture to almost 1/10 of what it would be otherwise.

Although results reported in this way make interpreting the effects of variables on the relative hazard easy, these results do not show all the parameters of the baseline hazard function because one of those parameters is β_0. We can, however, redisplay the results and ask for the coefficients:

```
. streg, nohr

Weibull regression -- log relative-hazard form

No. of subjects =           48           Number of obs    =         106
No. of failures =           31
Time at risk    =          714
                                         LR chi2(2)       =       34.61
Log likelihood  =    -41.992704          Prob > chi2      =      0.0000
```

_t	Coef.	Std. Err.	z	P>|z|	[95% Conf. Interval]	
protect	-2.207628	.4076113	-5.42	0.000	-3.006532	-1.408725
age	.1108134	.0378734	2.93	0.003	.036583	.1850439
_cons	-11.67104	2.90919	-4.01	0.000	-17.37295	-5.969135
/ln_p	.5188694	.1376486	3.77	0.000	.2490831	.7886556
p	1.680127	.2312671			1.282849	2.200436
1/p	.5951931	.0819275			.4544553	.7795152

From these results, we see that

$$h_0(t) = pt^{p-1}\exp(\beta_0) \approx 1.68t^{0.68}\exp(-11.67)$$
$$= 0.0000144t^{0.68}$$

The baseline hazard is so small because it is being evaluated at `protect==0` and `age==0` and, as before, we could move the baseline to reflect a more reasonable group by estimating on, say, `age60 = age-60`. In any case, as demonstrated in section 13.1.1, we could graph the baseline hazard by typing

```
. gen h = 0.0000144 * _t^0.68
. line h _t, c(l) sort
```

This time, however, we will follow a different approach to estimating the baseline hazard and type

```
. replace protect=0
. replace age=0
. predict h, hazard
```

Here we let `predict` do all the work. `predict, hazard` will calculate the hazard at the recorded values of the covariates after any parametric estimation using `streg`. If we set all the covariates to zero, we obtain the baseline hazard. The full capabilities of `predict` when used after `streg` are covered in chapter 14.

If we perform the above and estimate the baseline hazard for our fit model, we can
graph it by typing

```
. line h _t, c(1) sort ytitle("baseline hazard")
```

which produces figure 13.7.

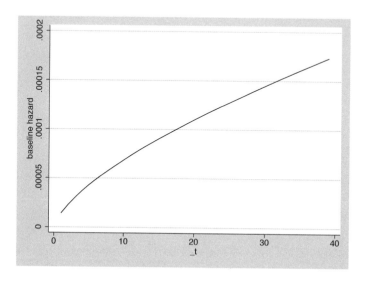

Figure 13.7. Estimated baseline hazard function for Weibull model

In section 13.1.1, we fit a step function for the hazard. We can compare that with
the baseline hazard produced by the Weibull model. In doing these kinds of comparisons
between classes of parametric models, it is important to make them at reasonable values
of the covariates. Thus we will make the comparison at `protect==0` and `age==70`. In
replicating the calculations for the step-function hazard, this time we let `predict` do
the work:

```
. use http://www.stata-press.com/data/cggm/hip2, clear
. stsplit myt, at(1(1)39)
. gen in_t1 = 0<=myt & myt<3
. gen in_t2 = 3<=myt & myt<8
. gen in_t3 = 8<=myt & myt<16
. gen in_t4 = 16<=myt
. streg protect age in_t2 in_t3 in_t4, dist(exp)
. gen hage = age
. gen hprotect = protect
. replace age = 70
. replace protect = 0
. predict hstep, hazard
. replace age = hage
. replace protect = hprotect
. streg protect age, dist(weib)
. replace age=70
```

```
. replace protect=0
. predict hweibull, hazard
. label variable hstep "Exponential/step"
. label variable hweibull "Weibull"
. line hstep hweibull _t, c(l l) sort ytitle("baseline hazard")
```

The result is shown in figure 13.8.

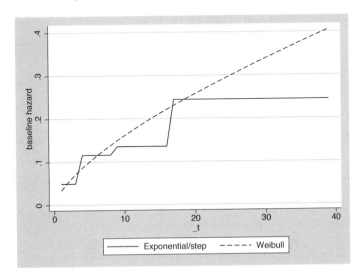

Figure 13.8. Comparison of exponential (step) and Weibull hazards

As we said in section 13.1.1, when commenting on figure 13.4, "Up until $t = 16$, we believe [the hazard] is not increasing faster than linearly ... and after that, we really do not know." We do not know because of the small amount of data after $t = 16$, so, in figure 13.8, we are not much bothered by the difference on the right between the two hazard functions.

Fitting null models

You can use `streg` to obtain estimates of the parameters of the parametric distribution when there are no covariates in the model. For example, if we wanted to find the Weibull distribution that fit our data by treating all subjects as alike (that is, ignoring `age` and `protect`), we could type

(Continued on next page)

```
. use http://www.stata-press.com/data/cggm/hip2, clear
(hip fracture study)

. streg, dist(weibull) time

        failure _d:  fracture
   analysis time _t:  time1
            id:  id

Fitting constant-only model:

Iteration 0:    log likelihood = -60.067085
Iteration 1:    log likelihood =  -59.30148
Iteration 2:    log likelihood = -59.298481
Iteration 3:    log likelihood = -59.298481

Fitting full model:
Iteration 0:    log likelihood = -59.298481

Weibull regression -- accelerated failure-time form

No. of subjects =           48            Number of obs    =        106
No. of failures =           31
Time at risk    =          714
                                          LR chi2(0)       =       0.00
Log likelihood  =   -59.298481            Prob > chi2      =          .
```

_t	Coef.	Std. Err.	z	P>|z|	[95% Conf. Interval]	
_cons	3.111177	.1489163	20.89	0.000	2.819306	3.403048
/ln_p	.1910097	.1476597	1.29	0.196	-.0983979	.4804174
p	1.210471	.1787378			.9062882	1.616749
1/p	.8261245	.1219853			.6185252	1.103402

and we obtain $\widehat{\beta}_0 = 3.11$ and $\widehat{p} = 1.21$, which completely specify a Weibull distribution with survivor function

$$\begin{aligned} S(t) &= \exp\{-\exp(\beta_0)t^p\} \\ &= \exp\{-\exp(3.11t^{1.21})\} \end{aligned}$$

Such estimation is useful to obtain the maximum likelihood estimates of the Weibull distribution for univariate data (or univariate data with censoring, as here).

Also by fitting a null model for each value of protect, we can graph and compare the hazard functions for the data when protect==0 and protect==1.

```
. streg if protect==0, dist(weib) nohr
  (output omitted)
. predict h0, hazard
. streg if protect==1, dist(weib) nohr
  (output omitted)
. predict h1, hazard
. label var h0 "protect==0"
. label var h1 "protect==1"
. line h0 h1 _t, c(l l) sort l1title("hazard")
```

This result is shown in figure 13.9.

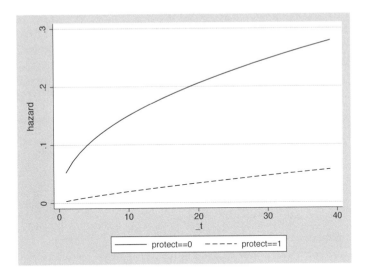

Figure 13.9. Estimated Weibull hazard functions over values of `protect`

Let us take a moment to reflect. In figure 8.2 of section 8.2.7, when we discussed nonparametric analysis, we parameterized neither the effect nor the underlying hazard function. Later, in section 8.6, we illustrated nonparametric tests, which similarly made no assumptions about the functional forms of the effect or the hazards.

In section 9.1, we used Cox regression. In that analysis, we parameterized the effect but not the underlying hazard.

Above, we just did a visual comparison, leaving the effect unparameterized but parameterizing the hazard function. And, of course, we could fit a Weibull model on `protect` and thus parameterize both.

All our tests and inspections have yielded the same result: there is a difference associated with wearing the hip-protection device. The results need not be the same, so which test should you use?

There is no simple answer, so let us instead understand how to determine the answer in particular cases.

The advantage of the modeling-the-effect approaches is that you can control for the effects of other variables. For instance, in our hip data, we know that patients vary in age, and we know age also affects outcome. In a carefully controlled experiment, we could ignore that effect because the average ages (and the distribution of age) of the control and experimental groups would be the same, but in our data that is not so.

The disadvantage of the modeling-the-effect approaches is that you could model the effect incorrectly in two ways. You could model incorrectly the effect of other variables, or you could mismodel the effect itself, for example, by stating its functional form incorrectly.

Effects of the form "apply the treatment and get an overall improvement" are often not simple. Effects can vary with other covariates (being perhaps larger for males than for females), and effects can vary with time, which is to say, aspects that change over time and that are not measured. For instance, a treatment might involve surgery, and then there may be a greater risk to be followed by a lesser risk in the future.

It is because of these concerns that looking at graphs such as figure 13.9 is useful, whether you are engaging in parametric or semiparametric modeling (although, when doing semiparametric modeling, you can only indirectly look at the hazard function by looking at the cumulative hazard or survivor function).

In most real circumstances, you will be forced into parametric or semiparametric analysis. Nonparametric analysis is useful when the experiment has been carefully controlled although even controlled experiments are sometimes not controlled adequately. Nonparametric analysis is always a useful starting point. In nonexperimental situations in the presence of covariates, you do this more as a data description technique rather than in hopes of producing any final analysis that you can believe. You, as a researcher, should be able to describe the survival experience, say, as reflected in a graph of the survivor function or cumulative hazard function for your data, ignoring the complications of confounding variables and the like. Before disentangling reality, you need to be able to describe the reality that you are starting with.

So, our position is that you will likely be forced into parameterizing the effect. This is perhaps due more to our past analysis experiences. In a well-designed, controlled experiment, however, there is nothing wrong with not parameterizing the effect and stopping at nonparametric analysis.

If you do need to continue, should you parameterize the hazard function? On this issue, different researchers feel differently. We are favorably disposed to parametric analysis when you have good reason to believe that the hazard function ought to follow a certain shape. Imposing a hazard function is an excellent way of improving the efficiency of your estimates and helping to avoid being misled by the fortuity of chance. On the other hand, when you do not have a good deductive reason to know the shape of the hazard, you should use semiparametric analysis.

When choosing between a semiparametric and parametric analysis, you must also take into consideration what information you are trying to obtain. If all you care about are hazard ratios (parameter effects) in a PH model, then you are probably better off with a semiparametric analysis. If you are interested in predicting the time to failure, however, some sort of parametric assumption as to the hazard is necessary. Here even if you do not have deductive knowledge as to the shape of the hazard, you can use `streg`, in all its implementations, to compare various functional forms of the hazard, and you can use the piecewise exponential model to "nonparametrically" check the validity of any parametric form you wish to posit.

13.2.2   Weibull regression in the AFT metric

In the AFT formulation of the Weibull, we have $\tau_j = \exp(-\mathbf{x}_j\boldsymbol{\beta}_x)t_j$, and in the Weibull regression model it is assumed that

$$\tau_j \sim \text{Weibull}(\beta_0, p)$$

τ_j is distributed as Weibull with parameters (β_0, p) with the cumulative distribution function

$$F(\tau) = 1 - \exp[-\{\exp(-\beta_0)\tau\}^p] \tag{13.5}$$

This implies then that

$$
\begin{aligned}
\ln(t_j) &= \mathbf{x}_j\boldsymbol{\beta}_x + \ln(\tau_j) \\
&= \beta_0 + \mathbf{x}_j\boldsymbol{\beta}_x + u_j
\end{aligned}
$$

where u_j follows the extreme value (Gumbel) distribution with shape parameter p, the result of which is that

$$E\{\ln(t_j)|\mathbf{x}_j\} = \beta_0 + \mathbf{x}_j\boldsymbol{\beta}_x + \frac{\Gamma'(1)}{p}$$

where $\Gamma'(1)$ is the negative of Euler's constant.   We can also derive the AFT formulation by accelerating the effect of time on survival experience. At baseline values of $\mathbf{x}$, $\tau_j = t_j$ because all covariates are equal to zero. Thus the baseline survivor function of t_j is obtained from (13.5) to be

$$S_0(t_j) = \exp[-\{\exp(-\beta_0)t_j\}^p]$$

In an AFT model, the effect of the covariates is to accelerate time by a factor of $\exp(-\beta_x\mathbf{x}_j)$. Thus for the AFT model,

$$
\begin{aligned}
S(t_j|\mathbf{x}_j) &= S_0\{\exp(-\mathbf{x}_j\boldsymbol{\beta}_x)t_j\} \\
&= \exp[-\{\exp(-\beta_0)\exp(-\mathbf{x}_j\boldsymbol{\beta}_x)t_j\}^p] \\
&= \exp[-\{\exp(-\beta_0 - \mathbf{x}_j\boldsymbol{\beta}_x)t_j\}^p] \tag{13.6}
\end{aligned}
$$

Comparison with (13.4) shows that one may transform the regression coefficients from one metric to the other by using the following relationship:

$$\boldsymbol{\beta}_{AFT} = \frac{-\boldsymbol{\beta}_{PH}}{p} \tag{13.7}$$

Also some authors cast the Weibull shape parameter p (which is common to both parameterizations) in $\sigma = 1/p$, where σ is known as the dispersion parameter. For convenience, Stata reports both the estimates and standard errors of p and $1/p$ in the output from Weibull regression. You fit Weibull models in the AFT metric by specifying the `time` option to `streg`:

```
. use http://www.stata-press.com/data/cggm/hip2, clear
(hip fracture study)

. streg protect age, dist(weib) time

        failure _d:  fracture
  analysis time _t:  time1
              id:  id

Fitting constant-only model:

Iteration 0:   log likelihood = -60.067085
Iteration 1:   log likelihood =  -59.30148
Iteration 2:   log likelihood = -59.298481
Iteration 3:   log likelihood = -59.298481

Fitting full model:

Iteration 0:   log likelihood = -59.298481
Iteration 1:   log likelihood = -54.887563
Iteration 2:   log likelihood = -42.123875
Iteration 3:   log likelihood = -41.993012
Iteration 4:   log likelihood = -41.992704
Iteration 5:   log likelihood = -41.992704

Weibull regression -- accelerated failure-time form
```

No. of subjects =	48		Number of obs	=	106
No. of failures =	31				
Time at risk =	714				
			LR chi2(2)	=	34.61
Log likelihood =	-41.992704		Prob > chi2	=	0.0000

_t	Coef.	Std. Err.	z	P>|z|	[95% Conf. Interval]
protect	1.313965	.2366229	5.55	0.000	.8501928 1.777737
age	-.0659554	.0221171	-2.98	0.003	-.1093041 -.0226067
_cons	6.946524	1.575708	4.41	0.000	3.858192 10.03486
/ln_p	.5188694	.1376486	3.77	0.000	.2490831 .7886556
p	1.680127	.2312671			1.282849 2.200436
1/p	.5951931	.0819275			.4544553 .7795152

The estimate of p is identical to that from when we fit this Weibull model in the PH metric, and the regression coefficients obey the transformation given in (13.7).

13.3 Gompertz regression (PH metric)

The Gompertz model is available only in PH metric and assumes a baseline hazard

$$h_0(t) = \exp(\gamma t)\exp(\beta_0)$$

so that in the PH model

$$
\begin{aligned}
h(t|\mathbf{x}_j) &= h_0(t)\exp(\mathbf{x}_j\boldsymbol{\beta}_x) \\
&= \exp(\gamma t)\exp(\beta_0 + \mathbf{x}_j\boldsymbol{\beta}_x)
\end{aligned}
$$

and thus,

$$H(t|\mathbf{x}_j) = \gamma^{-1} \exp(\beta_0 + \mathbf{x}_j \boldsymbol{\beta}_x)\{\exp(\gamma t) - 1\}$$
$$S(t|\mathbf{x}_j) = \exp[-\gamma^{-1} \exp(\beta_0 + \mathbf{x}_j \boldsymbol{\beta}_x)\{\exp(\gamma t) - 1\}]$$

The Gompertz distribution is an old distribution that has been extensively used by medical researchers and biologists modeling mortality data. This distribution is suitable for modeling data with monotone hazard rates that either increase or decrease exponentially with time, and the ancillary parameter γ controls the shape of the baseline hazard.

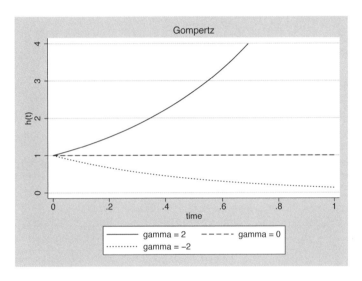

Figure 13.10. Gompertz hazard functions

As depicted in figure 13.10, if γ is positive, the hazard function increases with time; if γ is negative, the hazard decreases with time; if $\gamma = 0$, the hazard function is $\exp(\beta_0)$ for all t—it reduces to the hazard from the exponential model.

Some recent survival analysis texts such as Klein and Moeschberger (2003) restrict γ to be strictly positive. When $\gamma < 0$, $S(t|\mathbf{x})$ decreases to a nonzero constant as $t \to \infty$, implying that there is a nonzero probability of never failing (living forever). That is, the hazard remains positive but decreases to zero at an exponential rate, which is too rapid a decline to guarantee eventual failure. By restricting γ to be positive, one is assured that $S(t|\mathbf{x})$ tends to zero as $t \to \infty$.

Although the above argument may be desirable from a mathematical perspective, Stata's implementation takes the more traditional approach of not restricting γ. In survival studies, subjects are not monitored forever—there is a date when the study ends—and in many investigations (specifically in medical research), an exponentially decreasing hazard rate is clinically appealing.

Using our hip-fracture data, we type

```
. use http://www.stata-press.com/data/cggm/hip2, clear
(hip fracture study)
. streg protect age, dist(gompertz) nohr

        failure _d:  fracture
   analysis time _t:  time1
              id:  id

Fitting constant-only model:

Iteration 0:   log likelihood = -60.067085
Iteration 1:   log likelihood = -59.734027
Iteration 2:   log likelihood = -59.730841
Iteration 3:   log likelihood =  -59.73084

Fitting full model:

Iteration 0:   log likelihood =  -59.73084
Iteration 1:   log likelihood = -53.743223
Iteration 2:   log likelihood = -42.693409
Iteration 3:   log likelihood = -42.609467
Iteration 4:   log likelihood = -42.609352
Iteration 5:   log likelihood = -42.609352

Gompertz regression -- log relative-hazard form

No. of subjects =           48              Number of obs   =         106
No. of failures =           31
Time at risk    =          714
                                            LR chi2(2)      =       34.24
Log likelihood  =   -42.609352             Prob > chi2     =      0.0000
```

_t	Coef.	Std. Err.	z	P>|z|	[95% Conf. Interval]	
protect	-2.311232	.4435779	-5.21	0.000	-3.180629	-1.441835
age	.1055445	.0371977	2.84	0.005	.0326383	.1784508
_cons	-10.19274	2.721731	-3.74	0.000	-15.52723	-4.858245
gamma	.0752767	.0233602	3.22	0.001	.0294915	.1210618

and find that $\hat{\gamma} > 0$, meaning that we estimate a hazard that is increasing exponentially. We can plot the baseline hazard using

```
. replace protect = 0
(72 real changes made)
. replace age = 0
(106 real changes made)
. predict h0, hazard
. line h0 _t, c(l) sort l1title("baseline hazard")
```

which produces figure 13.11.

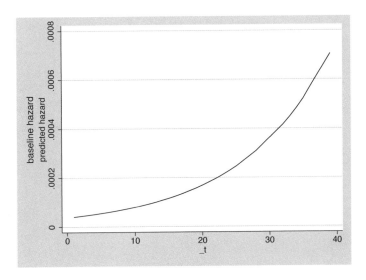

Figure 13.11. Estimated baseline hazard for the Gompertz model

Given our previous discussions on the exponential and Weibull models, this is a poor model for our data. We know that the hazard is increasing, and we have argued that it is probably increasing at a *decreasing* rate, something not allowed under a Gompertz specification where the hazard increases or decreases exponentially.

13.4 Lognormal regression (AFT metric)

In the AFT formulation, we have $\tau_j = \exp(-\mathbf{x}_j\boldsymbol{\beta}_x)t_j$, and for the lognormal regression model, it is assumed that

$$\tau_j \sim \text{Lognormal}(\beta_0, \sigma)$$

τ_j is distributed as lognormal with parameters (β_0, σ) with cumulative distribution function

$$F(\tau) = \Phi\left(\frac{\ln \tau - \beta_0}{\sigma}\right) \tag{13.8}$$

where $\Phi()$ is the cumulative distribution function for the standard Gaussian (normal) distribution. Thus

$$\begin{aligned} \ln(t_j) &= \mathbf{x}_j\boldsymbol{\beta}_x + \ln(\tau_j) \\ &= \beta_0 + \mathbf{x}_j\boldsymbol{\beta}_x + u_j \end{aligned}$$

where u_j follows a standard normal distribution with mean 0 and standard deviation σ. That is, for the lognormal model, transforming time into ln(time) converts the problem into one of simple linear regression (with possible censoring). As a result,

$$E\{\ln(t_j)|\mathbf{x}_j\} = \beta_0 + \mathbf{x}_j\boldsymbol{\beta}_x$$

We can also derive the AFT formulation by accelerating the effect of time on survival experience at baseline, where all covariates are equal to zero. Thus the baseline survivor function of t_j is obtained from (13.8):

$$S_0(t_j) = 1 - \Phi\left(\frac{\ln t_j - \beta_0}{\sigma}\right)$$

In an AFT model, the effect of the covariates is to accelerate time by a factor of $\exp(-\mathbf{x}_j\boldsymbol{\beta}_x)$. Thus for the AFT model,

$$
\begin{aligned}
S(t_j|\mathbf{x}_j) &= S_0\{\exp(-\mathbf{x}_j\boldsymbol{\beta}_x)t_j\} \\
&= 1 - \Phi\left[\frac{\ln\{\exp(-\mathbf{x}_j\boldsymbol{\beta}_x)t_j\} - \beta_0}{\sigma}\right] \\
&= 1 - \Phi\left\{\frac{\ln t_j - (\beta_0 + \mathbf{x}_j\boldsymbol{\beta}_x)}{\sigma}\right\} \quad (13.9)
\end{aligned}
$$

The attractive feature (for some problems) of this distribution is that the hazard function is nonmonotonic—it increases and then decreases; see figure 13.12.

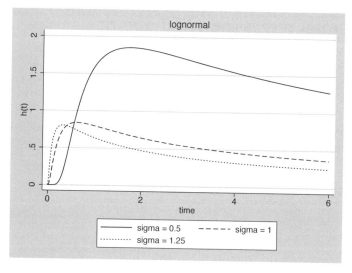

Figure 13.12. Examples of lognormal hazard functions ($\beta_0 = 0$)

This model has no natural PH interpretation. From (13.9), we could certainly derive the hazard as a function of $\mathbf{x}_j$ as

$$h(t|\mathbf{x}_j) = \frac{-\frac{d}{dt}S(t|\mathbf{x}_j)}{S(t|\mathbf{x}_j)}$$

but there is no choice of $h_0(t)$ for which this reduces to the form

$$h(t|\mathbf{x}_j) = h_0(t)\exp(\mathbf{x}_j\boldsymbol{\beta}_x)$$

and so this model is parameterized in the AFT metric only.

Although we do not think the lognormal model is appropriate for our hip-fracture data, by way of illustration we fit a lognormal model to it:

```
. use http://www.stata-press.com/data/cggm/hip2, clear
(hip fracture study)

. streg protect age, dist(lognormal)

        failure _d:  fracture
  analysis time _t:  time1
               id:  id

Fitting constant-only model:

Iteration 0:   log likelihood = -69.784524
Iteration 1:   log likelihood = -68.743214
Iteration 2:   log likelihood = -59.663856
Iteration 3:   log likelihood = -59.511256
Iteration 4:   log likelihood = -59.510919
Iteration 5:   log likelihood = -59.510919

Fitting full model:

Iteration 0:   log likelihood = -59.510919  (not concave)
Iteration 1:   log likelihood = -48.387777
Iteration 2:   log likelihood = -42.455265
Iteration 3:   log likelihood = -41.847689
Iteration 4:   log likelihood = -41.845325
Iteration 5:   log likelihood = -41.845325

Lognormal regression -- accelerated failure-time form

No. of subjects =           48            Number of obs    =        106
No. of failures =           31
Time at risk    =          714
                                          LR chi2(2)       =      35.33
Log likelihood  =    -41.845325           Prob > chi2      =     0.0000
```

_t	Coef.	Std. Err.	z	P>|z|	[95% Conf. Interval]	
protect	1.459569	.246986	5.91	0.000	.9754855	1.943653
age	-.0785641	.0222192	-3.54	0.000	-.1221129	-.0350152
_cons	7.45804	1.59032	4.69	0.000	4.341069	10.57501
/ln_sig	-.2952625	.1268218	-2.33	0.020	-.5438285	-.0466964
sigma	.7443362	.094398			.5805215	.9543771

Although the lognormal model does not fit into the PH framework, `predict` may
still be used after estimation to obtain the predicted hazard function. We can compare
the hazards for the `protect==0` and `protect==1` groups at age 70 by typing

```
. replace protect=0
(72 real changes made)
. replace age=70
(96 real changes made)
. predict h0, hazard
. replace protect=1
(106 real changes made)
. predict h1, hazard
. label var h0 "protect==0"
. label var h1 "protect==1"
. line h0 h1 _t, c(l l) sort l1title("h(t)")
```

which produces figure 13.13.

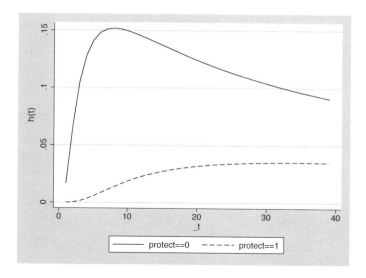

Figure 13.13. Comparison of hazards for a lognormal model

The non-PH nature of this model is now obvious. Given what we already know about
the hip data, we would argue that both of these hazard functions misspecify what is
occurring.

13.5   Loglogistic regression (AFT metric)

In the AFT metric, $\tau_j = \exp(-\mathbf{x}_j\boldsymbol{\beta}_x)t_j$, and for the loglogistic regression model, we assume that

$$\tau_j \sim \text{Loglogistic}(\beta_0, \gamma)$$

τ_j is distributed as loglogistic with parameters (β_0, γ) with cumulative distribution function

$$F(\tau) = 1 - \left[1 + \{\exp(-\beta_0)\tau\}^{\frac{1}{\gamma}}\right]^{-1} \tag{13.10}$$

Thus

$$
\begin{aligned}
\ln(t_j) &= \mathbf{x}_j\boldsymbol{\beta}_x + \ln(\tau_j) \\
&= \beta_0 + \mathbf{x}_j\boldsymbol{\beta}_x + u_j
\end{aligned}
$$

where u_j follows a logistic distribution with mean 0 and standard deviation $\pi\gamma/\sqrt{3}$. As a result,

$$E\{\ln(t_j)|\mathbf{x}_j\} = \beta_0 + \mathbf{x}_j\boldsymbol{\beta}_x$$

We can also derive the AFT formulation by accelerating the effect of time on survival experience. At baseline values of the covariates $\mathbf{x}$, $\tau_j = t_j$ because all covariates are equal to zero. Thus the baseline survivor function of t_j is obtained from (13.10) to be

$$S_0(t_j) = \left[1 + \{\exp(-\beta_0)t_j\}^{\frac{1}{\gamma}}\right]^{-1}$$

In an AFT model, the effect of the covariates is to accelerate time by a factor of $\exp(-\mathbf{x}_j\boldsymbol{\beta}_x)$. Thus for the AFT model,

$$
\begin{aligned}
S(t_j|\mathbf{x}_j) &= S_0\{\exp(-\mathbf{x}_j\boldsymbol{\beta}_x)t_j\} \\
&= \left[1 + \{\exp(-\beta_0)\exp(-\mathbf{x}_j\boldsymbol{\beta}_x)t_j\}^{\frac{1}{\gamma}}\right]^{-1} \\
&= \left[1 + \{\exp(-\beta_0 - \mathbf{x}_j\boldsymbol{\beta}_x)t_j\}^{\frac{1}{\gamma}}\right]^{-1}
\end{aligned}
$$

The loglogistic distribution closely resembles the lognormal distribution, and some examples of loglogistic hazards are given in figure 13.14.

(Continued on next page)

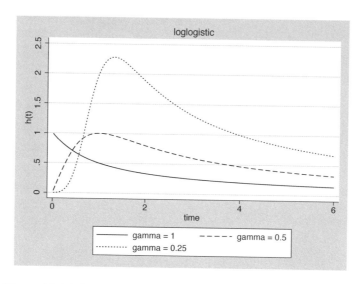

Figure 13.14. Examples of loglogistic hazard functions ($\beta_0 = 0$)

Like the lognormal model, the loglogistic model has no natural PH interpretation. One advantage of the loglogistic model over the lognormal model is that the loglogistic model has simpler mathematical expressions of the hazard and survivor functions, expressions that do not include the normal cumulative distribution function. If $\gamma < 1$, the loglogistic hazard increases and then decreases. If $\gamma \geq 1$, then the hazard is monotone decreasing.

Using our hip-fracture data, type

```
. use http://www.stata-press.com/data/cggm/hip2, clear
(hip fracture study)

. streg protect age, dist(llogistic)

        failure _d:  fracture
   analysis time _t:  time1
            id:  id

Fitting constant-only model:

Iteration 0:    log likelihood = -69.090852
Iteration 1:    log likelihood = -60.599923
Iteration 2:    log likelihood =   -59.3027
Iteration 3:    log likelihood = -59.298893
Iteration 4:    log likelihood = -59.298892

Fitting full model:

Iteration 0:    log likelihood = -59.298892  (not concave)
Iteration 1:    log likelihood = -48.506502
Iteration 2:    log likelihood = -42.884806
Iteration 3:    log likelihood = -42.242365
Iteration 4:    log likelihood = -42.240535
Iteration 5:    log likelihood = -42.240534
```

```
Loglogistic regression -- accelerated failure-time form
No. of subjects =           48              Number of obs    =         106
No. of failures =           31
Time at risk    =          714
                                            LR chi2(2)       =       34.12
Log likelihood  =    -42.240534             Prob > chi2      =      0.0000
```

_t	Coef.	Std. Err.	z	P>|z|	[95% Conf. Interval]	
protect	1.434467	.2483224	5.78	0.000	.9477643	1.92117
age	-.0755823	.0221193	-3.42	0.001	-.1189352	-.0322293
_cons	7.284475	1.562057	4.66	0.000	4.2229	10.34605
/ln_gam	-.8565429	.1476461	-5.80	0.000	-1.145924	-.5671619
gamma	.4246275	.0626946			.3179301	.5671327

We obtain results nearly identical to those produced under the lognormal model and every bit as inappropriate for these data. In fact, a comparison of the hazards for protect==0 and protect==1 with age held at 70 years,

```
. replace protect=0
(72 real changes made)
. replace age=70
(96 real changes made)
. predict h0, hazard
. replace protect=1
(106 real changes made)
. predict h1, hazard
. label var h0 "protect==0"
. label var h1 "protect==1"
. line h0 h1 _t, c(l l) sort l1title("h(t)")
```

will produce a graph nearly identical to that shown in figure 13.13.

The loglogistic and lognormal models are similar, and for most purposes are indistinguishable, much like probit and logistic regression models for binary data. Returning to our linear model for $\ln(t_j)$,

$$\ln(t_j) = \beta_0 + \mathbf{x}_j \boldsymbol{\beta}_x + u_j$$

For the lognormal model, u_j is normal with mean 0 and standard deviation σ; for the loglogistic model, u_j is logistic with mean 0 and standard deviation $\pi\gamma/\sqrt{3}$. For the above data, $\widehat{\gamma} = 0.425$, and thus $\pi\widehat{\gamma}/\sqrt{3} = 0.771$, which nearly equals $\widehat{\sigma} = 0.744$, estimated in section 13.4.

13.6   Generalized gamma regression (AFT metric)

In the AFT metric, $\tau_j = \exp(-\mathbf{x}_j\boldsymbol{\beta}_x)t_j$, and for gamma regression models, we assume that

$$\tau_j \sim \mathrm{Gamma}(\beta_0, \kappa, \sigma)$$

τ_j is distributed as generalized gamma with parameters $(\beta_0, \kappa, \sigma)$ with cumulative distribution function

$$F(\tau) = \begin{cases} I(\gamma, u), & \text{if } \kappa > 0 \\ \Phi(z), & \text{if } \kappa = 0 \\ 1 - I(\gamma, u), & \text{if } \kappa < 0 \end{cases} \tag{13.11}$$

where $\gamma = |\kappa|^{-2}$, $z_0 = \mathrm{sign}(\kappa)\{\ln(\tau) - \beta_0\}/\sigma$, $u = \gamma\exp(\sqrt{\gamma}z_0)$, $\Phi()$ is the standard normal cumulative distribution function, and $I(a, x)$ is the incomplete gamma function

$$I(a, x) = \frac{1}{\Gamma(a)} \int_0^x e^{-v} v^{a-1} dv$$

Thus

$$\begin{aligned} \ln(t_j) &= \mathbf{x}_j\boldsymbol{\beta}_x + \ln(\tau_j) \\ &= \beta_0 + \mathbf{x}_j\boldsymbol{\beta}_x + u_j \end{aligned}$$

where u_j has mean

$$E(u_j) = \frac{\sigma\Gamma(\gamma)}{\sqrt{\gamma}\Gamma'(\gamma)} + \ln(\gamma)$$

As a result,

$$E\{\ln(t_j)|\mathbf{x}_j\} = \beta_0 + \mathbf{x}_j\boldsymbol{\beta}_x + E(u_j)$$

We can also derive the AFT formulation by accelerating the effect of time on survival experience. At baseline values of the covariates $\mathbf{x}$, $\tau_j = t_j$ because all covariates are equal to zero. Thus the baseline survivor function of t_j is

$$S_0(t_j) = 1 - F(t_j)$$

where $F()$ is given in (13.11). Thus

$$\begin{aligned} S(t_j|\mathbf{x}_j) &= S_0\{\exp(-\mathbf{x}_j\boldsymbol{\beta}_x)t_j\} \\ &= 1 - F\{\exp(-\mathbf{x}_j\boldsymbol{\beta}_x)t_j\} \\ &= 1 - F^*(t_j) \end{aligned}$$

where $F^*()$ is $F()$ from (13.11) with z_0 replaced by

$$z = \text{sign}(\kappa) \frac{\ln(\tau) - (\beta_0 + \mathbf{x}_j \boldsymbol{\beta}_x)}{\sigma}$$

The generalized gamma distribution is a three-parameter distribution $(\beta_0, \kappa, \sigma)$ possessing a highly flexible hazard function that allows for many possible shapes. The gamma distribution includes as special cases the Weibull if $\kappa = 1$, in which case $p = 1/\sigma$; the exponential distribution if $\kappa = \sigma = 1$; and the lognormal distribution if $\kappa = 0$. As such, the generalized gamma model is commonly used for evaluating and selecting an appropriate parametric model for the data. For example,

```
. use http://www.stata-press.com/data/cggm/hip2, clear
(hip fracture study)

. streg protect age, dist(gamma)

        failure _d:  fracture
   analysis time _t:  time1
             id:  id

Fitting constant-only model:

Iteration 0:   log likelihood = -63.838211
Iteration 1:   log likelihood = -62.421374  (not concave)
Iteration 2:   log likelihood = -59.568933
Iteration 3:   log likelihood =   -59.0852
Iteration 4:   log likelihood = -59.074282
Iteration 5:   log likelihood = -59.074271
Iteration 6:   log likelihood = -59.074271

Fitting full model:

Iteration 0:   log likelihood = -59.074271  (not concave)
Iteration 1:   log likelihood = -46.552165
Iteration 2:   log likelihood = -44.237189  (not concave)
Iteration 3:   log likelihood = -42.025693
Iteration 4:   log likelihood = -41.514713
Iteration 5:   log likelihood = -41.483559
Iteration 6:   log likelihood = -41.483472
Iteration 7:   log likelihood = -41.483472

Gamma regression -- accelerated failure-time form
```

No. of subjects =	48			Number of obs	=	106
No. of failures =	31					
Time at risk =	714					
				LR chi2(2)	=	35.18
Log likelihood =	-41.483472			Prob > chi2	=	0.0000

_t	Coef.	Std. Err.	z	P>|z|	[95% Conf. Interval]	
protect	1.406694	.2551249	5.51	0.000	.9066582	1.906729
age	-.0727843	.0230869	-3.15	0.002	-.1180337	-.0275348
_cons	7.223303	1.597739	4.52	0.000	4.091793	10.35481
/ln_sig	-.3836532	.1768555	-2.17	0.030	-.7302836	-.0370228
/kappa	.4643881	.5287763	0.88	0.380	-.5719944	1.500771
sigma	.6813677	.1205036			.4817724	.9636541

and we note that the 95% confidence interval for κ $(-0.572, 1.501)$ includes both 0 and 1; thus, it does not rule out a Weibull model ($\kappa = 1$) or a lognormal model ($\kappa = 0$), which is not surprising given what we know about these data.

As such, a graph of the comparison of hazards for `protect==0` versus `protect==1` via

```
. replace protect=0
(72 real changes made)
. replace age=70
(96 real changes made)
. predict h0, hazard
. replace protect=1
(106 real changes made)
. predict h1, hazard
. label var h0 "protect==0"
. label var h1 "protect==1"
. line h0 h1 _t, c(l l) sort l1title("h(t)")
```

will produce a graph similar to that given for the lognormal (figure 13.13).

13.7 Choosing among parametric models

We may ask ourselves, given that we have several possible parametric models to choose from, how can we select one? The preferred answer is that the science for the problem at hand suggests an appropriate parametric model. If we think carefully about the underlying process that generated the failure times in our data and specifically about the possible shape of the hazard function, we can get a good idea of which parametric model or models we should evaluate.

From a purely statistical view, several strategies are available for selecting a parametric model, of which we will consider two:

1. When models are nested, the likelihood-ratio test or the Wald test can be used to discriminate between them. This can certainly be done for Weibull versus exponential, or gamma versus Weibull or lognormal.

2. When models are not nested, the likelihood-ratio and Wald tests are unsuitable, and we can use, for instance, the Akaike (1974) information criterion (AIC).

13.7.1 Nested models

Using our hip-fracture data, we first fit a generalized gamma model (the most general of the models available in `streg`) and test the following hypotheses:

1. H_o: $\kappa = 0$, in which case if H_o is true then the model is lognormal.
2. H_o: $\kappa = 1$, in which case if H_o is true then the model is Weibull.

3. $H_o: \kappa = 1, \sigma = 1$, in which case if H_o is true then the model is exponential (constant baseline hazard function).

We start by fitting the gamma model:

```
. use http://www.stata-press.com/data/cggm/hip2, clear
(hip fracture study)
. streg protect age, dist(gamma) nolog

       failure _d:  fracture
  analysis time _t:  time1
              id:  id

Gamma regression -- accelerated failure-time form

No. of subjects =          48          Number of obs   =        106
No. of failures =          31
Time at risk    =         714
                                        LR chi2(2)      =      35.18
Log likelihood  =   -41.483472          Prob > chi2     =     0.0000
```

_t	Coef.	Std. Err.	z	P>|z|	[95% Conf. Interval]
protect	1.406694	.2551249	5.51	0.000	.9066582 1.906729
age	-.0727843	.0230869	-3.15	0.002	-.1180337 -.0275348
_cons	7.223303	1.597739	4.52	0.000	4.091793 10.35481
/ln_sig	-.3836532	.1768555	-2.17	0.030	-.7302836 -.0370228
/kappa	.4643881	.5287763	0.88	0.380	-.5719944 1.500771
sigma	.6813677	.1205036			.4817724 .9636541

We can now perform our three tests:

(1) $H_o: \kappa = 0$

We can read the Wald test directly from the output since **streg** reports that the test statistic for /kappa is $z = 0.88$ with significance level 0.380. We could also perform the Wald test ourselves using **test**:

```
. test [kappa]_b[_cons] = 0
 ( 1)  [kappa]_cons = 0
           chi2(  1) =     0.77
         Prob > chi2 =   0.3798
```

The test statistic looks different because here a chi-squared value is reported and **streg** reported a normal value, but these tests are identical, as illustrated by the identical levels of significance.

We could also perform a likelihood-ratio test, which in this context is asymptotically equivalent to the Wald test, and the choice between the two is a matter of personal taste. Wald tests have the advantage that they can be used with sampling weights and robust estimates of the variance–covariance matrix of the parameters, whereas likelihood-ratio tests cannot. Because we specify neither sampling weights nor robust standard errors in this model, however, the likelihood-ratio test is ap-

propriate. For nested models (here the lognormal nested within the gamma), the
procedure for performing a likelihood-ratio test is as follows:

```
. streg protect age, dist(gamma)        /* Fit the saturated model */
  (output omitted )
. estimates store sat                    /* Save the results */
. streg protect age, dist(lnormal)       /* Fit the nested model */
  (output omitted )
. lrtest sat, force                      /* Likelihood-ratio test */
Likelihood-ratio test                                LR chi2(1)   =      0.72
(Assumption: . nested in sat)                        Prob > chi2 =    0.3949
```

This compares favorably with the asymptotically equivalent Wald test.

❏ **Technical note**

To perform the likelihood-ratio test, we needed to specify the `force` option to
`lrtest`; otherwise, Stata would refuse to perform this test. Normally, using `force`
is not a good idea because you are overriding Stata's judgment that the test has
no statistical validity. By default, Stata conservatively treats `streg` models of
differing distributions as not nested because not all combinations of distributions
result in nested models.

❏

(2) $H_o: \kappa = 1$

Here we will obtain the Wald test only, but first we need to refit our gamma model:

```
. quietly streg protect age, dist(gamma) nolog
. test [kappa]_b[_cons] = 1
 ( 1)  [kappa]_cons = 1
           chi2(  1) =     1.03
         Prob > chi2 =   0.3111
```

The test results do not preclude the use of the Weibull model for these data; these
results agree with what we already know about the nature of the hazard for hip
fracture.

(3) $H_o: \kappa = 1, \sigma = 1$

As in the first two tests, we can perform either a likelihood-ratio test or a Wald
test, but we opt for a Wald test so that we can demonstrate how to use `test` to
test two parameters simultaneously:

```
. test [kappa]_b[_cons] = 1, notest
 ( 1)  [kappa]_cons = 1
. test [ln_sig]_b[_cons] = 0, accum
 ( 1)  [kappa]_cons = 1
 ( 2)  [ln_sig]_cons = 0
           chi2(  2) =    15.86
         Prob > chi2 =   0.0004
```

streg, dist(gamma) estimated $\ln(\sigma)$ instead of σ, so to test $\sigma = 1$, we needed to test $\ln(\sigma) = 0$. The test results strongly reinforce what we already know: the risk of hip fracture is certainly not constant over time.

In section 13.1.1, (1) we used exponential regression to fit a step function and argued that the hazard was increasing and perhaps at a decreasing rate, (2) we found that the Weibull fit this exponentially estimated step function well, and (3) when we produced lognormal results, we made comments that the increasing and then decreasing hazard were inappropriate for these data. However, the statistical test we just performed cannot reject the lognormal specification. Furthermore, the Weibull and lognormal are not nested models, so a direct test such as the above to determine which is more suitable is not feasible.

When we fit our exponential step function, we did not put standard errors around our estimates of the steps (although streg reported them). If we had more data and if the pattern we think we saw persists, then the statistical test would reject the lognormal. The problem is that we do not have more data and cannot, on the basis of these data, justify our supposition that the hazard is monotonically increasing. Nevertheless, we would have no difficulty rejecting that assumption on the basis of our science, or at least our prior knowledge of hip fracture among elderly women.

13.7.2 Nonnested models

For nonnested models, Akaike (1974) proposed penalizing each model's log likelihood to reflect the number of parameters being estimated and then comparing them. Although the best-fitting model is the one with the largest log likelihood, the preferred model is the one with the lowest value of the Akaike Information Criterion (AIC). For parametric survival models, the AIC is defined as

$$\text{AIC} = -2\ln L + 2(k + c)$$

where k is the number of model covariates and c the number of model-specific distributional parameters.

For the models fit by Stata's streg, the values of c are given in table 13.1.

Table 13.1. streg models

Distribution	Metric	Hazard shape	c
Exponential	PH, AFT	constant	1
Weibull	PH, AFT	monotone	2
Gompertz	PH	monotone	2
Lognormal	AFT	variable	2
Loglogistic	AFT	variable	2
Generalized gamma	AFT	variable	3

To illustrate the AIC, we review the six parametric models as fit on the hip-fracture data with covariates `age` and `protect`. Table 13.2 gives the log likelihoods and AIC values from each model:

Table 13.2. AIC values for `streg` models

Distribution	Log likelihood	k	c	AIC
Exponential	−47.534656	2	1	101.06931
Weibull	−41.992704	2	2	91.98541
Gompertz	−42.609352	2	2	93.21870
Lognormal	−41.845345	2	2	91.69065
Loglogistic	−42.240534	2	2	92.48107
Generalized gamma	−41.483472	2	3	92.96694

Per the AIC criterion, the lognormal model is selected. The more reasonable (based on the science of hip fractures) Weibull model has virtually the same AIC score, 91.99 versus 91.69, and on that combined basis, we would choose the Weibull.

The information in table 13.2 can be obtained easily within Stata by using repeated calls to `streg` with `foreach` and then summarizing the model information using the `estimates stats` command:

```
. clear all

. use http://www.stata-press.com/data/cggm/hip2
(hip fracture study)

. foreach model in exponential weibull gompertz lognormal loglogistic gamma {
  2.          quietly streg age protect, dist(`model')
  3.          estimates store `model'
  4. }

. estimates stats _all
```

Model	Obs	ll(null)	ll(model)	df	AIC	BIC
exponential	106	−60.06709	−47.53466	3	101.0693	109.0596
weibull	106	−59.29848	−41.9927	4	91.98541	102.6392
gompertz	106	−59.73084	−42.60935	4	93.2187	103.8725
lognormal	106	−59.51092	−41.84533	4	91.69065	102.3444
loglogistic	106	−59.29889	−42.24053	4	92.48107	103.1348
gamma	106	−59.07427	−41.48347	5	92.96694	106.2841

```
Note:  N=Obs used in calculating BIC; see [R] BIC note
```

BIC stands for Bayesian information criterion, an alternative to AIC that is interpreted similarly—smaller values correspond to better-fitting models. For more information on BIC, see [R] **BIC note**.

14 Postestimation commands for parametric models

14.1 Use of predict after streg

predict after streg is used to generate a new variable containing predicted values or residuals, and what is calculated is determined by the option you specify; see table 14.1 for a list of available options.

Table 14.1. Options for predict after streg

Option	Contents
xb	$\widehat{\beta}_0 + \mathbf{x}_j \widehat{\boldsymbol{\beta}}_x$
stdp	Standard error of $\widehat{\beta}_0 + \mathbf{x}_j \widehat{\boldsymbol{\beta}}_x$
median time	Median survival time
median lntime	Median ln(survival time)
mean time	Mean survival time
mean lntime	Mean ln(survival time)
hazard	$h(\_t)$
hr	Hazard ratio
surv	$S(\_t|\_t0)$
csurv	$S(\_t|$ earliest $\_t0$ for the subject)
csnell	Cox–Snell residuals (partial)
ccsnell	Cumulative Cox–Snell residuals
mgale	Martingale residuals (partial)
cmgale	Cumulative martingale residuals
deviance	Deviance residuals

Calculations are obtained from the data currently in memory and do not need to correspond to the data used in fitting the model. This is true for all the calculations predict can make following streg. Below we give a more detailed list:

- xb—linear prediction
  xb calculates $\widehat{\beta}_0 + \mathbf{x}_j \widehat{\boldsymbol{\beta}}_x$, the linear prediction.

- stdp—standard error of the linear prediction
  stdp calculates the estimated standard error of $\widehat{\beta}_0 + \mathbf{x}_j\widehat{\boldsymbol{\beta}}_x$ based on the estimated variance–covariance matrix of $(\widehat{\beta}_0, \widehat{\boldsymbol{\beta}}_x)$.

- median time—predicted median survival time
  This is the default prediction; if you do not specify any options, predict will calculate the predicted median survival time, $\widehat{Q}(0.5|\mathbf{x}_j)$, reported in analysis-time units. The prediction is made from $t = 0$ conditional on constant covariates; thus if you have multiple-record-per-subject data with time-varying covariates, each record will produce a distinct predicted median survival time.

- median lntime—predicted median ln(survival time)
  The predicted median ln(survival time) is reported in ln(analysis time) units. As with median time, predictions are from $t = 0$ conditional on constant covariates, even in multiple-record data.

- mean time—predicted mean survival time
  The predicted mean survival time is reported in analysis-time units and is calculated as the integral of the survivor function from zero to infinity, given each observation's covariate values and the estimated model parameters.

- mean lntime—predicted mean ln(survival time)
  The predicted mean ln(survival time) is reported in analysis-time units and is calculated as the expected value of the distribution of ln(survival time), given each observation's covariate values and the estimated model parameters.

- hazard—predicted hazard function
  Calculates $h(\_t|\mathbf{x}_j)$ given the estimated model parameters. This calculation is not restricted to models fit in the proportional hazards (PH) metric.

- hr—predicted hazard ratio
  Calculates the relative hazard, $\exp(\mathbf{x}_j\widehat{\boldsymbol{\beta}}_x)$, for all models that may be parameterized as PH. This option is not allowed with accelerated failure-time (AFT) models. For the exponential and Weibull models (which may be cast in both metrics), the predicted hazard ratio is calculated regardless of the metric under which the model was fit; yet, the parameter estimates used in the above expression are from the PH version of the model.

- surv—predicted conditional survivor function
  Calculates $S(\_t|\_t0, \mathbf{x}_j)$, which is the probability of survival past time $\_t$ given survival to time $\_t0$, given $\mathbf{x}_j$ for each observation and the estimated model parameters. $\_t0$ and $\_t$ are specific to each observation.

- csurv—predicted survivor function
  Calculates a running product within each subject of the predictions produced by predict, surv. This amounts to the probability of survival past $\_t$ for each subject, given that the subject survives any gaps for which it is left unobserved.

- csnell—Cox–Snell residuals (partial)
  If you have single observations per subject, then csnell calculates the usual Cox–

Snell residuals from (11.1). Otherwise, it calculates the additive contribution to this observation to the subject's overall Cox–Snell residual.

- ccsnell—cumulative Cox–Snell residuals
  If you have single observations per subject, then ccsnell is equivalent to csnell. Otherwise, in the last observation (with respect to analysis time) on each subject, ccsnell records the sum of the partial Cox–Snell residuals and sets the other observations to missing. Only one value per subject is recorded—the overall sum—and it is placed on the last record for the subject.

- mgale—martingale-like residuals (partial)
  If you have single observations per subject, then mgale calculates the usual martingale residuals. Otherwise, it calculates the additive contribution of this observation to the subject's overall martingale residual. We use the term "martingale-like" because, although these residuals do not arise naturally from martingale theory for parametric survival models as they do for the Cox proportional hazard model, they do share similar properties.

- cmgale—cumulative martingale-like residuals
  If you have single observations per subject, then cmgale is equivalent to mgale. Otherwise, in the last observation on each subject, cmgale records the sum of mgale and sets the other observations to missing. Only one value per subject is recorded—the overall sum—and it is placed on the last record for the subject.

- deviance—deviance residuals
  deviance calculates the deviance residuals, which are a scaling of the martingale residuals to make them symmetric about zero. For multiple-record-per-subject data, only one value per subject is calculated, and it is placed on the last record for the subject.

14.1.1 Predicting the time of failure

After estimation with streg, you can type

1. predict *newvar*, time to obtain the predicted time of failure given $\mathbf{x}_j$
2. predict *newvar*, lntime to obtain the predicted ln(time of failure) given $\mathbf{x}_j$

If neither median nor mean is specified, the option median is assumed. However, specifying one of these options removes any ambiguity, and so we highly recommend it.

Combining these options gives four (potentially) distinct predictions. To demonstrate, we fit a Weibull model to our hip-fracture data, predict using the four combinations of these options, and list these predictions (along with the covariates) for the last observation:

```
. use http://www.stata-press.com/data/cggm/hip2
(hip fracture study)
```

```
. streg protect age, dist(weibull)
(output omitted)
. predict t_mean, time mean
. predict t_median, time median
. predict lnt_mean, lntime mean
. predict lnt_median, lntime median
. list _t protect age t_mean t_median lnt_mean lnt_median in L, abbrev(10)
```

	_t	protect	age	t_mean	t_median	lnt_mean	lnt_median
106.	39	1	67	41.60789	37.46217	3.497923	3.623332

Before we proceed with analyzing each of these predictions, let's fit the model again, this time in the AFT metric:

```
. use http://www.stata-press.com/data/cggm/hip2, clear
(hip fracture study)
. streg protect age, dist(weibull) time nolog
        failure _d:  fracture
   analysis time _t:  time1
              id:  id
Weibull regression -- accelerated failure-time form
No. of subjects =          48              Number of obs   =          106
No. of failures =          31
Time at risk    =         714
                                           LR chi2(2)      =        34.61
Log likelihood  =    -41.992704            Prob > chi2     =       0.0000
```

_t	Coef.	Std. Err.	z	P>|z|	[95% Conf. Interval]
protect	1.313965	.2366229	5.55	0.000	.8501928 1.777737
age	-.0659554	.0221171	-2.98	0.003	-.1093041 -.0226067
_cons	6.946524	1.575708	4.41	0.000	3.858192 10.03486
/ln_p	.5188694	.1376486	3.77	0.000	.2490831 .7886556
p	1.680127	.2312671			1.282849 2.200436
1/p	.5951931	.0819275			.4544553 .7795152

```
. predict t_mean, time mean
. predict t_median, time median
. predict lnt_mean, lntime mean
. predict lnt_median, lntime median
. list _t protect age t_mean t_median lnt_mean lnt_median in L, abbrev(10)
```

	_t	protect	age	t_mean	t_median	lnt_mean	lnt_median
106.	39	1	67	41.60789	37.46217	3.497923	3.623332

We did that just to demonstrate that nothing changes—it really is the same model—and predict was smart enough to take parameter estimates in either metric and apply them correctly when calculating the prediction. When we take a closer look at how these calculations are performed, we can switch between thinking in the PH and AFT metrics freely.

Let's look at our predict commands one at a time:

1. predict t_mean, time mean
   This gives the expected value of survival time, otherwise known as the first moment of the distribution of survival time. Letting T_j be the time to failure for a random observation with covariate values $\mathbf{x}_j$, the expected value of T_j is given by

$$\mu_{T_j} = E(T_j|\mathbf{x}_j) = \int_0^\infty tf(t|\mathbf{x}_j)dt \qquad (14.1)$$

$$= \int_0^\infty S(t|\mathbf{x}_j)dt$$

where $f()$ is the probability density function of T and $S()$ is the survivor function, both determined by the choice of parametric model. The "prediction" aspect of the calculation has to do with the model parameters that characterize $f()$ and $S()$ being replaced by their maximum likelihood estimates.

For the Weibull AFT model above, from (13.6) we have

$$S(t|\mathbf{x}_j) = \exp[-\{\exp(-\beta_0 - \mathbf{x}_j\boldsymbol{\beta}_x)t\}^p] \qquad (14.2)$$

Equivalently, we could use the Weibull PH survivor function, but we will stick with the AFT parameterization because those are the parameter estimates that we currently have handy. Performing the integration in (14.1) gives

$$\mu_{T_j} = \exp\{p(\beta_0 + \mathbf{x}_j\boldsymbol{\beta}_x)\}\Gamma\left(1 + \frac{1}{p}\right)$$

where $\Gamma()$ is the gamma function. Thus, for our model, the predicted mean survival time becomes

$$\widehat{\mu}_{T_j} = \exp\left\{\widehat{p}\left(\widehat{\beta}_0 + \widehat{\beta}_1\text{protect}_j + \widehat{\beta}_2\text{age}_j\right)\right\}\Gamma\left(1 + \frac{1}{\widehat{p}}\right) .$$

Plugging in the estimated model parameters and the values of protect and age from the last observation in our data yields $\widehat{\mu}_{T_{106}} = 41.60789$, which is what is given by predict.

2. predict t_median, time median
   This gives the median, or 50th percentile, of survival time. From (14.2) we can derive the quantile function $Q(u|\mathbf{x}_j)$ for the Weibull AFT model by noting that, because T_j is continuous, $Q(u|\mathbf{x}_j) = t$ if and only if $1 - S(t|\mathbf{x}_j) = u$. Thus

$$Q(u|\mathbf{x}_j) = \exp(\beta_0 + \mathbf{x}_j\boldsymbol{\beta}_x)\left\{-\ln(1 - u)\right\}^{1/p}$$

The median of T_j is

$$\widetilde{\mu}_{T_j} = Q(0.5|\mathbf{x}_j)$$

and is estimated by plugging in the covariate values $\mathbf{x}_j$ and the parameter estimates. For our model,

$$\widehat{\widetilde{\mu}}_{T_j} = \exp\left(\widehat{\beta}_0 + \widehat{\beta}_1 \texttt{protect}_j + \widehat{\beta}_2 \texttt{age}_j\right) \{\ln(2)\}^{1/\widehat{p}} \tag{14.3}$$

Plugging in the estimated model parameters and the values of `protect` and `age` from the last observation in our data yields $\widehat{\widetilde{\mu}}_{T_{106}} = 37.46217$, which is what is given by `predict`.

3. `predict lnt_mean, lntime mean`
   This predicts the expected value of $\ln(T_j)$ and can be derived in many ways. The most direct way would be a generalization of (14.1), which, instead of integrating t times the density, integrates $\ln(t)$ times the density:

$$E\{\ln(T_j)\} = \int_0^\infty \ln(t) f(t|\mathbf{x}_j) dt$$

For models that are parameterized as AFT, however, a better approach would be to derive the distribution of $\ln(T_j)$ via a change of variable and to directly calculate the mean of this new distribution. With AFT models, the change of variable recasts the model as one that is linear in the regressors, and thus the calculation of the mean reduces to calculating the mean of a residual term.

For example, our Weibull AFT model can alternatively be expressed as

$$\ln(T_j) = \beta_0 + \mathbf{x}_j \boldsymbol{\beta}_x + u_j \tag{14.4}$$

where the residual term u_j follows the extreme value distribution with shape parameter, p, such that $E(u_j) = \Gamma'(1)/p$, where $\Gamma'(1)$ is the negative of Euler's constant. Therefore,

$$
\begin{aligned}
E\{\ln(t_j)\} &= \beta_0 + \mathbf{x}_j \boldsymbol{\beta}_x + E(u_j) \\
&= \beta_0 + \mathbf{x}_j \boldsymbol{\beta}_x + \Gamma'(1)/p
\end{aligned}
$$

and for our model, this becomes

$$\widehat{E}\{\ln(T_j)\} = \widehat{\beta}_0 + \widehat{\beta}_1 \texttt{protect}_j + \widehat{\beta}_2 \texttt{age}_j + \Gamma'(1)/\widehat{p}$$

Plugging in our estimates and covariate values [$\Gamma'(1)$ is obtained in Stata as `digamma(1)` and $\widehat{p}$ as `e(aux_p)`] yields $\widehat{E}\{\ln(T_{106})\} = 3.497923$, which is obtained by `predict`.

For other AFT models such as the lognormal and the loglogistic, $E(u_j) = 0$ when the model is recast in the log-linear form, and thus $E\{\ln(T_j)\}$ is simply the linear predictor. For the Weibull (and exponential as a special case) model, however,

$E(u_j)$ is nonzero, and researchers often choose to ignore this fact and predict ln(time) using only the linear predictor. Even though this prediction is biased, it is widely used. In any case, if this alternate predicted ln(time) is desired, it is easily obtained: `predict` *newvar*, `xb` will give the linear predictor.

If you exponentiate a predicted mean of $\ln(T_j)$ to predict survival time, what you get is not equal to the mean of T_j but is another prediction of time to failure altogether. Also you can exponentiate the linear predictor (i.e., ignore the nonzero mean residual) and produce yet another predicted time to failure:

```
. gen e_lmean = exp(lnt_mean)
. predict xb, xb
. gen e_xb = exp(xb)
. list _t protect age t_mean t_median e_lmean e_xb in 1
```

	_t	protect	age	t_mean	t_median	e_lmean	e_xb
106.	39	1	67	41.60789	37.46217	33.04674	46.59427

The four predicted survival times are comparable but not equal. `t_mean` is the predicted mean failure time, `t_median` is the predicted median failure time, and `e_lmean` and `e_xb` are other predictions of failure time obtained via the alternate methods we have described. Each of these four predictions has been used by researchers, and our aim is not to debate the relative merits of each but to point out how they are calculated and that they generally do not coincide.

If you want to predict survival time, be aware of what the software gives you by default and how you can change the default to obtain the prediction you want. In Stata, these four predictions are easily obtained, and given the commands and options required for each, the context is clear.

❏ **Technical note**

Exponentiating a linear predictor to predict survival time (`e_xb` in the above) makes sense in an AFT model but not for a PH model when what you obtain are hazard ratios, which most likely have nothing to do with predicting a survival time.

❏

4. `predict lnt_median, lntime median`
   The direct calculation of the median of $\ln(T_j)$ would involve the derivation of the distribution of $\ln(T_j)$ from that of T_j via a change of variable. Once the distribution of $\ln(T_j)$ is obtained, you can derive the quantile function for this distribution and use it to evaluate the 50th percentile.

In an AFT model, when the model is recast as one that is log-linear in time such as (14.4), the calculation reduces to taking the median of the distribution of the residual, u_j, and adding $\beta_0 + \mathbf{x}_j\boldsymbol{\beta}_x$ to it. For many AFT models (such as the lognormal and loglogistic), the distribution of u_j is symmetric about zero, and thus $E(u_j) = \text{median}(u_j) = 0$. So, the median of $\ln(T_j)$ equals the mean of $\ln(T_j)$ and is simply the linear predictor in both cases.

The Weibull, however, is not such a model. Given the log-linear form (14.4), for the Weibull, u_j follows an extreme value distribution with shape parameter p. As such,

$$\text{median}(u_j) = \ln\{\ln(2)\}/p$$

and thus our estimate of the median log survival-time for our model becomes

$$\widehat{\text{median}}\{\ln(T_j)\} = \hat{\beta}_0 + \hat{\beta}_1\texttt{protect}_j + \hat{\beta}_2\texttt{age}_j + \ln\{\ln(2)\}/\hat{p}$$

which equals $\ln(\widehat{\tilde{\mu}}_{T_j})$, where $\widehat{\tilde{\mu}}_{T_j}$ is given in (14.3).

This is generally true because $\ln(T_j)$ is a monotone transformation of T_j, and the calculation of the median is invariant to such a transformation. Thus for our covariates and parameter estimates,

$$
\begin{aligned}
\widehat{\text{median}}\{\ln(T_{106})\} &= \ln\left(\widehat{\tilde{\mu}}_{T_{106}}\right) \\
&= \ln(37.46217) \\
&= 3.623332
\end{aligned}
$$

which agrees with what `predict` calculates.

Because medians are invariant to the log transformation, exponentiating the predicted median of ln(survival time) does not result in an alternate predicted survival time—it merely reproduces the predicted median survival time, something already considered.

Finally, be careful using `predict, time` and `predict, lntime` with multiple-record-per-subject data. Remember that each prediction is made for each record in isolation: the predicted failure time is from time 0 under the assumption of constant covariates. For example,

```
. use http://www.stata-press.com/data/cggm/hip2, clear
(hip fracture study)
. streg protect calcium, dist(weibull) time
  (output omitted)
. predict t_hat, mean time
. list id _t0 _t protect calcium _d t_hat if id==10
```

	id	_t0	_t	protect	calcium	_d	t_hat
11.	10	0	5	0	9.69	0	8.384427
12.	10	5	8	0	9.47	0	7.956587

The interpretation of `t_hat` in the first observation displayed is that if we had a subject who had `protect==0` and `calcium==9.69`, and those covariates were fixed, then the predicted (mean) time of failure would be 8.38. In the second observation, the prediction is that if we had a subject who was under continual observation from time 0 forward and the subject had fixed covariates of `protect==0` and `calcium==9.47`, the predicted time of failure would be 7.96. Neither prediction really has much to do with subject 10 in our data, given the nature of the time-varying covariate `calcium`.

14.1.2   Predicting the hazard and related functions

After estimation with `streg`, you can type

1. `predict` *newvar*, `hazard`
   to obtain the predicted hazard.

2. `predict` *newvar*, `hr`
   to obtain the predicted relative hazard (hazard ratio), assuming that the model has a PH implementation. Otherwise, this option is not allowed.

3. `predict` *newvar*, `surv`
   to obtain the predicted survivor function conditional on the beginning of an interval (determined by _t0).

4. `predict` *newvar*, `csurv`
   to obtain the (unconditional) predicted survivor function.

All these calculations (except `hr`) can be made, even for models fit using an AFT parameterization. All survival models may be characterized by their hazard function $h(t|\mathbf{x})$; it just may be that $h(t|\mathbf{x})$ cannot be written according to the proportional hazards decomposition $h_0(t)\exp(\mathbf{x}\boldsymbol{\beta}_x)$. `predict`, however, calculates $h()$ and not $h_0()$, and if you want $h_0()$ because it is appropriate given the model you have fit, you can obtain it by first setting $\mathbf{x} = \mathbf{0}$ and then predicting $h()$.

The values that `predict`, `hazard` calculates are $h(\_t)$. For example,

```
. use http://www.stata-press.com/data/cggm/hip2, clear
(hip fracture study)
. streg protect age, dist(weibull)
(output omitted )
. predict h, hazard
. list id _t0 _t protect age _d h if id==10 | id==21
```

	id	_t0	_t	protect	age	_d	h
11.	10	0	5	0	73	0	.1397123
12.	10	5	8	0	73	0	.1923367
35.	21	0	5	1	82	0	.041649
36.	21	5	6	1	82	1	.0471474

The above output reports

11. $\widehat{h}(5|\text{protect} == 0, \text{age} == 73) = 0.1397$

12. $\widehat{h}(8|\text{protect} == 0, \text{age} == 73) = 0.1923$

35. $\widehat{h}(5|\text{protect} == 1, \text{age} == 82) = 0.0416$

36. $\widehat{h}(6|\text{protect} == 1, \text{age} == 82) = 0.0471$

If we change the data after estimation, `predict` calculates results based on the previously estimated model parameters and the new data in memory:

```
. drop h
. replace age = 70
(96 real changes made)
. replace protect = 0
(72 real changes made)
. predict h, hazard
. list id _t0 _t protect age _d h if id==10 | id==21
```

	id	_t0	_t	protect	age	_d	h
11.	10	0	5	0	70	0	.1001977
12.	10	5	8	0	70	0	.1379384
35.	21	0	5	0	70	0	.1001977
36.	21	5	6	0	70	1	.1134256

The interpretation of this output is

11. $\widehat{h}(5|\text{protect} == 0, \text{age} == 70) = 0.1002$

12. $\widehat{h}(8|\text{protect} == 0, \text{age} == 70) = 0.1379$

35. $\widehat{h}(5|\text{protect} == 0, \text{age} == 70) = 0.1002$

36. $\widehat{h}(6|\text{protect} == 0, \text{age} == 70) = 0.1134$

`predict` always works like this, regardless of the option specified. `predict` calculates results; it does not retrieve results stored at the time of estimation. Changing the data in memory after estimation is a favorite trick to obtain desired calculations. Just remember to save your dataset before you change it.

`predict, surv` and `predict, csurv` calculate current and cumulative survivor probabilities using the data currently in memory:

```
. predict s, surv

. predict cs, csurv

. list id _t0 _t protect age _d s cs if id==10 | id==21
```

	id	_t0	_t	protect	age	_d	s	cs
11.	10	0	5	0	70	0	.7421641	.7421641
12.	10	5	8	0	70	0	.6986433	.518508
35.	21	0	5	0	70	0	.7421641	.7421641
36.	21	5	6	0	70	1	.8986372	.6669363

The interpretation of the previous output (for T = survival time) is given in table 14.2.

Table 14.2. Use of `predict, surv` and `predict, csurv`

Obs.	Var.	Calculation	Value
11	s	$\Pr(T > 5 \mid$ `protect` $== 0,$ `age` $== 70)$	0.7422
12	s	$\Pr(T > 8 \mid$ `protect` $== 0,$ `age` $== 70, T > 5)$	0.6986
12	cs	$\Pr(T > 8 \mid$ `history`$)$	$0.7422 \times 0.6986 = 0.5185$
35	s	$\Pr(T > 5 \mid$ `protect` $== 0,$ `age` $== 70)$	0.7422
35	s	$\Pr(T > 6 \mid$ `protect` $== 0,$ `age` $== 70, T > 5)$	0.8986
36	cs	$\Pr(T > 6 \mid$ `history`$)$	$0.7422 \times 0.8986 = 0.6669$

The cumulative probabilities calculated by `predict, csurv` are obtained by multiplying (within subject) the `predict, surv` results. This means that, if the covariates change during subspans, the change would be reflected in the calculated cumulative probabilities. We labeled the cumulative results as those given the "`history`" to emphasize that fact, and we did that even though, in our example, the covariates do not change.

❏ **Technical note**

This method of calculating cumulative probabilities via multiplication across the covariate history has its issues. When there are gaps or delayed entry, the net effect of the technique is to assume the probability of survival over the gaps is 1, just as for the likelihood function. This empirically based notion of survivor probabilities can make comparison of subjects with similar covariates more difficult if one of the subjects exhibits gaps and the other does not. (We will see another manifestation of the interpretation problem caused by gaps in section 14.2.)

❏

Finally, you may have noticed that there is no `cumhazard` option to `predict` after `streg` in current Stata. There is really no good reason for this, except to say that obtaining an estimate of the cumulative hazard function at time _t is easy given the one-to-one relationship between the survivor function and the cumulative hazard function. One types

```
. predict cs, csurv          /* We want cumulative survival here */
. gen chaz = -ln(cs)         /* H(t) = -ln{S(t)}   */
```

14.1.3   Calculating residuals

The residuals calculated by `predict` can be used in the same way as discussed in chapter 11, where we discussed regression diagnostics for the Cox model. The interpretation is always the same. There is, however, a difference in how (and when) you request the residuals. With the parametric models, the residuals are calculated after estimation using `predict`, whereas with the Cox model most of the residuals must be requested at the time of estimation via options to the `stcox` command. Requests after the fact are more convenient, and `stcox` requires that you request the residuals at the time of estimation because of computer efficiency. For Cox regression, the residuals (like the likelihood) are based on calculations involving summations across risk pools, and it is in the identification of those risk pools that the computer spends substantial time. Thus, rather than reforming the risk pools postestimation, it is more efficient to use the risk pools that were already formed during the estimation.

In any case, the intricacies of diagnostics for Cox models carry over to parametric models for survival data and thus are not something on which we elaborate. These diagnostics are important in parametric models; it is just that we think the application of those topics covered in chapter 11 can be applied straightforwardly to this context.

As an example, we fit a Weibull model to our hip data and calculate the (cumulative) Cox–Snell residuals to look for outliers:

```
. use http://www.stata-press.com/data/cggm/hip2, clear
(hip fracture study)
. streg protect age, dist(weib)
  (output omitted)
. predict cc, ccsnell
(58 missing values generated)
. scatter cc _t, mlabel(id) msymbol(i)
```

The result is shown in figure 14.1.

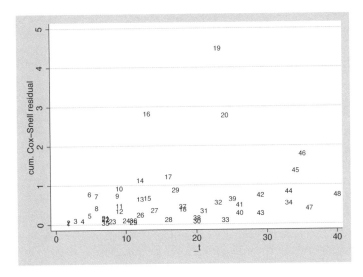

Figure 14.1. Cumulative Cox–Snell residuals for a Weibull model

We see that id==16, 19, & 20 are probable outliers. In looking at outliers, we find it useful to also have the predicted time of failure (the mean is fine) based on the model:

```
. predict pred_t, time mean
. list _t0 _t pred_t age protect _d cc if id==16
```

	_t0	_t	pred_t	age	protect	_d	cc
23.	0	5	5.782123	77	0	0	.
24.	5	12	5.782123	77	0	1	2.819476

We see that this 77-year-old woman failed at time 12, which was a long time after the predicted (mean) time of 5.78. Note also that cc is missing in observation 23. We asked for the cumulative Cox–Snell residuals because we had multiple-observation-per-subject data, and when we did that, the summed residual was filled in only on the last record of each subject. That is why predict in the previous log said "58 missing values generated".

Variable pred_t has the same value in both observations. As we warned, predicted failure times are made as if covariates are constant. In our model of age and protect, the covariates are indeed constant, so that is a correct result, but in general, this may not be what we want.

You may be asking yourself, Why not just find outliers based on pred_t? There are two answers:

1. You can correctly calculate "residual time" = _t − pred_t only for those who fail; censored observations can also be outliers.

2. The distribution of pred_t can be nonsymmetric, which makes finding outliers based on residual time difficult.

14.2 Using stcurve

stcurve is a wonderfully handy command for graphing after estimation the estimated survivor functions, hazard functions, and cumulative hazard functions. stcurve does nothing you cannot do for yourself using the predict and graph commands, but stcurve makes doing certain things easier.

Remember that predict calculates predicted values, for each observation in the data, at that observation's recorded values of the covariates. If you were to type, say, predict h, hazard and then graph h versus _t, what you would see is a hopeless mishmash of predicted hazards, each at different values of the covariates. The solution is to replace, in each observation, the values of the covariates with constant values (such as protect==0 and age==70) before making the prediction and then graph that. Remember, however, to save your dataset first so that you can get your real data back. In chapter 13, replacing the values of the covariates with the constant values was the approach we followed for making graphs of the functions, and that approach is well worth learning because it will let you draw any graph of which you can conceive, such as that in figure 13.8 where we compared Weibull and piecewise exponential hazards.

Usually, however, what we want are "simple" graphs, and stcurve can draw those for us. Moreover, stcurve can draw graphs comparing groups, and these graph might otherwise overtax our knowledge of Stata syntax.

stcurve can graph

1. The survivor function. Type stcurve, survival ...

2. The cumulative hazard function. Type stcurve, cumhaz ...

3. The hazard function. Type stcurve, hazard ...

stcurve can graph any of those functions at the values for the covariates you specify. The syntax is as follows:

  stcurve, ...at(*varname*=# *varname*=# ...)

If you do not specify a variable's value, the average value is used. Thus, if the at() option is omitted altogether, a graph is produced for all the covariates held at their average values. The at() option can also be generalized to graph the function evaluated at different values of the covariates on the same graph. The syntax is

  stcurve, ...at1(*varname*=# *varname*=# ...) at2(...) at3(...) ...

For example, we can fit a generalized gamma model to our hip data and plot the estimated hazard function for several covariate patterns:

```
. use http://www.stata-press.com/data/cggm/hip2, clear
(hip fracture study)
. streg protect age, dist(gamma)
  (output omitted )
. stcurve, hazard
>             at1(protect=0 age=70)
>             at2(protect=0 age=75)
>             at3(protect=0 age=80)
>             at4(protect=0 age=85)
```

The results are shown in figure 14.2.

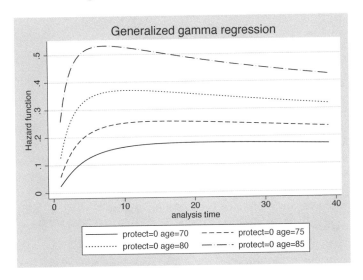

Figure 14.2. Hazard curves for the gamma model (`protect==0`)

For hazard curves for models that are not constrained to fit the PH assumption (such as the gamma), we emphasize that there is not one hazard function but one function per pattern of covariates: $h(t|\mathbf{x})$. For PH models, these functions can be summarized as $h(t|\mathbf{x}) = h_0(t) \exp(\mathbf{x}\boldsymbol{\beta}_x)$. Here we are being flexible by fitting a gamma model, and we can check the PH assumption graphically.

These graphs only gently violate the PH assumption; the top graph for the oldest group actually turns down, whereas the bottom graph for the youngest group does not.

(Continued on next page)

The corresponding graph for those with `protect==1` is

```
. stcurve, hazard
>               at1(protect=1 age=70)
>               at2(protect=1 age=75)
>               at3(protect=1 age=80)
>               at4(protect=1 age=85)
```

and the result is shown in figure 14.3.

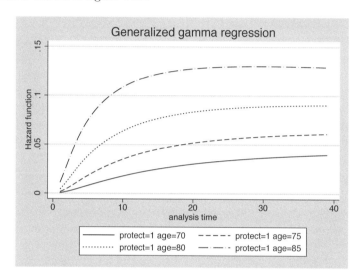

Figure 14.3. Hazard curves for the gamma model (`protect==1`)

In comparing this graph with the previous one, note carefully the different scales of the y axis.

In a technical note in section 14.1.2, we mentioned that `predict, csurv` calculates the cumulative probability of survival and that, given the way this prediction is calculated, special care must be taken when comparing subjects with similar covariate patterns where one subject has gaps and the other does not. Because of gaps, even if two subjects have the same covariate values, they may have different predictions of cumulative survival for any given time. We can see this by plotting the cumulative survivor function as calculated using `predict` after setting `age==70` and `protect==0`:

```
. use http://www.stata-press.com/data/cggm/hip2, clear
(hip fracture study)
. streg protect age, dist(weibull)
  (output omitted)
. replace age = 70
(96 real changes made)
. replace protect = 0
(72 real changes made)
```

```
. predict cs, csurv
. line cs _t, c(l) sort l1title("Cumulative survival probability")
```

The result is shown in figure 14.4.

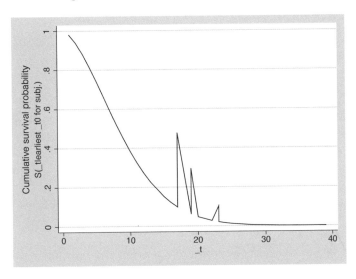

Figure 14.4. Cumulative survivor probability as calculated by `predict`

The spikes in the plot indicate time gaps in the data, and even though all observations have the same covariate values (we set them that way), the predicted cumulative survivor function is not a smooth function of _t.

When covariates are held constant, the cumulative survivor probability as calculated by `predict, csurv` will equal $S(\_t|\mathbf{x})$ only if there are no time gaps in the data. When there are gaps, `stcurve` comes to the rescue. `stcurve` calculates $S(\_t|\mathbf{x})$ directly from the parametric form of this function, ignoring the gaps. A superior plot of the predicted $S(\_t|\mathbf{x})$ is obtained using

```
. use http://www.stata-press.com/data/cggm/hip2, clear
(hip fracture study)
. streg protect age, dist(weibull)
  (output omitted )
. stcurve, survival at(protect=0 age=70)
```

The result is shown in figure 14.5.

(Continued on next page)

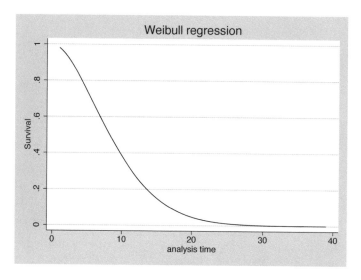

Figure 14.5. Survivor function as calculated by `stcurve`

Given what we have seen, our advice is to use `predict, surv` if you want individual-specific survivor probabilities that take gaps and time-varying covariates into account. Use `stcurve, survival` if you want to plot the estimated overall survivor function.

15 Generalizing the parametric regression model

In chapter 13, we discussed the six parametric models for survival data available in Stata's `streg` command: exponential, Weibull, lognormal, loglogistic, Gompertz, and generalized gamma. These models represent the basis of parametric survival models available in Stata. Here we discuss other aspects and options to `streg` to expand and generalize these models.

15.1 Using the ancillary() option

The `ancillary()` option is how you specify linear predictors for the "other" parameters of the assumed distribution. For instance, the Weibull distribution has one ancillary parameter, p, which is usually fit as a constant. p has been constant in every example we have considered. By specifying `ancillary(male)`, for instance, you could specify that $p = p_0 + p_1 \mathtt{male}$; if `male` were an indicator variable that assumed the value 0 for females and 1 for males, then you would effectively be saying that p has one constant value for males and another constant value for females. We will explore below why you might want to do that.

The exponential model has no ancillary parameters (thus `ancillary()` is not relevant); the Weibull, Gompertz, lognormal, and loglogistic models each have one ancillary parameter, and the generalized gamma model has two (for which there is a corresponding `anc2()` option to go along with `ancillary()`).

To understand why we might want to parameterize ancillary parameters, let us reconsider the expanded hip-fracture data first discussed in chapter 9. This dataset, in addition to the data on female patients, contains data on males and includes the additional variable `male` specifying the patient's sex.

Let us begin by fitting a combined male and female Weibull proportional hazards (PH) model with `protect` and `age` as covariates.

```
. use http://www.stata-press.com/data/cggm/hip3
(hip fracture study)

. streg protect age, dist(weibull) nohr nolog

        failure _d:  fracture
   analysis time _t:  time1
              id:  id

Weibull regression -- log relative-hazard form

No. of subjects =          148              Number of obs   =        206
No. of failures =           37
Time at risk    =         1703
                                            LR chi2(2)      =      49.97
Log likelihood  =   -77.446477             Prob > chi2     =     0.0000
```

_t	Coef.	Std. Err.	z	P>|z|	[95% Conf. Interval]	
protect	-2.383745	.3489219	-6.83	0.000	-3.067619	-1.699871
age	.0962561	.03467	2.78	0.005	.0283042	.1642079
_cons	-10.63722	2.597762	-4.09	0.000	-15.72874	-5.545695
/ln_p	.4513032	.1265975	3.56	0.000	.2031767	.6994297
p	1.570357	.1988033			1.225289	2.012605
1/p	.6367977	.080617			.4968686	.816134

The hazard function for the Weibull PH model is

$$h(t_j|\mathbf{x}_j) = \exp(\beta_0 + \mathbf{x}_j\boldsymbol{\beta}_x)pt_j^{p-1}$$

which for our model translates into the estimated hazard

$$\widehat{h}(t_j|\mathbf{x}_j) = \exp\left(-10.64 - 2.38\text{protect}_j + 0.10\text{age}_j\right)1.57t_j^{0.57}$$

If we wanted to compare males with females, we could add the variable `male` to our covariate list.

```
. streg protect age male, dist(weibull) nohr nolog

        failure _d:  fracture
  analysis time _t:  time1
              id:  id

Weibull regression -- log relative-hazard form

No. of subjects =        148              Number of obs    =        206
No. of failures =         37
Time at risk    =       1703
                                          LR chi2(3)       =      62.42
Log likelihood  =    -71.224402           Prob > chi2      =     0.0000
```

_t	Coef.	Std. Err.	z	P>|z|	[95% Conf. Interval]	
protect	-2.095949	.3632392	-5.77	0.000	-2.807885	-1.384013
age	.0907551	.0339681	2.67	0.008	.0241788	.1573314
male	-1.413811	.4555996	-3.10	0.002	-2.30677	-.5208524
_cons	-9.667022	2.562691	-3.77	0.000	-14.6898	-4.64424
/ln_p	.3784328	.1302032	2.91	0.004	.1232392	.6336264
p	1.459995	.190096			1.131155	1.884432
1/p	.684934	.0891806			.5306639	.8840521

Because `male` is in our covariate list, we assume that the hazards for males and females are proportional (given `age` and `protect`). The net effect is that only the scale of the distribution of t_j changes according to the patient's sex. Because `male==0` for females and `male==1` for males, the above output gives the estimated hazards

$$\widehat{h}(t_j|\mathbf{x}_j) = \begin{cases} \exp\left(-9.67 - 2.10\texttt{protect}_j + 0.09\texttt{age}_j\right) 1.46t_j^{0.46}, & \text{if female} \\ \exp\left(-11.08 - 2.10\texttt{protect}_j + 0.09\texttt{age}_j\right) 1.46t_j^{0.46}, & \text{if male} \end{cases}$$

Because the Wald test for `male` is $z = -3.10$ with significance level 0.002, we are confident that gender does have some effect on the risk of hip fracture. We are not, however, convinced that the effect is proportional. Perhaps gender's effect is to alter the shape of the hazard instead, and the above output is just a best approximation as to the effect of gender given the constraints of the PH assumption.

The Weibull distribution has a scale parameter and a shape parameter. The scale parameter is the exponentiated linear predictor and is thus modeled using the covariate list in `streg`. The other parameter is the shape, p, and if we want this parameter to be modeled, we use the `ancillary()` option to `streg`:

```
. streg protect age, dist(weibull) ancillary(male) nohr nolog

        failure _d:  fracture
  analysis time _t:  time1
               id:  id

Weibull regression -- log relative-hazard form

No. of subjects =           148                Number of obs   =        206
No. of failures =            37
Time at risk    =          1703
                                               LR chi2(2)      =      39.80
Log likelihood  =    -69.323532                Prob > chi2     =     0.0000
```

_t	Coef.	Std. Err.	z	P>|z|	[95% Conf. Interval]
_t					
protect	-2.130058	.3567005	-5.97	0.000	-2.829178 -1.430938
age	.0939131	.0341107	2.75	0.006	.0270573 .1607689
_cons	-10.17575	2.551821	-3.99	0.000	-15.17722 -5.174269
ln_p					
male	-.4887189	.185608	-2.63	0.008	-.8525039 -.1249339
_cons	.4540139	.1157915	3.92	0.000	.2270667 .6809611

Here we are assuming that $\ln(p)$ is not constant over the data but instead is a linear function of some covariates $\mathbf{z}$, $\ln(p) = \alpha_0 + \mathbf{z}\boldsymbol{\alpha}_z$, where $\mathbf{z}$ need not be distinct from $\mathbf{x}$. We parameterize in $\ln(p)$ instead of p itself because p is constrained to be positive, whereas a linear predictor can take on any values on the real line.

From our estimation results, we see that $\widehat{\ln(p)} = 0.454$ for females and $\widehat{\ln(p)} = 0.454 - 0.489 = -0.035$ for males. Thus $\hat{p} = \exp(0.454) = 1.57$ for females and $\hat{p} = \exp(-0.035) = 0.97$ for males. The estimated hazards are then

$$\widehat{h}(t_j|\mathbf{x}_j) = \begin{cases} \exp\left(-10.18 - 2.13\texttt{protect}_j + 0.09\texttt{age}_j\right) 1.57 t_j^{0.57}, & \text{if female} \\ \exp\left(-10.18 - 2.13\texttt{protect}_j + 0.09\texttt{age}_j\right) 0.97 t_j^{-0.03}, & \text{if male} \end{cases}$$

and if we believe this model, we would say that the hazard for males given `age` and `protect` is almost constant.

The Wald test for `male` in the above output is $z = -2.63$ with significance level 0.008, and thus we are still convinced of the effect of gender on the risk of hip fracture. However, we are no better off than we were before. Here it may be that the effect of gender is actually proportional, and by not allowing a shift in the hazard to take place, the model actually had to change the shape of the hazard to accommodate this effect.

When we included `male` in the covariate list, we assumed that the effect of gender was on the scale of the hazard and that the shape of the hazard was the same for both sexes. When we specified `ancillary(male)`, we assumed that the shape of the hazard changed according to gender but that the scale remained unaffected. We can be fully general and allow both the scale and shape to vary by gender by using

```
. streg protect age if male, dist(weib)
. streg protect age if !male, dist(weib)
```

but if we did that we would be assuming that the effect of protect and age differed according to gender as well. There is certainly nothing wrong with this, but let us assume that what we want is to allow the scale and shape of the hazard to vary with gender but to constrain the effect of protect and age to remain the same. This is achieved by specifying male both in the covariate list and in ancillary():

```
. streg protect age male, dist(weibull) ancillary(male) nohr nolog

         failure _d:  fracture
   analysis time _t:  time1
               id:  id

Weibull regression -- log relative-hazard form

No. of subjects =            148          Number of obs   =         206
No. of failures =             37
Time at risk    =           1703
                                          LR chi2(3)      =       40.28
Log likelihood  =      -69.082313         Prob > chi2     =      0.0000
```

_t	Coef.	Std. Err.	z	P>|z|	[95% Conf. Interval]	
_t						
protect	-2.185115	.3645006	-5.99	0.000	-2.899523	-1.470707
age	.0966629	.0345663	2.80	0.005	.0289142	.1644117
male	.7382003	1.036608	0.71	0.476	-1.293514	2.769915
_cons	-10.61465	2.646846	-4.01	0.000	-15.80238	-5.42693
ln_p						
male	-.7116757	.3834735	-1.86	0.063	-1.46327	.0399185
_cons	.5079011	.1358255	3.74	0.000	.2416881	.7741142

Comparing the Wald tests for both instances of male in this model favors our second version of this story: the effect of gender is on the shape of the hazard, and given that the shape changes, the effect on the scale is not significant.

Satisfied that we need to include male only in the estimation of the ancillary parameter, we can now use ancillary() to test the PH assumption on the other two covariates, protect and age. For example,

(Continued on next page)

```
. streg protect age, dist(weibull) ancillary(male protect) nohr nolog
         failure _d:  fracture
   analysis time _t:  time1
                id:  id
Weibull regression -- log relative-hazard form
No. of subjects =         148                Number of obs   =        206
No. of failures =          37
Time at risk    =        1703
                                             LR chi2(2)      =      13.69
Log likelihood  =  -69.264251               Prob > chi2     =     0.0011
```

_t	Coef.	Std. Err.	z	P>|z|	[95% Conf. Interval]	
_t						
protect	-2.456007	1.023378	-2.40	0.016	-4.46179	-.4502238
age	.0934049	.0340489	2.74	0.006	.0266704	.1601394
_cons	-10.02396	2.580379	-3.88	0.000	-15.08141	-4.966508
ln_p						
male	-.4801077	.1850198	-2.59	0.009	-.8427398	-.1174756
protect	.0763409	.2215628	0.34	0.730	-.3579142	.510596
_cons	.4220745	.1505132	2.80	0.005	.1270741	.7170749

We see no evidence to contradict the PH assumption for `protect`. A similar analysis performed on `age` would also fail to reject the PH assumption. Because the above test of proportional hazards is based on a comparison of nested models, we could also perform a likelihood-ratio test,

```
. streg protect age, dist(weibull) ancillary(male protect) nohr nolog
  (output omitted)
. estimates store protect
. streg protect age, dist(weibull) ancillary(male) nohr nolog
  (output omitted)
. lrtest protect ., force
Likelihood-ratio test                        LR chi2(1)  =       0.12
(Assumption: . nested in protect)            Prob > chi2 =      0.7306
```

with similar results. To perform this test, we had to use the `force` option to `lrtest`. We used this option because we are omitting a parameter from the ancillary parameter list and not the main parameter list. This omission causes `lrtest` to think the models are not nested because they have differing log likelihoods under the null hypothesis that all main parameters are zero. In any case, the use of `force` is justified because we know that the models are nested.

The use of the `ancillary()` option is not restricted to PH models; the option may be used with AFT models, and the interpretation of results is a straightforward extension of the above.

15.2 Stratified models

Stratified models, as you may remember from section 9.3, concern estimation when the baseline hazard is allowed to differ for different groups. Males and females, for instance, might each have their own hazard function rather than that of males being constrained to be a proportional replica of that for females. In section 9.3, we allowed this in the context of semiparametric estimation.

In parametric stratified estimation, each group is similarly allowed to have its own baseline hazard function, but the hazard functions are constrained to be of the same family. If the hazard for females is Weibull, then so must be the hazard for males.

There is an obvious connection of this idea with the `ancillary()` option. In the previous section, when we typed

```
. streg protect age male, dist(weibull) ancillary(male) nohr nolog
```

we allowed both the scale and shape of the hazard to vary with `male`, yet we constrained the effects of `protect` and `age` to be the same for both sexes. Examining the output from that estimation, we can construct the estimated hazards for males and females as

$$\widehat{h}(t_j|\mathbf{x}_j) = \begin{cases} \exp\left(-10.61 - 2.19\texttt{protect}_j + 0.10\texttt{age}_j\right) 1.66 t_j^{0.66}, & \text{if female} \\ \exp\left(-9.88 - 2.19\texttt{protect}_j + 0.10\texttt{age}_j\right) 0.82 t_j^{-0.18}, & \text{if male} \end{cases}$$

Because this is a PH model, we can write the estimated hazard as

$$\widehat{h}(t_j|\mathbf{x}_j) = \widehat{h}_0(t_j) \exp(-2.19\texttt{protect}_j + 0.10\texttt{age}_j)$$

for the estimated baseline hazard $\widehat{h}_0(t_j)$ such that

$$\widehat{h}_0(t_j) = \begin{cases} \exp\left(-10.61\right) 1.66 t_j^{0.66}, & \text{if female} \\ \exp\left(-9.88\right) 0.82 t_j^{-0.18}, & \text{if male} \end{cases}$$

and we can graph the comparison of these baseline hazards by typing

```
. gen h0female = exp(-10.61)*1.66*_t^(0.66)
. gen h0male = exp(-9.88)*0.82*_t^(-0.18)
. line h0female h0male _t, sort l1title("Baseline hazard")
```

This produces figure 15.1 and graphically verifies what we already know: the shape of the hazard is different for males and females; the hazards are not proportional.

(Continued on next page)

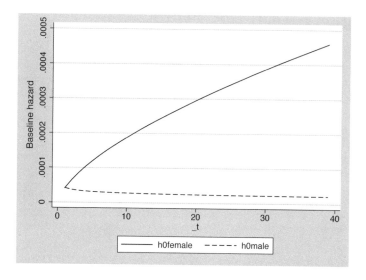

Figure 15.1. Comparison of baseline hazards for males and females

Thus we have a model that assumes proportional hazards with respect to `protect` and `age`, yet we have different baseline hazard functions for males and females. Said differently, we have a proportional hazards model on `protect` and `age` that is stratified on `male` and is analogous to the stratified Cox model.

`streg`'s `strata(`*varname*`)` option provides an easier way to fit these models:

```
. streg protect age, dist(weibull) strata(male) nohr nolog
        failure _d:  fracture
   analysis time _t:  time1
               id:  id

Weibull regression -- log relative-hazard form
No. of subjects =          148                Number of obs   =         206
No. of failures =           37
Time at risk    =         1703
                                              LR chi2(3)      =       40.28
Log likelihood  =   -69.082313               Prob > chi2     =      0.0000
```

_t	Coef.	Std. Err.	z	P>|z|	[95% Conf.	Interval]
_t						
protect	-2.185115	.3645006	-5.99	0.000	-2.899523	-1.470707
age	.0966629	.0345663	2.80	0.005	.0289142	.1644117
_Smale_1	.7382003	1.036608	0.71	0.476	-1.293514	2.769915
_cons	-10.61465	2.646846	-4.01	0.000	-15.80238	-5.42693
ln_p						
_Smale_1	-.7116757	.3834735	-1.86	0.063	-1.46327	.0399185
_cons	.5079011	.1358255	3.74	0.000	.2416881	.7741142

These are precisely the same estimates we obtained previously, the only difference being that the effect of gender is now labeled as _Smale_1. The prefix _S is used to denote variables created by streg.

When you specify strata(*varname*), *varname* is used to identify a categorical variable that identifies the strata. streg takes this categorical variable and from it constructs indicator variables that uniquely identify the strata. Then streg puts those indicator variables in the model everywhere they need to appear so that the baseline hazards are allowed to differ. For the model fit above, the variable male was already an indicator variable, so this amounted to putting male in the two places required: first in the main equation (scale) and second in the ancillary equation (shape).

Suppose that instead of age, we have the variable agecat that equals 1 if age is less than 65, 2 if age is between 65 and 74 inclusive, and 3 if age is greater than 74.

```
. gen agecat = 1
. replace agecat = 2 if age >=65
(166 real changes made)
. replace agecat = 3 if age > 74
(52 real changes made)
```

We can then hypothesize a model where the hazard is proportional with respect to protect and male but stratified on agecat:

```
. streg protect male, dist(weibull) strata(agecat) nolog nohr

        failure _d:  fracture
   analysis time _t:  time1
                id:  id

Weibull regression -- log relative-hazard form

No. of subjects =          148          Number of obs   =        206
No. of failures =           37
Time at risk    =         1703
                                         LR chi2(4)      =      57.86
Log likelihood  =    -72.948131          Prob > chi2     =     0.0000
```

_t	Coef.	Std. Err.	z	P>|z|	[95% Conf. Interval]	
_t						
protect	-2.082556	.3676436	-5.66	0.000	-2.803125	-1.361988
male	-1.501734	.4554929	-3.30	0.001	-2.394483	-.6089838
_Sagecat_2	.3400986	1.610656	0.21	0.833	-2.816729	3.496927
_Sagecat_3	1.132034	1.670177	0.68	0.498	-2.141454	4.405521
_cons	-3.646434	1.453472	-2.51	0.012	-6.495187	-.7976804
ln_p						
_Sagecat_2	-.0683576	.370067	-0.18	0.853	-.7936756	.6569604
_Sagecat_3	-.0875935	.3915644	-0.22	0.823	-.8550456	.6798586
_cons	.4174532	.3334877	1.25	0.211	-.2361707	1.071077

We see that, for this model, the scale and shape of the hazard can vary with respect to agecat. It is evident from the labeling of the output that agecat==1 is used as the base category. That is, `agecat==1` when both indicators `_Sagecat_2` and `_Sagecat_3` are equal to zero. For this model, the estimated baseline hazards are

$$
\widehat{h}_0(t_j) = \begin{cases}
\exp\left(-3.65\right) 1.52 t_j^{0.52}, & \text{if } \texttt{agecat==1} \\[2mm]
\exp\left(-3.31\right) 1.42 t_j^{0.42}, & \text{if } \texttt{agecat==2} \\[2mm]
\exp\left(-2.51\right) 1.39 t_j^{0.39}, & \text{if } \texttt{agecat==3}
\end{cases}
$$

and given the Wald tests displayed, none of these baseline hazards is significantly different from any other.

We demonstrated the `strata()` option using a PH model, but there is nothing stopping us from applying it to an AFT model. For an AFT model, when we stratify, we not only allow time to accelerate or decelerate with respect to the strata but also allow the actual shape of the baseline survivor function to vary with the strata as well.

The `strata()` option is really just a convenience option that attaches the term "stratification" to something that we could have done manually without this option. Except for variable labels in the output,

        . streg ..., strata(*varname*) ...

is synonymous with

        . xi: streg ... i.*varname*, ancillary(i.*varname*) ...

for models with one ancillary parameter, synonymous with

        . xi: streg ... i.*varname*, ancillary(i.*varname*) anc2(i.*varname*) ...

for the generalized gamma model, and synonymous with

        . xi: streg ... i.*varname*, ...

for the exponential. See [R] **xi** for more information about the creation of indicator variables.

15.3   Frailty models

Frailty models are a further generalization of the parametric regression models available to users of `streg`. A *frailty* model can take the form of an overdispersion/heterogeneity model or a random-effects model. Random-effects models are referred to as *shared frailty* models to make the distinction between the two, and here we define the term *unshared frailty* model to mean an overdispersion/heterogeneity model. What follows is a brief discussion of these models; for a more detailed treatment (and a more complete bibliography), see Gutierrez (2002).

15.3.1 Unshared frailty models

Each parametric regression model available in `streg` can be characterized by the hazard function $h(t_j|\mathbf{x}_j)$, regardless of whether the model is parameterized in the PH or the AFT metric. For PH models, the hazard function is conveniently represented as a baseline hazard that is multiplicatively affected by the covariates, but such a representation is not necessary in the development that follows.

A frailty model in the unshared case defines the hazard to be

$$h(t_j|\mathbf{x}_j, \alpha_j) = \alpha_j h(t_j|\mathbf{x}_j) \tag{15.1}$$

where α_j is some unobserved observation-specific effect. The effect, α_j, is known as a frailty and represents that individuals in the population are heterogeneous because of factors that remain unobserved. The frailties are positive quantities not estimated from the data but instead assumed to have mean 1 (for purposes of identifiability) and variance θ, and θ is estimated from the data. If $\alpha_j < 1$, then the effect is to decrease the hazard, and thus such subjects are known to be less frail than their counterparts. If $\alpha_j > 1$, then these frailer subjects face an increased risk.

The term *frailty* was first suggested by Vaupel, Manton, and Stallard (1979) in the context of mortality studies and by Lancaster (1979) in the context of duration of unemployment. Much of the initial work on frailty models is due to Hougaard (1984, 1986a, 1986b).

Given the relationship between the hazard and survivor functions, it can be shown from (15.1) that

$$S(t_j|\mathbf{x}_j, \alpha_j) = \{S(t_j|\mathbf{x}_j)\}^{\alpha_j}$$

where $S(t_j|\mathbf{x}_j)$ is the survivor function for a standard parametric model, such as the ones described in chapter 13. The unconditional survivor function is obtained by integrating out the unobservable α_j, and for this we need to assume a distribution for α_j. Two popular choices (popular for many reasons, one of which is mathematical tractability) are the gamma distribution and the inverse-Gaussian distribution, both available in Stata.

If α_j has probability density function $g(\alpha_j)$, then the unconditional survivor function is obtained by integrating out the frailty,

$$S_\theta(t_j|\mathbf{x}_j) = \int_0^\infty \{S(t_j|\mathbf{x}_j)\}^{\alpha_j} g(\alpha_j) d\alpha_j \tag{15.2}$$

where we use the subscript θ to emphasize the dependence on the frailty variance θ.

When α_j follows a gamma distribution with mean 1 and variance θ,

$$g(\alpha_j) = \frac{\alpha_j^{1/\theta - 1} \exp(-\alpha_j/\theta)}{\Gamma(1/\theta)\theta^{1/\theta}}$$

and (15.2) becomes

$$S_\theta(t_j|\mathbf{x}_j) = [1 - \theta \ln \{S(t_j|\mathbf{x}_j)\}]^{-1/\theta}$$

When α_j follows an inverse-Gaussian distribution with mean 1 and variance θ,

$$g(\alpha_j) = \left(\frac{1}{2\pi\theta\alpha_j^3}\right)^{1/2} \exp\left\{-\frac{1}{2\theta}\left(\alpha_j - 2 + \frac{1}{\alpha_j}\right)\right\}$$

and (15.2) becomes

$$S_\theta(t_j|\mathbf{x}_j) = \exp\left\{\frac{1}{\theta}\left(1 - [1 - 2\theta \ln\{S(t_j|\mathbf{x}_j)\}]^{1/2}\right)\right\}$$

As such, when modeling individual heterogeneity, a frailty model is just the standard parametric model with the addition of one new parameter, θ, and a new definition of the survivor function from which all the likelihood calculations derive in the standard way. For example, for the Weibull model in the PH metric,

$$S(t_j|\mathbf{x}_j) \quad = \quad \exp\{-\exp(\beta_0 + \mathbf{x}_j\boldsymbol{\beta}_x)t_j^p\}$$

For the Weibull–PH model with gamma-distributed heterogeneity,

$$S_\theta(t_j|\mathbf{x}_j) \quad = \quad \left\{1 + \theta \exp(\beta_0 + \mathbf{x}_j\boldsymbol{\beta}_x)t_j^p\right\}^{-1/\theta}$$

and both here and in general, $S_\theta(t_j|\mathbf{x}_j)$ reduces to $S(t_j|\mathbf{x}_j)$ as θ goes to zero.

15.3.2   Example: Kidney data

To illustrate frailty models, we will be analyzing a dataset from a study of 38 kidney dialysis patients. Originally considered in McGilchrist and Aisbett (1991), this study is concerned with the prevalence of infection at the catheter insertion point. Two recurrence times are measured for each patient. A catheter is inserted, and the first time to infection (in days) is measured. If the catheter is removed for reasons other than infection, then the first recurrence time is censored. Should infection occur, the catheter is removed, the infection cleared, and then after some predetermined period the catheter is reinserted. The second time to infection is measured as time elapsed between the second insertion and the second infection or censoring. The second recurrence time is censored if either the catheter is removed for reasons other than infection or the follow-up period for the patient ends before infection occurs.

```
. use http://www.stata-press.com/data/cggm/kidney, clear
(Kidney data, McGilchrist and Aisbett, Biometrics, 1991)

. describe
Contains data from http://www.stata-press.com/data/cggm/kidney.dta
  obs:            38                        Kidney data, McGilchrist and
                                            Aisbett, Biometrics, 1991
  vars:            7                        21 Feb 2002 15:37
  size:          722 (99.9% of memory free)
```

variable name	storage type	display format	value label	variable label
patient	float	%7.0g		Patient ID
time1	int	%9.0g		recurrence time to first infection
fail1	byte	%4.0g		equals 1 if infection, 0 if censored
time2	int	%9.0g		recurrence time to second infection
fail2	byte	%4.0g		equals 1 if infection, 0 if censored
age	float	%6.0g		Patient age
gender	byte	%6.0g		Patient gender (0=male, 1=female)

```
Sorted by:
. list in 21/30
```

	patient	time1	fail1	time2	fail2	age	gender
21.	21	562	1	152	1	46.5	0
22.	22	24	0	402	1	30	1
23.	23	66	1	13	1	62.5	1
24.	24	39	1	46	0	42.5	1
25.	25	40	1	12	1	43	0
26.	26	113	0	201	1	57.5	1
27.	27	132	1	156	1	10	1
28.	28	34	1	30	1	52	1
29.	29	2	1	25	1	53	0
30.	30	26	1	130	1	54	1

The variables time1 and time2 record the two times to infection or censoring, and the variables fail1 and fail2 are the associated failure indicator variables (1 = failure, 0 = censored). Other covariates that were measured for each patient were patient age and gender (0 = male, 1 = female).

Before we begin analyzing these data, we note that the data are in wide form and must first be reshaped into long form before we stset the dataset.

```
. reshape long time fail, i(patient) j(pat_insert)
(note: j = 1 2)
Data                                    wide    ->   long

Number of obs.                            38    ->     76
Number of variables                        7    ->      6
j variable (2 values)                           ->   pat_insert
xij variables:
                                  time1 time2   ->   time
                                  fail1 fail2   ->   fail
```

```
. list in 1/14, sepby(patient)
```

	patient	pat_in~t	time	fail	age	gender
1.	1	1	16	1	28	0
2.	1	2	8	1	28	0
3.	2	1	13	0	48	1
4.	2	2	23	1	48	1
5.	3	1	22	1	32	0
6.	3	2	28	1	32	0
7.	4	1	318	1	31.5	1
8.	4	2	447	1	31.5	1
9.	5	1	30	1	10	0
10.	5	2	12	1	10	0
11.	6	1	24	1	16.5	1
12.	6	2	245	1	16.5	1
13.	7	1	9	1	51	0
14.	7	2	7	1	51	0

```
. summarize, sep(0)
```

Variable	Obs	Mean	Std. Dev.	Min	Max
patient	76	19.5	11.03872	1	38
pat_insert	76	1.5	.5033223	1	2
time	76	97.68421	128.3424	2	562
fail	76	.7631579	.4279695	0	1
age	76	43.69737	14.73795	10	69
gender	76	.7368421	.4432733	0	1

Our data now consist of 76 observations, with each observation chronicling one catheter insertion and with two observations per patient.

Before we stset the dataset, we need to consider carefully how we want to define analysis time. Looking at the first patient, we find that the recurrence times are 16 and 8. Given what we know about these data, we realize that these times represent two separate observations that took place on different time lines; they do not represent one observation of a two-failure process for which failure occurred at $t = 16$ and then 8 days later at $t = 24$ (for instance). We know this from our description of how the data were collected: the time to the first infection is measured, and then the infection is cleared and more time (10 weeks) is allowed to pass, which essentially resets the "risk clock" back to zero. The time at which the second catheter insertion takes place marks the onset of risk for the second infection, and the second recurrence time is measured as the time that has elapsed from this second onset of risk.

Thus analysis time is just recurrence time as measured, and because the time at which the catheter is inserted marks the onset of risk, _t0 will equal zero for each observation. The way we have chosen to measure analysis time is an assumption we are making. We assume that 5 days after the first catheter insertion is, for accumulated risk, indistinguishable from 5 days after the second insertion. Statistical considerations do not force us to make that particular assumption. Statistically, we must make some assumption to measure analysis time, but we are free to make any assumption we please. We make this particular assumption because of the substantive arguments given above; this seems the appropriate way to measure analysis time.

Our assumption does not imply that we must treat each insertion as an *independent* observation, and we would not be at all surprised if the two recurrence times for each patient were correlated. The two recurrence times are distinct observations on the same failure process, and whether these times are correlated is an issue left to be resolved by our analysis, not one we need to concern ourselves with when stsetting the data.

Therefore, for stset, we define an "observation" (or synonymously a "subject") to be one catheter insertion and not the aggregated data for one patient. Using this definition of a subject, we have single-record-per-subject data, and thus we are not required to stset an ID variable that uniquely identifies each subject. However, it is always good practice to define one anyway because we may want to stsplit our records later, and stsplit does require that an ID variable be stset.

```
. gen insert_id = _n
. stset time, failure(fail) id(insert_id)
  (output omitted)
. format _t0 %4.0g
. format _t %5.0g
. format _d %4.0g
. format gender %6.0g
```

(*Continued on next page*)

```
. list insert_id patient pat_insert _t0 _t _d gender age in 1/10, sepby(patient)
```

	insert~d	patient	pat_in~t	_t0	_t	_d	gender	age
1.	1	1	1	0	16	1	0	28
2.	2	1	2	0	8	1	0	28
3.	3	2	1	0	13	0	1	48
4.	4	2	2	0	23	1	1	48
5.	5	3	1	0	22	1	0	32
6.	6	3	2	0	28	1	0	32
7.	7	4	1	0	318	1	1	31.5
8.	8	4	2	0	447	1	1	31.5
9.	9	5	1	0	30	1	0	10
10.	10	5	2	0	12	1	0	10

Satisfied at how our data are stset, we save the data as kidney2.dta.

❏ **Technical note**

Determining what is an "observation" can be confusing. Say that you agree with us that analysis time ought to be measured as time between insertion and infection. That definition, we have just argued, will lead us to stset the data by typing

```
. stset ..., id(insert_id) ...
```

But you may wonder why we could not type

```
. stset ..., id(patient) exit(time .)
```

where we would make an "observation" a patient and include exit(time .) to allow for multiple failures (two) per patient.

Here is a test that will help you dismiss inconsistent ways of defining "observations": determine the maximum number of potential failures in the dataset at the earliest possible instance at which failure could occur. In our dataset we have 38 patients, but in theory, after a short time, there could have been up to 76 failures because each catheter insertion for each patient could have developed an infection immediately.

The maximum number of potential failures must equal the number of "observations", and so you are led to a contradiction.

An observation, given our definition of analysis time, is a catheter insertion.

For cases where the observations do represent multiple failures per subject, there are many methods for analysis; however, these methods are beyond the scope of this introductory text. See Cleves (1999) for a survey of the methodology in this area.

❏

15.3.3   Testing for heterogeneity

Continuing our study of the kidney data, we avoid (only for now) the issue of correlation between observations within each patient by temporarily confining our analysis to consider only the first catheter insertion for each patient (`pat_insert==1`). We begin by fitting the most general of our parametric models, the generalized gamma:

```
. use http://www.stata-press.com/data/cggm/kidney2, clear
(Kidney data, McGilchrist and Aisbett, Biometrics, 1991)

. keep if pat_insert==1
(38 observations deleted)

. streg age gender, dist(gamma) nolog

        failure _d:  fail
   analysis time _t:  time
               id:  insert_id

Gamma regression -- accelerated failure-time form

No. of subjects =          38                  Number of obs   =          38
No. of failures =          27
Time at risk    =        3630
                                               LR chi2(2)      =        4.16
Log likelihood  =   -48.376399                 Prob > chi2     =      0.1247
```

_t	Coef.	Std. Err.	z	P>|z|	[95% Conf. Interval]	
age	-.0094449	.0147097	-0.64	0.521	-.0382754	.0193856
gender	1.219718	.5438289	2.24	0.025	.1538334	2.285603
_cons	3.640631	.7896671	4.61	0.000	2.092912	5.18835
/ln_sig	.2113217	.1347486	1.57	0.117	-.0527807	.4754242
/kappa	-.2071306	.5150676	-0.40	0.688	-1.216644	.8023833
sigma	1.23531	.1664563			.9485881	1.608696

The Wald test of H_o: $\kappa = 0$ has significance level 0.688, and thus there is insufficient evidence to reject H_o, which in essence says that a lognormal model would serve our purposes equally well. Out of `streg`'s six parametric models, the lognormal gives the highest Akaike information criterion (AIC) index (see sec. 13.7.2 for a description of AIC).

(Continued on next page)

```
. streg age gender, dist(lnormal) nolog

        failure _d:  fail
   analysis time _t:  time
             id:  insert_id

Lognormal regression -- accelerated failure-time form

No. of subjects =           38          Number of obs   =           38
No. of failures =           27
Time at risk    =         3630
                                        LR chi2(2)      =         4.81
Log likelihood  =   -48.455243          Prob > chi2     =       0.0902
```

_t	Coef.	Std. Err.	z	P>|z|	[95% Conf. Interval]	
age	-.0079733	.0143949	-0.55	0.580	-.0361867	.0202402
gender	1.113414	.4948153	2.25	0.024	.1435937	2.083234
_cons	3.779857	.7215091	5.24	0.000	2.365725	5.193989
/ln_sig	.2090131	.1336573	1.56	0.118	-.0529503	.4709766
sigma	1.232461	.1647274			.9484272	1.601557

With these regression results, we can use stcurve to graph the comparative estimated hazards for males and females with age held at its mean value:

```
. stcurve, hazard at1(gender=0) at2(gender=1)
```

The result is shown in figure 15.2.

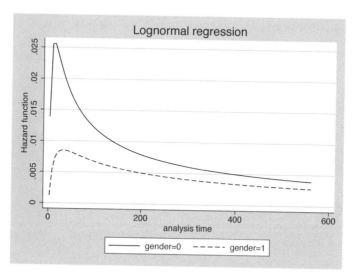

Figure 15.2. Comparison of lognormal hazards for males and females

The hazard for males is much greater than that for females. The hazard, in any case, is estimated from the lognormal model to rise and then fall; if we choose to adhere to the AIC criterion, this is as good an estimate of the hazard as we can get by using `streg` in the standard way.

If we interpreted this hazard as the hazard each individual faces, then as time passes the instantaneous risk of infection falls. If this were indeed the individual hazard, then physicians should tell their patients that if infection hasn't taken place by a certain time after catheter insertion, then it is unlikely to happen at all and they need not worry. On the other hand, we obtained these estimates by assuming that all patients are identical other than in age and gender. If subjects differ in unobserved ways in their inherent probability of failure—if some are more robust and others more frail—then we must be more careful about our interpretation of results and about how we estimate them.

In frailty models, there is a distinction between the hazard individuals face and the population hazard that arises by averaging over all the survivors. In a heterogeneous population, the population hazard can fall while the individual hazards all rise because, over time, the population becomes populated by more and more robust individuals as the frailer members fail. This is known as the frailty effect, and it virtually ensures that population hazards decline over time, regardless of the shape of the hazards that individuals face. We will provide evidence that this is indeed occurring in these data.

Under the assumption of a heterogeneous population, it could actually be the case that each individual's risk rises the longer the catheter remains inserted, and so the clinical recommendation is to remove the catheter as soon as possible, even though for the population as a whole, the infection rate falls.

Under the frailty model, the individual hazard function is written $h(t_j|\mathbf{x}_j, \alpha_j) = \alpha_j h(t_j|\mathbf{x}_j)$, and the resulting population hazard is written as $h_\theta(t_j|\mathbf{x}_j)$, which is obtained from (15.2) in the standard way:

$$h_\theta(t_j|\mathbf{x}_j) = -\left\{\frac{d}{dt}S_\theta(t_j|\mathbf{x}_j)\right\}\left\{S_\theta(t_j|\mathbf{x}_j)\right\}^{-1}$$

As θ tends to 0, the population and individual hazard functions coincide and $\lim_{\theta \to 0} h_\theta() = h()$.

As when fitting a model without frailty, you assume a parametric form for the hazard function, but in models with frailty, you make that assumption about the hazard function individuals face. The population hazard function is just whatever it turns out to be given the estimate of θ and the assumed distribution of α_j, which in Stata may be gamma or inverse-Gaussian.

We can fit a model with lognormal individual hazard and gamma-distributed frailty by specifying the additional option `frailty(gamma)` to `streg`:

```
. streg age gender, dist(lnormal) frailty(gamma) nolog

        failure _d:  fail
   analysis time _t:  time
              id:  insert_id

Lognormal regression -- accelerated failure-time form
                    Gamma frailty
No. of subjects =         38                  Number of obs   =         38
No. of failures =         27
Time at risk     =       3630
                                             LR chi2(2)      =       5.02
Log likelihood  =  -48.351289                Prob > chi2     =     0.0813
```

_t	Coef.	Std. Err.	z	P>|z|	[95% Conf. Interval]	
age	−.0097803	.0144447	−0.68	0.498	−.0380913	.0185307
gender	1.260102	.5566201	2.26	0.024	.1691466	2.351057
_cons	3.653898	.7309858	5.00	0.000	2.221192	5.086604
/ln_sig	.1257008	.2199912	0.57	0.568	−.305474	.5568757
/ln_the	−1.965189	2.298354	−0.86	0.393	−6.46988	2.539502
sigma	1.133943	.2494575			.736774	1.745211
theta	.1401294	.3220669			.0015494	12.67336

```
Likelihood-ratio test of theta=0: chibar2(01) =      0.21 Prob>=chibar2 = 0.324
```

Or if you prefer, we can fit a model with inverse-Gaussian distributed frailties:

```
. streg age gender, dist(lnormal) frailty(invgauss) nolog

        failure _d:  fail
   analysis time _t:  time
              id:  insert_id

Lognormal regression -- accelerated failure-time form
                 Inverse-Gaussian frailty
No. of subjects =         38                  Number of obs   =         38
No. of failures =         27
Time at risk     =       3630
                                             LR chi2(2)      =       5.03
Log likelihood  =  -48.347311                Prob > chi2     =     0.0810
```

_t	Coef.	Std. Err.	z	P>|z|	[95% Conf. Interval]	
age	−.009801	.0144568	−0.68	0.498	−.0381357	.0185337
gender	1.255811	.5422437	2.32	0.021	.1930326	2.318589
_cons	3.646516	.7389398	4.93	0.000	2.19822	5.094811
/ln_sig	.1186605	.2374315	0.50	0.617	−.3466967	.5840176
/ln_the	−1.727569	2.700183	−0.64	0.522	−7.01983	3.564692
sigma	1.125988	.2673449			.7070198	1.793228
theta	.1777159	.4798655			.000894	35.32857

```
Likelihood-ratio test of theta=0: chibar2(01) =      0.22 Prob>=chibar2 = 0.321
```

The choice of frailty distribution has implications in the interpretation of how the relative hazard (with respect to the covariates) changes with time; see the technical note below. Regardless of the choice of frailty distribution, however, from examining the likelihood-ratio test of $H_o: \theta = 0$ at the bottom of each output, we realize that if we are willing to accept that the individual hazards are indeed lognormal, then there is not much evidence pointing toward a population that is heterogeneous.

❏ **Technical note**

In hazard-metric frailty models, exponentiated coefficients have the interpretation of hazard ratios at $t = 0$ only. After that, the effects of covariate differences become muted as the more frail experience failure and so are removed from the surviving population. In gamma frailty models, the effect of covariate differences eventually diminishes completely in favor of the frailty effect. In inverse-Gaussian frailty models, the effect of covariate differences never vanishes, going instead to the square root of the hazard ratio effect at $t = 0$. For more information, see Gutierrez (2002).

❏

Suppose, however, that the science underlying infection due to catheter insertions dictated that individual hazards must increase with time. Our seemingly good results with the lognormal model would then be an artifact caused by the heterogeneity in the data. We could examine this possibility by fitting a Weibull individual hazard model with gamma frailty, and then ask (1) is the Weibull parameter $p > 1$ (which is to say, do individual hazard functions monotonically rise) and (2) does this alternate model fit the data about as well or better?

Let us begin by fitting the Weibull model with gamma frailty:

```
. streg age gender, dist(weibull) frailty(gamma) nolog time

           failure _d:  fail
     analysis time _t:  time
                   id:  insert_id

Weibull regression -- accelerated failure-time form
                  Gamma frailty

No. of subjects =          38                 Number of obs   =          38
No. of failures =          27
Time at risk    =        3630
                                              LR chi2(2)      =        4.78
Log likelihood  =  -48.410832                 Prob > chi2     =      0.0917
```

_t	Coef.	Std. Err.	z	P>|z|	[95% Conf. Interval]	
age	-.0109489	.013751	-0.80	0.426	-.0379004	.0160026
gender	1.338382	.5071762	2.64	0.008	.3443351	2.332429
_cons	3.570254	.6510318	5.48	0.000	2.294255	4.846253
/ln_p	.4965479	.3129399	1.59	0.113	-.1168032	1.109899
/ln_the	.3610338	.7067754	0.51	0.609	-1.02422	1.746288
p	1.643039	.5141727			.8897603	3.034052
1/p	.6086281	.190464			.3295923	1.123898
theta	1.434812	1.01409			.3590763	5.733282

```
Likelihood-ratio test of theta=0: chibar2(01) =      4.20 Prob>=chibar2 = 0.020
```

We chose the AFT parameterization of the Weibull simply to ease the comparison with the lognormal, which is available only in the AFT metric.

From these results, we see that our alternative theory has some merit:

(1) The point estimate of the Weibull parameter p is greater than 1 (although the 95% confidence interval includes values below 1).

(2) The fit of this model, as reflected in log-likelihood values, is about the same as that of the lognormal model (being about -48.4 in both cases).

We will not feel completely comfortable until we see graphs of the population and individual hazards, and we will be discussing how to obtain those graphs. In the meantime, you can look ahead to figures 15.3 and 15.4. Individual hazards rise monotonically, yet the model reproduces the rise-and-then-fall shape of the population hazard previously exhibited by the lognormal model. (Later, we will also explore what might be the true shape of the individual hazard.)

Individual hazards rise, yet population hazards ultimately fall because of heterogeneity. So we are not surprised to see that in the reported results above, the likelihood-ratio test for H_o: $\theta = 0$ would be rejected, here at the 0.02 level.

Let us now return to the issue of obtaining graphs of the hazard functions. After fitting standard (nonfrailty) parametric models, you can type `predict ... , hazard` to obtain the estimated hazard function. After fitting a parametric model with frailty, you type

        . predict ..., hazard unconditional

or

        . predict ..., hazard alpha1

to obtain the estimated hazard functions. With the `unconditional` option, you obtain the population hazard function $h_\theta(t_j|\mathbf{x}_j)$, and with the `alpha1` option, you obtain the mean individual hazard $h(t_j|\mathbf{x}_j)$. The options `unconditional` and `alpha1` can also be used when using `predict` to obtain other predictions after fitting parametric frailty models.

❑ Technical note

For univariate (nonshared) frailty models, the default prediction (that is, if neither `unconditional` nor `alpha1` is specified) is the unconditional (population) function. For shared frailty models, which we discuss later, the default prediction is the individual ($\alpha_j = 1$) function. The reason for the change in default behavior has to do with the interpretation of the results. In the shared frailty case, careful consideration must be given to the interpretation of the unconditional functions. See Gutierrez (2002) for more details. In any case, we recommend always specifying either `unconditional` or `alpha1` when predicting the hazard or survivor function after fitting a frailty model.

❑

The `alpha1` and `unconditional` options may also be specified in `stcurve` so that we can easily graph either implementation (conditional or unconditional) of the hazard, cumulative hazard, and survivor functions. From our Weibull/gamma model, we can graph the comparative population hazards for males and females,

```
. stcurve, hazard unconditional at1(gender=0) at2(gender=1)
```

which produces figure 15.3, and the similarity to figure 15.2 is striking.

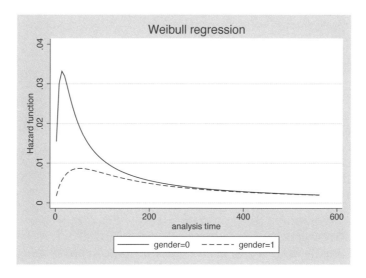

Figure 15.3. Comparison of Weibull/gamma population hazards

Regardless of which story we believe, whether the lognormal individual hazard with no frailty (figure 15.2) or the Weibull individual hazard with significant frailty (figure 15.3), the estimate of the population hazard is virtually the same.

We can also graph the individual hazards for the Weibull/gamma model,

```
. stcurve, hazard alpha1 at1(gender=0) at2(gender=1)
```

which produces figure 15.4.

(Continued on next page)

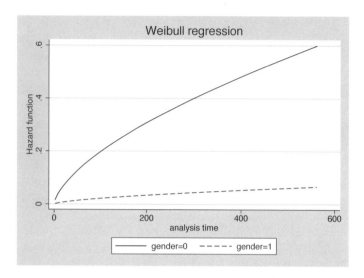

Figure 15.4. Comparison of Weibull/gamma individual ($\alpha_j = 1$) hazards

As expected, the individual hazards are monotone increasing. Thus if we believe this model—that is, if we believe that the science dictates a monotone increasing hazard—then we must also believe that there is some patient-level effect that we are not measuring. A physician looking at this model would realize that some individuals are just more susceptible to infection than others for reasons unmeasured and, when consulting with a patient, would actually warn that the risk of infection increases (for that patient) the longer the catheter remains inserted.

Which model better fits the data? Do individual hazards monotonically rise, and are hazards falling in the data merely artifacts of the heterogeneity? Or do individual hazards rise and then fall in lock step with the population hazard? Clearly, patients are heterogeneous—we would never argue that age and sex fully describe the differences. In the two extremes we have posed, we are really asking whether the heterogeneity is so great that it dominates the production of population hazards from individual hazards, or here, whether heterogeneity can be ignored.

Which of our two alternate theories is better supported by the data? Throughout this section, we have confined the analysis to the first infection time for each patient, and thus we wish to consider the full dataset (both infection times) to answer this question.

15.3.4 Shared frailty models

In section 15.3.3, we confined our analysis to the first time to infection for each patient because we wished to investigate the existence of a latent patient frailty in the univariate case. We now consider both infection times for each patient:

```
. use http://www.stata-press.com/data/cggm/kidney2, clear
(Kidney data, McGilchrist and Aisbett, Biometrics, 1991)
. list insert_id patient pat_insert _t0 _t _d gender age in 1/6, separator(0)
```

	insert~d	patient	pat_in~t	_t0	_t	_d	gender	age
1.	1	1	1	0	16	1	0	28
2.	2	1	2	0	8	1	0	28
3.	3	2	1	0	13	0	1	48
4.	4	2	2	0	23	1	1	48
5.	5	3	1	0	22	1	0	32
6.	6	3	2	0	28	1	0	32

Realizing that the two recurrence times for each patient are probably correlated, we could take one of our standard parametric survival models and adjust the standard errors to take into account the intrapatient correlation.

```
. streg age gender, dist(weibull) vce(cluster patient) time nohr nolog

         failure _d:  fail
   analysis time _t:  time
                id:  insert_id

Weibull regression -- accelerated failure-time form

No. of subjects    =         76            Number of obs   =         76
No. of failures    =         58
Time at risk       =       7424
                                            Wald chi2(2)    =       3.38
Log pseudolikelihood =   -103.44362         Prob > chi2     =     0.1848

                   (Std. Err. adjusted for 38 clusters in patient)
```

_t	Coef.	Robust Std. Err.	z	P>|z|	[95% Conf. Interval]	
age	−.004559	.0097206	−0.47	0.639	−.0236111	.0144932
gender	.9194971	.5413907	1.70	0.089	−.1416092	1.980603
_cons	4.30243	.7075207	6.08	0.000	2.915715	5.689145
/ln_p	−.1028083	.0798087	−1.29	0.198	−.2592305	.053614
p	.9023	.0720114			.7716451	1.055077
1/p	1.108279	.0884503			.9477979	1.295933

By specifying vce(cluster patient), we are saying that we do not believe these to be 76 independent observations but rather 38 independent "clusters" of observations. When you specify vce(cluster *varname*), you get robust standard errors, those obtained via the Huber/White/sandwich estimator of variance. Should there be intracluster correlation, the robust standard errors are better indicators of the sample-to-sample variability of the parameter estimates and thus produce more accurate tests of the effects of covariates. For more details on the sandwich estimator of variance, see [U] **20.15 Obtaining robust variance estimates**. We do not want to make much of such an analysis here because, for most situations we can imagine where observations are correlated within groups, they are usually correlated because of some overall group characteristic (a frailty) that is not being measured. What we instead want to do is to model the correlation using a shared frailty model.

Shared frailty models are the survival-data analogue to random-effects models. A *shared* frailty is a frailty model where the frailties are no longer observation specific but instead are shared across groups of observations, thus causing those observations within the same group to be correlated. Here the two infection times for each patient share the same frailty because we assume the frailty to be a characteristic of the patient, not of the catheter insertion.

The generalization of (15.1) to a case where frailty is assumed to be shared across groups of observations is

$$h(t_{ij}|\mathbf{x}_{ij}, \alpha_i) = \alpha_i h(t_{ij}|\mathbf{x}_{ij})$$

for data consisting of n groups with the ith group comprising n_i observations. The index i denotes the group ($i = 1, \ldots, n$), and j denotes the observation within group ($j = 1, \ldots, n_i$). The frailties, α_i, are shared within each group and are assumed to follow either a gamma or inverse-Gaussian distribution (as before). The frailty variance, θ, is estimated from the data and measures the variability of the frailty across groups.

To fit a shared frailty model, one need add only the option shared(*varname*) to streg, frailty(), where *varname* is an ID variable describing those groups wherein frailties are shared. Returning to our example, we again compare the lognormal and Weibull models. Because this time we are considering both catheter insertions for each patient, we assume that the frailty is shared at the patient level.

We first fit a lognormal with gamma-shared frailty:

```
. streg age gender, dist(lnormal) frailty(gamma) shared(patient) nolog
       failure _d:  fail
  analysis time _t:  time
             id:  insert_id

Lognormal regression --
       accelerated failure-time form            Number of obs     =        76
       Gamma shared frailty                     Number of groups  =        38
Group variable: patient
No. of subjects =         76                    Obs per group: min =         2
No. of failures =         58                                  avg =         2
Time at risk    =       7424                                  max =         2
                                                LR chi2(2)        =     16.65
Log likelihood  =   -97.594575                  Prob > chi2       =    0.0002
```

_t	Coef.	Std. Err.	z	P>|z|	[95% Conf. Interval]	
age	-.0067002	.0098278	-0.68	0.495	-.0259623	.0125619
gender	1.422234	.3343544	4.25	0.000	.7669111	2.077556
_cons	3.331102	.4911662	6.78	0.000	2.368434	4.29377
/ln_sig	.0670784	.1187767	0.56	0.572	-.1657197	.2998764
/ln_the	-1.823592	.995803	-1.83	0.067	-3.77533	.1281455
sigma	1.069379	.1270173			.8472837	1.349692
theta	.1614447	.1607671			.0229295	1.136718

```
Likelihood-ratio test of theta=0: chibar2(01) =      1.57 Prob>=chibar2 = 0.105
```

From the likelihood-ratio test of H_o: $\theta = 0$, we find the frailty effect to be insignificant at the 10% level. This is not the overwhelming lack of evidence of heterogeneity that we had before in the univariate case, but the effect of frailty is still small in this model. If the individual hazards are indeed lognormal, then there is little evidence of a heterogeneous population of patients.

(Continued on next page)

If we fit a Weibull with gamma shared frailty,

```
. streg age gender, dist(weibull) fr(gamma) shared(patient) time nohr nolog

        failure _d:  fail
  analysis time _t:  time
               id:   insert_id

Weibull regression --
        accelerated failure-time form        Number of obs      =        76
        Gamma shared frailty                 Number of groups   =        38
Group variable: patient

No. of subjects =            76              Obs per group: min =         2
No. of failures =            58                           avg =         2
Time at risk    =          7424                           max =         2

                                            LR chi2(2)         =     14.81
Log likelihood  =   -98.006931              Prob > chi2        =    0.0006
```

_t	Coef.	Std. Err.	z	P>|z|	[95% Conf. Interval]	
age	-.0067052	.0102377	-0.65	0.512	-.0267707	.0133602
gender	1.506616	.3659291	4.12	0.000	.7894085	2.223824
_cons	3.557985	.5224117	6.81	0.000	2.534077	4.581894
/ln_p	.2410369	.1336503	1.80	0.071	-.0209129	.5029866
/ln_the	-.4546298	.4747326	-0.96	0.338	-1.385089	.475829
p	1.272568	.1700791			.9793043	1.653653
1/p	.7858127	.1050241			.6047219	1.021133
theta	.6346829	.3013047			.2503016	1.609348

```
Likelihood-ratio test of theta=0: chibar2(01) =     10.87 Prob>=chibar2 = 0.000
```

we see a significant frailty effect. Thus, if we believe the individual hazard to be Weibull, then we must also be willing to believe in an unobserved patient-level effect. Also we estimated $\hat{p} = 1.27$; that is, the estimated individual hazard for this model is monotone increasing.

This is the same situation as before in the unshared case; if we used stcurve to plot and compare the hazard functions as we did before, then we would see the same phenomena.

To help us settle the issue of which of the two competing models is preferable, we will use our favorite trick of fitting a piecewise-exponential model (piecewise-constant hazard) with a gamma-shared frailty model. By splitting time into several discrete intervals and fitting a separate hazard on each, we are being more flexible in our specification of the individual hazard. By being more flexible, we are allowing the data if not entirely to speak for themselves then at least to get a word in as to what the individual hazard and level of heterogeneity are.

To fit the piecewise-exponential model, we will generate a set of indicator variables that comprise a grid of 10 intervals spanning $[0, 60), [60, 120), \dots, [540, \infty)$ (_t==562 is the largest value in our data). Before we define our intervals, however, we need to stsplit our data at the appropriate times:

```
. stsplit my_t, at(60(60)562)
(92 observations (episodes) created)

. forvalues k = 1/9 {
  2.        gen in_`k´ = ((`k´-1)*60 <= my_t) & (my_t < `k´*60)
  3. }
```

The indicators are constructed so that the baseline interval (that for which all the indicators equal zero) is the interval $[540, \infty)$.

We now fit the piecewise-exponential model with gamma-shared frailty by including the indicator variables in the covariate list:

```
. streg age gender in_*, dist(exp) fr(gamma) shared(patient) nolog nohr time

       failure _d:  fail
  analysis time _t:  time
              id:  insert_id

Exponential regression --
         accelerated failure-time form        Number of obs      =        168
         Gamma shared frailty                 Number of groups   =         38
Group variable: patient

No. of subjects =           76                Obs per group: min =          2
No. of failures =           58                              avg =   4.421053
Time at risk    =         7424                              max =         14

                                              LR chi2(11)        =      20.90
Log likelihood  =    -95.241838               Prob > chi2        =     0.0344
```

_t	Coef.	Std. Err.	z	P>|z|	[95% Conf. Interval]	
age	-.0060301	.0119445	-0.50	0.614	-.0294409	.0173807
gender	1.416426	.4525747	3.13	0.002	.5293963	2.303456
in_1	2.566041	1.350038	1.90	0.057	-.0799854	5.212067
in_2	3.229574	1.339194	2.41	0.016	.6048016	5.854347
in_3	2.366064	1.29691	1.82	0.068	-.1758328	4.907961
in_4	2.1349	1.305189	1.64	0.102	-.4232243	4.693024
in_5	2.44805	1.373428	1.78	0.075	-.2438189	5.139919
in_6	2.788362	1.52622	1.83	0.068	-.2029743	5.779697
in_7	2.659777	1.518699	1.75	0.080	-.3168189	5.636373
in_8	2.319784	1.501516	1.54	0.122	-.6231325	5.262701
in_9	1.233349	1.296951	0.95	0.342	-1.308628	3.775326
_cons	1.057467	1.488013	0.71	0.477	-1.858986	3.973919
/ln_the	-.8007126	.5527555	-1.45	0.147	-1.884094	.2826683
theta	.4490089	.2481921			.1519667	1.326665

```
Likelihood-ratio test of theta=0: chibar2(01) =      6.62 Prob>=chibar2 = 0.005
```

The first thing we notice is the likelihood-ratio test of $H_o: \theta = 0$. By not specifying a strict parametric form for the individual hazard but instead leaving the model to its own devices with a more flexible specification, we still find evidence of a heterogeneous population of patients. From our previous debate over the form of the individual hazard, this evidence leads us to initially favor the Weibull model over the lognormal model because, for the Weibull, we also have significant heterogeneity.

We can see the similarity in estimation results between the piecewise-exponential shared frailty model and the Cox shared frailty model (which makes no assumption about the functional form of the hazard) as fit to the same data in section 9.4. In the Cox shared frailty model, $\widehat{\theta} = 0.475$ versus $\widehat{\theta} = 0.449$ for the piecewise exponential. The magnitudes of the coefficients on `age` and `gender` are also similar across both models, yet the signs are reversed. The sign reversal is because we fit the piecewise exponential model in the AFT metric whereas the Cox model is inherently a PH model.

We can use `predict` followed by `line` to examine the estimated individual hazards for males and females from the piecewise exponential model. `stcurve` will not work here because we have indicator variables that vary over time, and `stcurve` likes to hold things constant. Because we are generating our own graph, and because of the sparseness of the data in the right tail of the distribution, we generate our own plotting grid rather than use the existing `_t` in our data:

```
. drop _all
. set obs 400
obs was 0, now 400
. gen time = _n
. stset time

     failure event:  (assumed to fail at time=time)
obs. time interval:  (0, time]
 exit on or before:  failure

       400  total obs.
         0  exclusions

       400  obs. remaining, representing
       400  failures in single record/single failure data
     80200  total analysis time at risk, at risk from t =         0
                          earliest observed entry t =         0
                           last observed exit t =       400
. forvalues k = 1/9 {
  2.          gen in_`k' = ((`k'-1)*60 <= _t) & (_t < `k'*60)
  3. }
. gen age = 43.7    /* mean age */
. gen gender = 0
. predict h_male, hazard alpha1
. label var h_male "males"
. replace gender = 1
(400 real changes made)
. predict h_female, hazard alpha1
. label var h_female "females"
. line h_female h_male _t, sort l1title("hazard")
```

This produces figure 15.5.

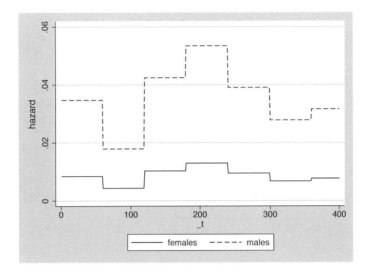

Figure 15.5. Comparison of piecewise-constant individual ($\alpha = 1$) hazards

Even though the piecewise-exponential model features a level of heterogeneity similar to that of the Weibull, the hazard is by no means monotone increasing and in fact begins to decrease steadily around $t = 200$. Admittedly, the data are pretty sparse past this point, so we should not make too much of the estimated hazards at the later time points. In any case, our model-agnostic piecewise-constant hazard offers little in the way of a scientific interpretation, but fitting this model at least served to further confirm that, whatever the form of the individual hazard, there is evidence of patient heterogeneity.

From the above example, the heterogeneity cannot simply be ignored for this problem. Concerning whether individual hazards are rising or falling, we have inconclusive evidence. In the absence of a strong theory, we would need more data.

16 Power and sample-size determination for survival analysis

In some cases, we are stuck with the data we have and, in others, we get to design a data-collection effort—a study—to produce the data that we will analyze. This chapter concerns the latter, namely, the design of studies for use in survival analysis. Our goal is not to discuss the nitty-gritty of data collection, although that is also important. Instead, our goal is to describe the data to be collected for the relevant statistical terms, which, for survival analysis, turn out to be size, allocation, and duration. How many subjects should be included? How should those subjects be allocated between experimentals and controls? How long should the study continue? Imagine how embarrassing it would be if, after discovering a statistically insignificant result, some smart statistician pointed out that you never had much hope of detecting the effect given its likely magnitude. Or, perhaps not quite as embarrassing, after successfully detecting the effect, the same statistician points out the resources you wasted because the study was much larger than what was needed, and points out you should have known that at the outset.

Recall that hip-fracture study introduced in section 7.6. To remind you, the goal was to investigate the performance of an inflatable device to protect the elderly from hip fractures as a result of falls. In section 7.6, we just took the data as given: 48 women over the age of 60, 28 randomly assigned to wear the device, the remaining 20 used as controls, and all 48 followed until the end of the study, which we will now tell you was 40 months. Why 48 women and not, say, 108 or 18? Why were more assigned to wear the new device than not? Why 40 months and not 50? Or 10?

Those questions were answered before the data were collected, and the answers were based on the following logic. The physician who invented the device believes that the true reduction in hip fractures because of the device is at least 60%. If he is right, he wants to obtain results showing that the device in fact yields a reduction, but he does not at this time need to measure exactly what the reduction is. For the pilot study, it will be sufficient to say that, at the 5% significance level, the device yields a reduction in hip fractures. So now the statistician advising him thought thusly: we will measure the effect of the hazard ratio of hip fracture of experimentals relative to controls. Sixty percent reduction means a hazard ratio of 0.4. How big a study do we need so that we can detect with, say, 80% probability (power) a smaller-than-1 hazard ratio at the 5% significance level, given a true hazard ratio of 0.4?

Big is a vague word and, in the case of survival studies, it has three components: sample size measured in number of subjects, duration of the study, and allocation of those subjects between experimentals (inflatable hip-protection device wearers) and controls. Obviously, more data produce more precise results, but in the survival study case, we need to consider carefully what we need more of. What we need are failure events. Lots of subjects, none of whom fail, produce no useful data. In the sample-size determination formulas that we will discuss, we will need a certain number of failures, and then we will have to guess how long to run the experiment to observe that number. On the other hand, if the length of our experiment is fixed, then we need to figure out how many subjects we need to ensure that a certain number of them fail by the end of the experiment. Of course, the required length of the experiment and the required number of subjects are both tied to the probability of a failure, but if we knew that, we would not need to perform the experiment at all.

The allocation between experimentals and controls is important. Until one thinks carefully about the survival analysis problem, one probably expects that the allocation between experimentals and controls should be fifty-fifty. So now think more carefully. What we often want is roughly an equal number of failures in each group. If the hip device really delays failure, and if we are going to follow both experimentals and controls over the same time period, then we are going to need more experimentals than controls to get close to an equal number of failures.

That is the subject of this chapter: how these questions may be answered using the `stpower` command.

In the case of the log-rank test for comparing two survivor curves (that of device wearers and that of controls), you will need to specify the underlying hazard ratio (0.4 in our case), the power (.8), the distribution of controls and experimentals (we will use 42% controls), and the expected censoring rate among controls (5% in our case). The result is

```
. stpower logrank 0.05, power(0.8) hratio(0.4) p1(0.42)
Estimated sample sizes for two-sample comparison of survivor functions
Log-rank test, Freedman method
Ho: S1(t) = S2(t)

Input parameters:
        alpha =    0.0500   (two sided)
           s1 =    0.0500
           s2 =    0.3017
       hratio =    0.4000
        power =    0.8000
           p1 =    0.4200

Estimated number of events and sample sizes:
            E =        39
            N =        48
           N1 =        20
           N2 =        28
```

where `E` in the output is the number of events (failures) required to be observed in the study. How did we decide to run the study for 40 months? We decided it at the outset and that determined the .05 rate of censoring we specified in the control group. We knew from prior experience that, over a 40-month period, 95% of patients such as those in the control group have a hip fracture. Taking into account the censoring rate in the control group, the anticipated censoring rate in the experimental group based on the hypothesized hazard ratio, and the unequal allocation of subjects, we found that we needed a total of 48 subjects in the study to observe 39 failures.

`stpower` has other capabilities. It can solve for power given the other characteristics so you can examine what the power would be under a given scenario. It can also solve for effect size (`hratio`), in effect turning the problem on its head, so you can examine what the minimal detectable underlying hazard ratio would be.

Say you decide on 48 subjects, with 28 in the treatment group, and 40 months, and the person in charge of running the experiment comes back and says, "There is no way we can run that long. Tell me something that can be done in one year." In that case, you will need to explore other solutions that still yield 80% power and 0.05 significance at the hypothesized 0.4 hazard ratio, and those solutions are going to involve larger sample sizes. To see this, note that shortening the study will increase the censoring rate, given as 0.05 in our original scenario. More censorings mean less failures, and thus we would need more subjects to achieve the required 39 failures.

Finally, let us say a few words about the 80% power. That is a number you are going to have to choose. Perhaps you will want to choose 90%, or 95%, or 75%. The hip-replacement study is fictional, but were it real, our thinking would have been that this was just a pilot study and that, in the meetings we attended, the physician was quite clear that he thought the device yielded at least a 60% reduction in hip fractures, with the emphasis on at least. Based on that information, we figured that a four out of five chance of detecting 0.4, the hazard-ratio equivalent of a 60% reduction, would be good enough. For different problems, perhaps you would choose different powers.

16.1 Estimating sample size

As we mentioned in the introduction, in survival data, measuring the amount of information contained in your data is not as simple as counting the physical observations—the notion of sample size is different. Inference (and therefore power) directly depends on the number of observed events (failures), and there is an indirect relationship between power and sample size, which arises because more subjects usually means more events. Calculating the required number of subjects based on the required number of failures requires assuming or estimating a value for the probability that a subject fails during the course of the study.

In the best of all worlds, that probability is one. Then the sample size is identical to the number of events. What that means is no censoring; all subjects are followed until the failure event occurs.

In practice, it may not be feasible to follow subjects until all of them fail because the study would take too long. Instead, a study is terminated after a predetermined period of time. This often leads to having subjects who did not experience an event of interest by the end of the study; instead, *administrative censoring* of subjects occurs.

In the presence of censoring, we additionally need to assume or estimate a value for the probability that a subject fails during the course of the study to obtain the required sample size. stpower provides ways, specific to the type of analysis to be performed, to specify or compute this probability.

16.1.1   Multiple-myeloma data

Before we proceed, let us first describe the dataset we will use throughout this chapter.

We consider data on 65 patients who were diagnosed with multiple myeloma and treated with alkylating agents, as described in Krall, Uthoff, and Harley (1975). Forty-eight patients died during the study. Nine covariates believed to be related to the survival of patients were recorded for each patient. The data on the failure times and the 9 covariates are recorded in myeloma.dta.

```
. use http://www.stata-press.com/data/cggm/myeloma
(Multiple myeloma patients)

. describe

Contains data from http://www.stata-press.com/data/cggm/myeloma.dta
  obs:            65                        Multiple myeloma patients
 vars:            11                        9 Nov 2007 16:17
 size:         1,950 (99.9% of memory free)
```

variable name	storage type	display format	value label	variable label
time	float	%9.0g		Survival time from diagnosis to nearest month + 1
died	byte	%9.0g		0 - Alive, 1 - Dead
lnbun	float	%9.0g		log BUN at diagnosis
hemo	float	%9.0g		Hemoglobin at diagnosis
platelet	byte	%9.0g	normal	Platelets at diagnosis
age	byte	%9.0g		Age (complete years)
lnwbc	float	%9.0g		Log WBC at diagnosis
fracture	byte	%9.0g	present	Fractures at diagnosis
lnbm	float	%9.0g		log % of plasma cells in bone marrow
protein	byte	%9.0g		Proteinuria at diagnosis
scalcium	byte	%9.0g		Serum calcium (mgm%)

```
Sorted by:
```

```
. stset time, failure(died)

     failure event:  died != 0 & died < .
  obs. time interval:  (0, time]
  exit on or before:  failure

        65  total obs.
         0  exclusions

        65  obs. remaining, representing
        48  failures in single record/single failure data
    1560.5  total analysis time at risk, at risk from t =          0
                             earliest observed entry t =           0
                                  last observed exit t =          92
```

The analysis time variable is `time`, recording time from diagnosis to death in months, and the failure variable is `died`. The rest of the variables described above are the covariates of interest from the original study.

In this chapter, our emphasis is on the design of a new study. We are not so much interested in the actual analysis of these multiple-myeloma data. Rather, we treat this study as a pilot study and use it to obtain the information needed to design a new study.

16.1.2   Comparing two survivor functions nonparametrically

Suppose that we want to design a study to compare a new treatment for multiple myeloma with the standard treatment (using alkylating agents). The analysis of the collected data will use the nonparametric log-rank test (see section 8.6) to compare the survivor functions of the control (standard treatment) and the experimental (new treatment) groups, given as $S_1(t)$ and $S_2(t)$, respectively. We need an estimate of the sample size sufficient to detect an effect of the new treatment if it exists. `stpower logrank` gives you this estimate.

We need to choose some a priori values for the components used in the computation. These include (1) the probability of a type I error α (or significance level) of the test, (2) the probability of a type II error β (or power $1 - \beta$) of the test, (3) the effect size of a treatment desired to be detected by the test, and (4) the proportion of subjects in the control group π_1.

The probability of a type I error (significance level) of a test is the probability of rejecting the null hypothesis of no effect when, in fact, there is no effect. The probability of a type II error is the probability of not rejecting the null hypothesis when, in fact, you should. Equivalently, the power is the probability of rejecting the null hypothesis when it is false. The goal of power analysis is to find a sample size that minimizes the probability of a type I error while maximizing the power. As such, we fix the type I error rate at some low value (usually 5%), fix power to some high value (80%, 90%, or some other), and then determine the sample size necessary to achieve both.

The effect size is chosen to represent the minimal clinically significant effect of a treatment. With survival data, specifying the effect size may be tricky without any assumptions about the survivor functions. In general, the effect size is a function of time. Under the proportional-hazards (PH) assumption, $S_2(t) = \{S_1(t)\}^\Delta$, the hazard ratio Δ is constant and so it is commonly used to represent the effect size in survival studies. If $\Delta < 1$, then the survival in the experimental group is higher relative to the survival in the control group; the standard treatment is inferior. If $\Delta > 1$, then the standard treatment is superior to the new treatment. We will be interested in, say, a 40% reduction in the hazard of the experimental group corresponding to a hazard ratio of 0.6.

The allocation of subjects between both groups also affects the required sample size. The simplest design is an equal-allocation design, that is, assigning an equal number of subjects to either protocol. In many nonsurvival settings, equal-allocation designs are optimal in the sense that they maximize power as a function of allocation.

Sometimes one may prefer (or be restricted to) using an unequal-allocation design, without regard to optimality. For example, ethical considerations may dictate giving the "superior" treatment to more subjects. Another consideration is cost.

As discussed previously, in survival studies the optimal designs tend to balance the number of failures in both groups by allocating more subjects to the experimental group, that is, the group with the lower failure rate. This leads to an unequal allocation of subjects. In practice, determining just the right allocation to achieve equal failures may be difficult, and so the more simple equal-allocation design is often used. In fact, Kalish and Harrington (1988) found that equal-allocation designs maintain high efficiency relative to equal-event designs under common survival-study scenarios. In any case, the decision is yours.

`stpower`, by default, assumes an equal-allocation design, and this is what we use in the examples that follow. You may specify unequal allocation by using either option `p1()` or option `nratio()`; see [ST] **stpower**.

As we mentioned earlier, the power of the test is directly related to the number of events observed in the study. The total number of events E required to be observed to ensure a power $1 - \beta$ of the log-rank test to detect an effect when the hazard ratio is Δ, according to Schoenfeld (1981), is

$$E = \frac{(z_{1-\alpha/k} + z_{1-\beta})^2}{\pi_1(1 - \pi_1)\ln^2(\Delta)} \tag{16.1}$$

where $z_{(1-\alpha/k)}$ and $z_{(1-\beta)}$ are the $(1 - \alpha/k)$th and the $(1 - \beta)$th quantiles, respectively, of the standard normal distribution, with $k = 1$ for the one-sided test and $k = 2$ for the two-sided test.

Returning to our study, we want to be able to detect a 40% reduction in the hazard of the experimental group ($\Delta = 0.6$) with a power of 90% ($1 - \beta = 0.9$) allowing only a 5% chance for a type I error (significance level $\alpha = 0.05$). To obtain the required number of events (sample size), we type

```
. stpower logrank, hratio(0.6) power(0.9) schoenfeld
Estimated sample sizes for two-sample comparison of survivor functions
Log-rank test, Schoenfeld method
Ho: S1(t) = S2(t)

Input parameters:
        alpha =    0.0500  (two sided)
    ln(hratio) =   -0.5108
        power =    0.9000
           p1 =    0.5000
Estimated number of events and sample sizes:
          E =       162
          N =       162
         N1 =        81
         N2 =        81
```

The specified study parameters and those used by default are reported in the output under `Input parameters`. Here `p1` denotes π_1, the proportion of subjects allocated to the control group, and equals 0.5 under the default 1:1 randomization. From the output, the required number of events and sample size is 162. We need to enroll a total of 162 patients (81 to each group) and follow each until they die.

❏ **Technical note**

In the above syntax, we also specified option `schoenfeld` to use the Schoenfeld formula, which calculates the log metric in the computation. Testing the equality of the two survivor functions is equivalent to testing H_o: $\Delta = 1$ or, alternatively, H_o: $\ln(\Delta) = 0$. Both these null hypotheses express an equality of survivor functions. Log-rank tests based on these two nulls are equivalent in large samples, but in small samples the choice of metric can make a difference. As such, both metrics are available with `stpower logrank`. The standard metric (H_o: $\Delta = 1$) is the default and uses methodology from Freedman (1982). The methodology for the log-metric computation is attributed to Schoenfeld (1981).

❏

The above computation assumed no censoring, which we consider now. Our pilot multiple-myeloma study indicates that roughly 4% of patients survive for 89 months. With a 40% decrease in the hazard of the experimental group, we expect roughly $100\% \times 0.04^{0.6} = 14\%$ to survive for 89 months. This means that, to observe all 162 patients die, we would need to continue our study for more than 89 months.

Instead, suppose we have enough resources to run the study for only 35 months. Then all patients who survive for 35 months will be (administratively) censored at the end of the study. In this case, the estimate of the sample size N must be adjusted for the reduction in the observed number of failures, caused by administrative censoring, in the fixed-length study. The required sample size is then

$$N = \frac{E}{p_E}$$

where the overall probability of a subject failing in a study, $p_E = 1 - \{S_1(T) + S_2(T)\}/2$, is estimated using the average of the survival probabilities in both groups at the end of the study (time T). Administrative censoring results in the increase of the required sample size to ensure that a certain number of events is observed.

Continuing our example, from our pilot study we find that 36% of subjects in the control group are expected to survive to the end of the study (35 months). To obtain the results in this case, we specify this proportion, 0.36, immediately following `stpower logrank`:

```
. stpower logrank 0.36, hratio(0.6) power(0.9) schoenfeld

Estimated sample sizes for two-sample comparison of survivor functions
Log-rank test, Schoenfeld method
Ho: S1(t) = S2(t)

Input parameters:
        alpha =      0.0500   (two sided)
           s1 =      0.3600
           s2 =      0.5417
   ln(hratio) =     -0.5108
        power =      0.9000
           p1 =      0.5000

Estimated number of events and sample sizes:
            E =         162
            N =         294
           N1 =         147
           N2 =         147
```

Here the proportion of subjects expected to survive in the experimental group was computed using the specified control-group survival and hazard ratio: $0.5417 = 0.36^{0.6}$. As an alternative to `hratio()`, we could have specified this proportion, 0.5417, as the second term following `stpower logrank`

```
. stpower logrank 0.36 0.5417, power(0.9) schoenfeld
  (output omitted)
```

thus obtaining the same results.

The estimated sample size increases from 162 to 294 after accounting for the expected 45% $[(0.36 + 0.5417)/2$ from the output] censoring in the study. The required number of events remains unchanged because censoring is assumed to occur independently of failure.

❏ Technical note

According to Hsieh (1999), in equal-allocation studies the Schoenfeld method performs better than the Freedman method. In equal-event studies, the Freedman method performed better.

❏

We assumed a two-sided test because, even though we expect the new treatment to be better than the standard, we are not 100% positive it will be. By allowing for the possibility that the new treatment is inferior, we are being conservative in our sample-size calculations. You can instead obtain results for a one-sided test by specifying option `onesided`.

16.1.3   Comparing two exponential survivor functions

Continuing with our multiple-myeloma example, we notice that the assumption of an exponential survivor function for the control group (treated with alkylating agents) is reasonable for these data; see figure 16.1 produced by the following:

```
. use http://www.stata-press.com/data/cggm/myeloma, clear
(Multiple myeloma patients)
. quietly stset time, failure(died) noshow
. streg, dist(exponential) nohr nolog
Exponential regression -- log relative-hazard form
No. of subjects =           65              Number of obs   =          65
No. of failures =           48
Time at risk    =       1560.5
                                            LR chi2(0)      =        0.00
Log likelihood  =   -87.648532             Prob > chi2     =         .
```

_t	Coef.	Std. Err.	z	P>|z|	[95% Conf. Interval]
_cons	-3.481561	.1443376	-24.12	0.000	-3.764457 -3.198664

```
. predict s0, surv
. sts graph, survival addplot(line s0 _t, sort title(Survival estimates)
> legend(order(1 "Kaplan-Meier" 2 "Exponential")))
```

(Continued on next page)

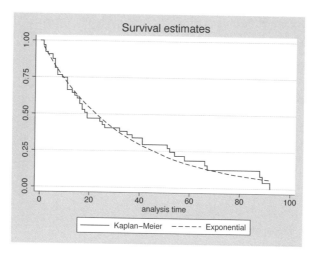

Figure 16.1. Kaplan–Meier and exponential survivor functions for multiple-myeloma data

From the above output, the estimated monthly hazard rate in the control group is $\lambda_1 = \exp(-3.482) = 0.03$. Then, assuming a 40% reduction in the hazard because of the new treatment, the anticipated monthly hazard rate in our experimental group is $\lambda_2 = (1 - 0.4)\lambda_1 = 0.6 \times 0.03 = 0.018$. The disparity in the two exponential survivor functions $S_1(t) = e^{-\lambda_1 t}$ and $S_2(t) = e^{-\lambda_2 t}$ can be tested by comparing the hazards λ_1 and λ_2. Let's use stpower exponential to estimate the required sample size, assuming exponential survivor functions:

```
. stpower exponential 0.03, hratio(0.6) power(0.9)
Note: input parameters are hazard rates.

Estimated sample sizes for two-sample comparison of survivor functions
Exponential test, hazard difference, conditional
Ho: h2-h1 = 0

Input parameters:
        alpha =    0.0500  (two sided)
           h1 =    0.0300
           h2 =    0.0180
        h2-h1 =   -0.0120
        power =    0.9000
           p1 =    0.5000

Estimated sample sizes:
            N =       174
           N1 =        87
           N2 =        87
```

Note that 0.03 above is the control-group hazard rate and not the censoring rate (as with stpower logrank). Alternatively, a hazard rate can be obtained from the exponential survival probability $S(t)$ as $\lambda = -\log\{S(t)\}/t$. This survival probability can be specified in place of the hazard rate if option t() is also specified and contains the time point t

corresponding to the specified survival probability; see [ST] **stpower exponential** for details.

From the output, we get a larger estimate of the sample size (174) than we obtained earlier from **stpower logrank** (162). Why did the sample size increase? By default, **stpower exponential** reports results for the hazard-difference test, the test with null hypothesis H_o: $\lambda_2 - \lambda_1 = 0$. The sample-size and power computations based on this test are conservative relative to those based on the (equivalent in large samples) log-hazard difference test with the null hypothesis H_o: $\ln(\lambda_2) - \ln(\lambda_1) = \ln(\lambda_2/\lambda_1) = 0$ (Lachin and Foulkes 1986). We can request this alternative test by specifying option **loghazard**:

```
. stpower exponential, hratio(0.6) power(0.9) loghazard

Estimated sample sizes for two-sample comparison of survivor functions
Exponential test, log-hazard difference, conditional
Ho: ln(h2/h1) = 0

Input parameters:
        alpha =    0.0500   (two sided)
    ln(h2/h1) =   -0.5108
        power =    0.9000
           p1 =    0.5000

Estimated sample sizes:
            N =       162
           N1 =        81
           N2 =        81
```

Here we obtain the earlier estimate of the sample size of 162. It being exactly the same as that from **stpower logrank** is mostly coincidence; in general, they will be close because they share a common metric. Notice that we omitted the hazard rate in the control group (0.03) in the above syntax. The log-hazard difference (log hazard-ratio) test requires only an estimate of the hazard ratio (0.6) for this computation. The hazard-difference test is based on the difference between the two hazards—the ratio of hazards is not sufficient to determine this difference and thus the control-group hazard rate must also be specified.

If we now restrict our study to a fixed length of 35 months, we need to know the proportion of subjects in the control group that survive to the end of the study. We know that, under the exponential model with hazard rate λ, the probability that a subject fails by time $t^\star$ is $P(T \le t^\star) = 1 - S(t^\star) = 1 - \exp(-\lambda t^\star)$, where T denotes time to failure. As such, the censoring and failure probabilities are completely determined (given hazard rates) by the length of the study, $t^\star$, which we specify using option **fperiod()**:

(Continued on next page)

```
. stpower exponential 0.03, hratio(0.6) power(0.9) fperiod(35)
Note: input parameters are hazard rates.

Estimated sample sizes for two-sample comparison of survivor functions
Exponential test, hazard difference, conditional
Ho: h2-h1 = 0

Input parameters:

          alpha =    0.0500   (two sided)
             h1 =    0.0300
             h2 =    0.0180
          h2-h1 =   -0.0120
          power =    0.9000
             p1 =    0.5000

Accrual and follow-up information:

       duration =   35.0000
      follow-up =   35.0000

Estimated sample sizes:
              N =       300
             N1 =       150
             N2 =       150
```

As expected, our estimate of the required sample size increases; the required sample size is now 300. For the log-hazard difference test, we must now specify the control-group hazard rate (in addition to the hazard ratio) necessary to estimate censoring probabilities.

```
. stpower exponential 0.03, hratio(0.6) power(0.9) fperiod(35) loghazard detail
Note: input parameters are hazard rates.

Estimated sample sizes for two-sample comparison of survivor functions
Exponential test, log-hazard difference, conditional
Ho: ln(h2/h1) = 0

Input parameters:

          alpha =    0.0500   (two sided)
             h1 =    0.0300
             h2 =    0.0180
      ln(h2/h1) =   -0.5108
          power =    0.9000
             p1 =    0.5000

Accrual and follow-up information:

       duration =   35.0000
      follow-up =   35.0000

Estimated sample sizes:
              N =       290
             N1 =       145
             N2 =       145

Estimated expected number of events:
          E|Ha =       162        E|Ho =       164
         E1|Ha =        94       E1|Ho =        82
         E2|Ha =        68       E2|Ho =        82
```

We obtain the sample-size estimate of 290, which is close to that obtained earlier by `stpower logrank` (294) for our fixed-length study of 35 months. This is not surprising because, in general, the powers of the log-rank and the exponential log hazard-ratio tests are approximately the same provided the censoring pattern is the same in both the experimental and control groups.

In addition, we specified option `detail` to report the expected number of events, given a sample size of 290, under both the alternative and null hypotheses. The expected number of events under the alternative, E|Ha, can be treated as analogous to the required number of events, E, from `stpower logrank`. Indeed, E|Ha = 162 from above matches E from `stpower logrank` (obtained previously).

Why do we need `stpower exponential` if we can get the same (or very similar) results by using `stpower logrank`? Well, as we demonstrate later, `stpower exponential` allows you to account for more flexible study designs because of its underlying distributional assumptions.

❏ **Technical note**

Estimates of sample size and power for more complex designs obtained under the assumption of the exponential model can be used with log-rank test and with the Cox PH model. Lachin (2000, 410) notes that the computations based on the hazard-difference test would be preferable if the subsequent analysis is to be done using the log-rank test. Computations based on the log-hazard difference test are preferable if the analysis will be performed using the Cox model.

❏

16.1.4 Cox regression models

In our previous examples, the goal of our new study was to investigate the effect of the new multiple-myeloma treatment. We chose to explore the effect by comparing the survivor functions in treatment versus control, without regard to any other factors. Suppose now that we want to adjust our analysis for other factors such as age, levels of hemoglobin, and so on, available from our pilot study. The inclusion of additional covariates in the model often makes the PH assumption more plausible. From previous discussions, we know that we can account for covariates, for example, by using a Cox regression model. We simply include our binary covariate of interest (the treatment indicator) along with other factors as explanatory variables.

Going back to our pilot multiple-myeloma study described in section 16.1.1, we believe that the amount of blood urea nitrogen (BUN) has a significant effect on survival of the patients (Krall, Uthoff, and Harley 1975; Hsieh and Lavori 2000). Suppose that we want to design a new study to investigate this effect. From the pilot study,

```
. use http://www.stata-press.com/data/cggm/myeloma
(Multiple myeloma patients)
. quietly stset time, failure(died)
```

```
. stcox lnbun hemo platelet age lnwbc fracture lnbm protein scalcium
          failure _d:  died
    analysis time _t:  time
Iteration 0:   log likelihood = -154.85799
Iteration 1:   log likelihood = -146.68114
Iteration 2:   log likelihood = -146.29446
Iteration 3:   log likelihood = -146.29404
Refining estimates:
Iteration 0:   log likelihood = -146.29404

Cox regression -- Breslow method for ties

No. of subjects =          65              Number of obs   =          65
No. of failures =          48
Time at risk    =      1560.5
                                           LR chi2(9)      =       17.13
Log likelihood  =   -146.29404             Prob > chi2     =      0.0468
```

_t	Haz. Ratio	Std. Err.	z	P>|z|	[95% Conf. Interval]	
lnbun	6.039698	3.915714	2.77	0.006	1.694948	21.52158
hemo	.88134	.0633096	-1.76	0.079	.7655945	1.014584
platelet	.7783403	.3949809	-0.49	0.621	.2878816	2.104384
age	.9872866	.0192274	-0.66	0.511	.9503117	1.0257
lnwbc	1.424365	1.015848	0.50	0.620	.3520058	5.763584
fracture	1.401968	.5709897	0.83	0.407	.6310487	3.114678
lnbm	1.431803	.6958989	0.74	0.460	.5522987	3.711869
protein	1.013153	.0265138	0.50	0.618	.9624971	1.066475
scalcium	1.134223	.1172804	1.22	0.223	.9261535	1.389038

The effect of lnbun (recording the log of the amount of BUN) on the survival is rather large. The increase of one unit of log(BUN) is associated with the sixfold increase in the hazard of a failure.

Let us pretend that we did not detect a clinically meaningful difference at the pilot stage. Instead, we want to design a new study to ensure that we have high power to detect the effect of log(BUN) corresponding to a twofold increase in a hazard (a hazard ratio of 2).

In the context of a Cox PH model, we are interested in the sample-size computation for the Wald test for the regression coefficient β_1 (log of the hazard ratio) with the null hypothesis H_o: $(\beta_1, \beta_2, \ldots, \beta_p) = (0, \beta_2, \ldots, \beta_p)$ against the alternative H_a: $(\beta_1, \beta_2, \ldots, \beta_p) = (\beta_{1a}, \beta_2, \ldots, \beta_p)$. β_1 represents the coefficient for the covariate of interest, x_1, and the rest of the coefficients are for the confounders $x_2, \ldots, x_p$ and are treated as nuisance parameters.

The required number of events for the test on a single coefficient from a Cox model in the presence of other covariates is

$$E = \frac{(z_{1-\alpha/k} + z_{1-\beta})^2}{\sigma^2 \beta_{1a}^2 (1 - R^2)}$$

where σ is the standard deviation of x_1, and R^2 is the squared multiple correlation coefficient of x_1 versus other predictors $x_2, \ldots, x_p$ (Hsieh and Lavori 2000). The term $\{\sigma^2(1-R^2)\}^{-1}$ accounts for the variability in the estimate of β_1—the larger the variance of the estimate, the larger the required number of events and sample size. A factor $1-R^2$ is the adjustment to the variance of the estimate of β_1 inflated by the presence of other covariates correlated with x_1. Large values of R^2 will lead to the increase in the number of events. You can see the resemblance with the respective formula for the log-rank test given in (16.1). In fact, for a single binary covariate, this formula reduces to the one given in Schoenfeld (1983) and in Schoenfeld (1981) for the log-rank test.

In the presence of censoring, the required sample size is estimated as $N = E/p_E$, where p_E is the overall probability of a subject failing during the study. `stpower cox` provides option `failprob()` to specify the estimate of this probability. You may already have this estimate available to you or, in the case of a binary covariate, you can use `stpower logrank` to get it. `stpower logrank` saves this result in `r(Pr_E)`.

Let's now compute the required sample size. First, we need estimates of the standard deviation σ and multiple correlation coefficient R^2. We can get those by using `summarize` followed by `regress`:

```
. summarize lnbun
```

Variable	Obs	Mean	Std. Dev.	Min	Max
lnbun	65	1.3929	.3126297	.7782	2.2355

```
. regress lnbun hemo platelet age lnwbc fracture lnbm protein scalcium
```

Source	SS	df	MS		
Model	1.15026278	8	.143782848		
Residual	5.10492732	56	.091159416		
Total	6.2551901	64	.097737345		

Number of obs = 65
F(8, 56) = 1.58
Prob > F = 0.1525
R-squared = 0.1839
Adj R-squared = 0.0673
Root MSE = .30193

lnbun	Coef.	Std. Err.	t	P>|t|	[95% Conf. Interval]	
hemo	-.0043198	.0182425	-0.24	0.814	-.0408639	.0322242
platelet	-.0468442	.1301531	-0.36	0.720	-.3075722	.2138838
age	.0075177	.0041724	1.80	0.077	-.0008405	.015876
lnwbc	.429355	.1675451	2.56	0.013	.0937219	.7649882
fracture	-.0874605	.0919659	-0.95	0.346	-.2716904	.0967694
lnbm	-.0078937	.1114837	-0.07	0.944	-.2312223	.2154349
protein	.010002	.0070262	1.42	0.160	-.0040731	.0240771
scalcium	.0193276	.022566	0.86	0.395	-.0258776	.0645328
_cons	-.7467908	.7684998	-0.97	0.335	-2.286281	.7926993

The effect size is a hazard ratio of 2 or, alternatively, a coefficient (a log hazard ratio) of $\ln(2) = 0.6931$. Either one may be specified with `stpower cox`, either by using the option `hratio(2)` or by specifying 0.6931 directly following `stpower cox`.

Recall that 17 of 65 total patients survived in the pilot study, so we can estimate the overall death rate to be $1 - 17/65 = 0.738$. We are quite sure that the increase in survival is associated with the increase in `lnbun`, and so we request a one-sided test.

We now can supply all the estimates to the respective options of `stpower cox` to obtain the estimates of the required number of events and the sample size:

```
. stpower cox, hratio(2) power(0.9) sd(0.3126) r2(0.1839) failprob(0.738) onesided
Estimated sample size for Cox PH regression
Wald test, log-hazard metric
Ho: [b1, b2, ..., bp] = [0, b2, ..., bp]

Input parameters:
          alpha =    0.0500   (one sided)
             b1 =    0.6931
             sd =    0.3126
          power =    0.9000
       Pr(event) =    0.7380
             R2 =    0.1839

Estimated number of events and sample size:
              E =       224
              N =       303
```

The estimated sample size is 303 and the required number of events is 224.

16.2   Accounting for withdrawal and accrual of subjects

Thus far our examples contained either no censoring or the administrative censoring associated with a fixed-length study. In this section, we describe two other components of the study design commonly encountered in practice: loss to follow-up (withdrawal) and accrual. The former is just another type of censoring. The latter is related to how subjects enroll in the study.

16.2.1   The effect of withdrawal or loss to follow-up

In most experiments, there is additional censoring before the end of the experiment. This is called withdrawal or loss to follow-up. Withdrawal arises because, for instance, some subjects move or experience side effects due to treatment. The former probably affects controls and experimentals equally. The latter, obviously, does not. `stpower` can handle both situations. Remember, of course, that just as in the analysis stage, we must assume that withdrawal (or any form of censoring) occurs for reasons unrelated to failure.

By default, `stpower` assumes no censoring. To account for withdrawal, both `stpower logrank` and `stpower cox` use a conservative adjustment: the sample size is simply divided by one minus the proportion of subjects anticipated to withdraw from the study. This proportion is specified in `wdprob()`. This approach is deemed conservative because it assumes all withdrawals occur at the outset, and it makes no attempt to incorporate additional data available from staggered withdrawals. In contrast, `stpower exponential` assumes that subjects withdraw at a constant exponential rate during the whole course of the study, and it provides options to specify the parameters of this exponential distribution separately for each treatment group.

Here is a summary of how `stpower` handles administrative censoring and loss to follow-up:

`stpower logrank`

- **Administrative censoring**. Specify the probability of surviving to the end of the study immediately following the command.

- **Withdrawal**. Use option `wdprob()` to specify the proportion of subjects anticipated to withdraw. In this case, the sample-size estimate is divided by one minus the specified probability, which amounts to assuming all withdrawals occurred at the outset. Thus, the adjustment is conservative. The same adjustment is made to both controls and experimentals. If withdrawal rates really differ, specify the rate as the larger of the two; this will make results even more conservative.

`stpower exponential`

- **Administrative censoring**. Use option `fperiod(d)`, where d equals the total length of the study.

- **Withdrawal**. Use either option `losshaz()` or option `lossprob()` plus `losstime()` to specify exponential withdrawal. Note that these options may take one or two arguments, which allows for different withdrawal rates for controls and experimentals.

`stpower cox`

- **Administrative censoring**. Use option `failprob()` to incorporate administrative censoring. `failprob()` specifies the overall failure rate for the study period.

- **Withdrawal**. Same as `stpower logrank`.

16.2.2 The effect of accrual

A study is often divided into two phases: an accrual phase of length R during which subjects are recruited, and a follow-up phase of length f during which subjects are followed. The total duration of a study is then $T = R + f$. At the study-design stage, we must account for accrual for the same reason that we must account for censoring. They both reduce the aggregate time at risk.

Recall our fixed-length study comparing survivor functions in two groups. Our previous sample-size computations assumed that subjects were under observation for all

35 months, that is $f = T = 35$ months and $R = 0$. In practice, this means that subjects were recruited simultaneously. Computationally, this is also equivalent to recruiting subjects at different times and following each of them for 35 months. The latter rarely happens in practice because such a design increases the length of the study without the benefit of increasing the aggregate time at risk.

What if we cannot recruit all patients at once and, instead, the recruitment is going to take place over $R = 20$ months after the study begins? In this case, patients who enter the study later will have a shorter follow-up time. As a worst case, the follow-up time decreases from 35 to 15 months for subjects who enter at the end of the recruitment period (20 months). Decreased follow-up times decrease the number of failures we will observe. There are two things we can do to ensure that the required number of events is observed in the presence of accrual: increase the duration of the study or enter more patients. We consider the latter in our examples and assume the more typical situation where the duration of the study is predetermined. How many more patients are required depends on the length of the accrual R and on how subjects enter the study, that is, the pattern of the accrual.

The assumption of uniform accrual, when subjects are entering a study at a constant rate, is commonly used in practice; `stpower logrank` and `stpower exponential` can account for this. If subjects enter a study uniformly over a period of $[0, R]$, the overall probability of failing in the study conditional on the entry times is

$$p_E = 1 - \frac{1}{R} \int_f^T \widetilde{S}(t) dt$$

where $\widetilde{S}(t) = \pi_1 S_1(t) + (1 - \pi_1) S_2(t)$ is the average survival for both groups. Without making any parametric assumptions about the survivor functions, the above integral can be computed numerically using pilot estimates of the survivor function in the control group in the range $[f, T]$. In the absence of these estimates at many time points, Schoenfeld (1983) suggests computing the probability from the survivor estimates at three specific time points by using Simpson's rule:

$$p_E \approx 1 - \left\{ \widetilde{S}(f) + 4\widetilde{S}(0.5R + f) + \widetilde{S}(T) \right\} / 6$$

`stpower logrank` provides options for either using Simpson's rule with three time points $(f, R/2 + f, T)$ or using standard numerical integration with a larger number of points. `stpower exponential` assumes exponential survivor functions, in which case the above integral has a closed-form expression. In this case, only the accrual period needs to be specified in addition to the previously discussed study parameters.

Sometimes the accrual pattern is not uniform. Accrual may be faster (or slower) at the beginning of the accrual period and die off (or accelerate) toward the end of the accrual period. This pattern may then be modeled, among other ways, by a truncated exponential model with distribution function: $G(t) = \{1 - \exp(-\gamma t)\} / \{1 - \exp(-\gamma R)\}$. Here the parameter γ governs the shape of the distribution: slow accrual at the outset when $\gamma < 0$ and fast accrual at the outset when $\gamma > 0$. `stpower exponential` provides options to adjust for this type of accrual pattern.

stpower cox does not provide options to account for accrual directly. For binary co-variates, accrual can be handled indirectly by obtaining the estimate of the probability of failing in the study (by using stpower logrank and stpower exponential), adjusted for accrual, and supplying this estimate to stpower cox via option failprob().

Accrual patterns are assumed to be the same for controls and experimentals and, by default, no accrual is assumed. Here is a summary:

stpower logrank

- **Uniform accrual**. Specify option simpson(s_1, s_2, s_3), where s_1, s_2, and s_3 are survival probabilities in the control group. s_1 is the probability at the mini-mum follow-up time f; s_3 is at the end of the study T; and s_2 is at the midpoint between the two, $(f+T)/2 = f + R/2$, which is called the *average follow-up time*. Alternatively, if you have an estimate of the survivor function for the control group at more points, use option st1(), which allows you to specify as many integration points as you wish.

- **Truncated exponential accrual**. Not available.

stpower exponential

- **Uniform accrual**. Use option aperiod() to specify the length of accrual period.

- **Truncated exponential accrual**. Use option aperiod() to specify the length of the accrual period, and use either (1) option ashape() to specify the shape parameter, or (2) option aprob() and either aptime() or atime() to specify the shape indirectly.

stpower cox

- **Uniform accrual**. There are no options to handle accrual directly. With binary covariates, use, for example, stpower logrank to obtain overall probability of failure saved in r(Pr_E), and then specify that number in option failprob().

- **Truncated exponential accrual**. Same as for stpower cox uniform accrual.

16.2.3   Examples

Returning to our example, let's see how accrual and withdrawals affect previously ob-tained sample sizes. We begin with accrual.

Suppose that we need to recruit subjects into our study, and we are planning to do that over some period of time. Although determining the length of the accrual may be of interest in itself, here we simply assume that we estimated it to be 20 months based on our pilot study. For a fixed 35-month duration study, this leaves 15 months for the follow-up period.

To account for uniform accrual with `stpower logrank`, we first estimate the survivor function in the control group from our pilot study by using `sts generate`. We then save only the estimates in the range $[f = 15, T = 35]$ in variable `surv1` and specify this variable and variable `_t` containing the time points to `stpower logrank`, via option `st1()`. Restricting the range of the survival estimates (or, equivalently, the time points) is important. `stpower logrank` determines the ranges of integration f and T as the respective minimum and maximum values of the specified time variable (`_t` here) among the nonmissing values of the two specified variables.

```
. use http://www.stata-press.com/data/cggm/myeloma
(Multiple myeloma patients)

. quietly stset time, failure(died) noshow

. sts gen surv1 = s

. replace surv1 = . if _t<15 | _t>35
(48 real changes made, 48 to missing)

. stpower logrank, hratio(0.6) power(0.9) sch st1(surv1 _t)
Note: probability of an event is computed using numerical
      integration over [15.00, 35.00].

Estimated sample sizes for two-sample comparison of survivor functions
Log-rank test, Schoenfeld method
Ho: S1(t) = S2(t)

Input parameters:

        alpha =    0.0500  (two sided)
   ln(hratio) =   -0.5108
        power =    0.9000
           p1 =    0.5000

Estimated number of events and sample sizes:
          E =        162
          N =        340
         N1 =        170
         N2 =        170
```

Recall that, in the absence of accrual (see the example on p. 332), we estimated a total of 294 subjects required for a study with a 35-month follow-up period. In this study, a 20-month uniform accrual period decreases the average follow-up time $(f + R/2)$ from 35 to 25 months. As a result, we need 46 more subjects to observe the required number of events (162) in this study.

Alternatively, if we only had estimates of the survivor function for the control group at three points $(f = 15, f + R/2 = 25, \text{ and } T = 35)$, we would specify them directly in option `simpson()`. For example, we use `sts list` to list these survival estimates from our pilot study,

```
. sts list, at(15 25 35)
```

Time	Beg. Total	Fail	Survivor Function	Std. Error	[95% Conf. Int.]	
15	33	24	0.6062	0.0633	0.4707	0.7171
25	21	9	0.4275	0.0672	0.2949	0.5536
35	17	3	0.3610	0.0669	0.2334	0.4900

```
Note: survivor function is calculated over full data and evaluated at
      indicated times; it is not calculated from aggregates shown at left.
```

and we then supply them to `stpower logrank`'s option `simpson()`:

```
. stpower logrank, hratio(0.6) power(0.9) schoenfeld simpson(0.6 0.43 0.36)
Note: probability of an event is computed using Simpson's rule with
      S1(f) = 0.60, S1(f+R/2) = 0.43, S1(T) = 0.36
      S2(f) = 0.74, S2(f+R/2) = 0.60, S2(T) = 0.54

Estimated sample sizes for two-sample comparison of survivor functions
Log-rank test, Schoenfeld method
Ho: S1(t) = S2(t)

Input parameters:
        alpha =    0.0500   (two sided)
    ln(hratio) =   -0.5108
        power =    0.9000
           p1 =    0.5000

Estimated number of events and sample sizes:
          E =        162
          N =        344
         N1 =        172
         N2 =        172
```

We obtain a similar estimate of the sample size, 344, versus 340 obtained previously.

In situations where nonparametric estimates of the survivor function in the control group are unavailable, we can use the assumption of exponential survival. We can then account for uniform accrual by specifying the accrual period in `stpower exponential`'s option `aperiod()`. For example, in the presence of a 20-month accrual and a follow-up period of 15 months,

```
. stpower exponential 0.03, hratio(0.6) power(0.9) aperiod(20) fperiod(15)
Note: input parameters are hazard rates.

Estimated sample sizes for two-sample comparison of survivor functions
Exponential test, hazard difference, conditional
Ho: h2-h1 = 0

Input parameters:
        alpha =    0.0500   (two sided)
           h1 =    0.0300
           h2 =    0.0180
        h2-h1 =   -0.0120
        power =    0.9000
           p1 =    0.5000
```

```
Accrual and follow-up information:
    duration =    35.0000
   follow-up =    15.0000
     accrual =    20.0000   (uniform)
Estimated sample sizes:
           N =        380
          N1 =        190
          N2 =        190
```

the required sample size increases from 300 (as obtained in section 16.1.3) to 380. Assuming exponential survival allows us to at least get a rough estimate of the sample size adjusted for accrual. To get a final estimate, we could also consider a more flexible survival model, for example, a piecewise exponential, to see how the sample size changes. This could be done via simulation or as described in Barthel et al. (2006) and Barthel, Royston, and Babiker (2005).

According to uniform accrual, to recruit 380 subjects during 20 months, we need to be recruiting $380/20 = 19$ subjects per month. What happens if, for instance, we recruit more subjects earlier and fewer later? In other words, how does a nonuniform accrual pattern affect our estimates? The answer depends on the functional form of the accrual pattern.

Suppose that the recruitment of subjects is faster at the beginning of the accrual period but slows down toward the end of the accrual period in such a way that 50% of the subjects are recruited with only 17% of the accrual period being elapsed. This entry distribution, under the assumption that it is exponential in shape, is as shown in figure 16.2.

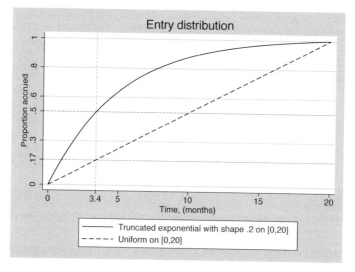

Figure 16.2. Accrual pattern of subjects entering a study over a period of 20 months

Also suppose that, in addition to being administratively censored at 35 months, subjects are expected to be withdrawing from the study at an exponential rate of 20% per year for reasons unrelated to death.

Although we could proceed as before using the command line, in such complex study designs it is convenient to specify the design parameters interactively by using `stpower exponential`'s dialog box. In this way, we can try out different values of study parameters and different study designs before deciding on the final sampling plan. Using the dialog box, you just fill in the appropriate values and click **Submit**, then change the design specification and click **Submit** again to get new results.

To invoke `stpower exponential`'s dialog box, select **Statistics > Power and sample size > Exponential test**. The dialog with default settings will appear. The following screenshots of the dialog boxes were obtained from Windows Vista.

For our example, we specify values for power and hazard rates on the *Main* tab, as shown in figure 16.3.

(*Continued on next page*)

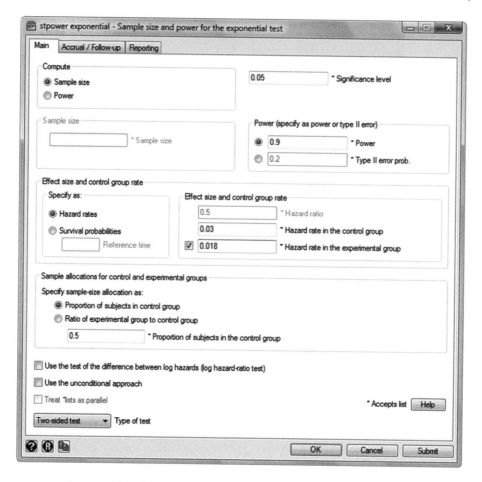

Figure 16.3. *Main* tab of `stpower exponential`'s dialog box

There are several choices of how to specify the failure distributions for the two groups. We can select specification of either hazard rates or survival probabilities at a reference time in the **Specify as:** group box. Then we can specify the failure parameter of the control group and the effect size expressed as a hazard ratio, or we can specify the failure parameters for both groups. In the latter case, the effect size is computed from the two failure parameters. In our example, we choose to specify the hazard rates for both groups.

Next we proceed to the *Accrual/Follow-up* tab. There we specify accrual and follow-up periods, choose the accrual pattern, and request adjustment for exponentially distributed losses to follow-up, as demonstrated in figure 16.4.

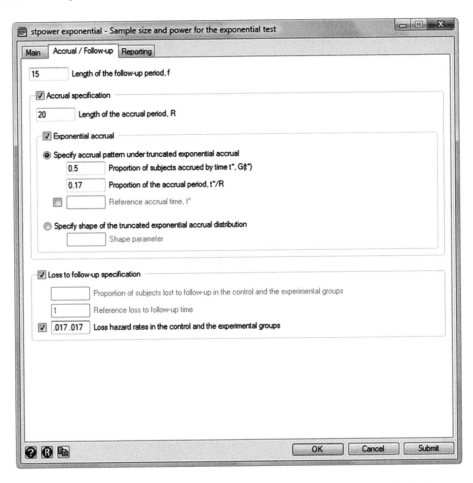

Figure 16.4. *Accrual/Follow-up* tab of `stpower exponential`'s dialog box

We specified the truncated exponential accrual distribution by providing the proportion of subjects (0.5) accrued by the proportion of the accrual period (0.17). Alternatively, we could have specified the shape parameter (0.2 from figure 16.2) in the respective edit box. Notice also that we specified the monthly loss to follow-up hazard rates of $0.2/12 = 0.017$ instead of yearly hazard rates of 0.2. All specified hazard rates must agree with the time scale specified for the accrual and follow-up periods (months in our example). Similar to the failure rates, we could have specified the proportion of subjects lost to follow-up at or prior to reference time in the respective edit boxes instead of specifying the loss to follow-up hazard rates.

By clicking the **Submit** button, we generate and execute the following command:

```
. stpower exponential 0.03 0.018, power(0.9) fperiod(15) aperiod(20)
> losshaz(.017 .017) aptime(0.17)
Note: input parameters are hazard rates.

Estimated sample sizes for two-sample comparison of survivor functions
Exponential test, hazard difference, conditional
Ho: h2-h1 = 0

Input parameters:
        alpha =     0.0500  (two sided)
           h1 =     0.0300
           h2 =     0.0180
        h2-h1 =    -0.0120
        power =     0.9000
           p1 =     0.5000

  Accrual and follow-up information:
     duration =    35.0000
    follow-up =    15.0000
      accrual =    20.0000  (exponential)
   accrued(%) =      50.00  (by time t*)
           t* =     3.4000  (17.00% of accrual)
          lh1 =     0.0170
          lh2 =     0.0170

Estimated sample sizes:
            N =        410
           N1 =        205
           N2 =        205
```

The required sample size increases from 380 (obtained earlier) to the final estimate of 410. The presence of loss to follow-up led to the increase in the required sample size. Ignoring loss to follow-up would lead to a decrease of the sample size from 380 to 330. We invite you to verify this by unchecking the **Loss to follow-up specification** box on the dialog and resubmitting the results. The convex shape of the accrual distribution results in a larger average follow-up time compared with the uniform accrual and thus decreases the required sample size. In the case of a concave accrual pattern (for example, exponential distribution with negative values of the shape parameter), the required sample size would increase. You can further play with the dialog by specifying other shapes of the accrual distribution and varying the loss to follow-up rates to see how these affect the estimates of sample size. To get results for uniform accrual, simply uncheck the **Exponential accrual** box.

In our example, we made the parametric assumption that our accrual and loss to follow-up distributions are exponential. When these assumptions are suspect or in the presence of other complexities associated with the nature of the trial (for example, lagged effect of a treatment, more than two treatment groups, or clustered data) and with the behavior of the participants (for example, noncompliance of subjects with the assigned treatment or competing risks), one may consider obtaining required sample size or power by simulation; for example, see Feiveson (2002) for a general discussion. Barthel et al. (2006); Barthel, Royston, and Babiker (2005); and Lakatos (1988) present sample-size computations and power computations for multiarm (multiple groups) trials under more flexible design conditions. They also relax the assumption of PH.

16.3   Estimating power and effect size

So far we have been interested in estimating sample size given power and effect size. Sometimes, we may want to turn the problem on its head and estimate power given sample size and effect size or, perhaps, the effect size given power and sample size. All of the issues discussed earlier apply here as well. The only difference is that we change what is to be treated as unknown: the formulas are now viewed as functions of sample size and effect size when computing power, and sample size and power when computing effect size. In the latter case, obtaining a solution requires using an iterative method.

Recall our example of a study investigating the effect of blood urea nitrogen (BUN) on the survival of patients described in section 16.1.4. We obtained a sample size of 303 required to detect a twofold increase in the hazard with 90% power using a one-sided Wald test.

Suppose that we could not recruit that many subjects. Also we do not have enough resources to extend the follow-up period to observe more events. We want to know by how much the power is decreased for the obtainable sample size of 65, the actual number of patients in the original multiple-myeloma dataset.

To answer this question, we leave everything specified as before in section 16.1.4 and replace the `power(0.9)` option with `n(65)`:

```
. stpower cox, hratio(2) n(65) sd(0.3126) r2(0.1839) failprob(0.738) onesided
Estimated power for Cox PH regression
Wald test, log-hazard metric
Ho: [b1, b2, ..., bp] = [0, b2, ..., bp]

Input parameters:
        alpha =     0.0500   (one sided)
           b1 =     0.6931
           sd =     0.3126
            N =         65
    Pr(event) =     0.7380
           R2 =     0.1839

Estimated number of events and power:
            E =         48
        power =     0.3862
```

We estimate the power of the test to be only 39%.

We also find that, for this sample size and 90% power, the minimal detectable increase in the coefficient is, disregarding the sign, 1.496 corresponding to a hazard ratio of $\exp(1.496) = 4.46$.

(Continued on next page)

```
. stpower cox, n(65) power(0.9) sd(0.3126) r2(0.1839) failprob(0.738) onesided
Estimated coefficient for Cox PH regression
Wald test, log-hazard metric
Ho: [b1, b2, ..., bp] = [0, b2, ..., bp]

Input parameters:
         alpha =    0.0500  (one sided)
            sd =    0.3126
             N =        65
         power =    0.9000
     Pr(event) =    0.7380
            R2 =    0.1839

Estimated number of events and coefficient:
             E =        48
            b1 =   -1.4962
```

Although our pilot study suggests a large effect of BUN, based on the above results we would want to find additional resources to extend our recruitment period and obtain more subjects for our new study.

In general, use

$$\text{stpower} \ldots, \text{n}(\#) \ldots$$

when you want to estimate power, and

$$\text{stpower} \ldots, \text{n}(\#) \text{ power}(\#) \ldots$$

or

$$\text{stpower} \ldots, \text{n}(\#) \text{ beta}(\#) \ldots$$

when you want to estimate effect size. When option n() is omitted with stpower, the sample size is estimated. At the time of publication of this text, effect-size determination is not available with stpower exponential.

16.4 Tabulating or graphing results

At the design stage, we are usually interested in obtaining results for a combination of different values of study parameters to get an idea of how these changes affect the estimates of interest. The results may be presented in a table or as a graph. To accommodate this interest, `stpower` allows specifying multiple values (*numlists*) within its main options and can save results in a dataset for later production of power or other curves.

Recall our first example of sample-size computation for the log-rank test from section 16.1.2. We can request results to be displayed in a table by specifying option `table`:

```
. stpower logrank, hratio(0.6) power(0.9) schoenfeld table
Estimated sample sizes for two-sample comparison of survivor functions
Log-rank test, Schoenfeld method
Ho: S1(t) = S2(t)
```

Power	N	N1	N2	E	ln(HR)	Alpha*
.9	162	81	81	162	-.51083	.05

```
* two sided
```

Suppose that we want to compare the required number of events (sample size) for three values of the hazard ratio: 0.5, 0.6, and 0.7.

```
. stpower logrank, hratio(0.5 0.6 0.7) power(0.9) schoenfeld
Estimated sample sizes for two-sample comparison of survivor functions
Log-rank test, Schoenfeld method
Ho: S1(t) = S2(t)
```

Power	N	N1	N2	E	ln(HR)	Alpha*
.9	88	44	44	88	-.69315	.05
.9	162	81	81	162	-.51083	.05
.9	332	166	166	332	-.35667	.05

```
* two sided
```

We did not specify `table` to obtain a table in this example. If multiple values are specified within its options, `stpower` will automatically display results in tabular form.

Now, in addition to varying hazard ratios, we would like to obtain results for 80% and 90% power. We also want to customize the look of our table by selecting the specific columns and their order. We can do this by using option `columns()`:

(Continued on next page)

```
. stpower logrank, hrat(0.5(0.1)0.7) p(0.8 0.9) sch columns(e power hr p1 alpha)
Estimated sample sizes for two-sample comparison of survivor functions
Log-rank test, Schoenfeld method
Ho: S1(t) = S2(t)
```

E	Power	HR	P1	Alpha*
66	.8	.5	.5	.05
88	.9	.5	.5	.05
122	.8	.6	.5	.05
162	.9	.6	.5	.05
248	.8	.7	.5	.05
332	.9	.7	.5	.05

```
* two sided
```

stpower's subcommands offer various quantities that can be displayed in the table as columns; see the manual entries specific to each subcommand for a list of available columns and other table-related options.

We can also choose which columns to include in a customized table by using the dialog box. Recall the example given in section 16.2.3. Repeat all the steps from that section for filling in the *Main* and *Accrual/Follow-up* tabs. Now fill in the *Reporting* tab, as shown in figure 16.5. In the dialog that appears when you click the **Custom columns** button, select the columns for power, sample size, hazard rates in the control and the experimental group, and total and group-specific numbers of events under the null and alternative hypotheses.

Figure 16.5. Column specification in `stpower exponential`'s dialog

Click the **Accept** button and then the **Submit** button to obtain the table below.

```
. stpower exponential 0.03 0.018, power(0.9) fperiod(15) aperiod(20)
> losshaz(.017 .017) aptime(0.17) table
> columns(power n h1 h2 eo eo1 eo2 ea ea1 ea2) colwidth(7)
Note: input parameters are hazard rates.

Estimated sample sizes for two-sample comparison of survivor functions
Exponential test, hazard difference, conditional
Ho: h2-h1 = 0
```

Power	N	H1	H2	E|Ho	E1|Ho	E2|Ho	E|Ha	E1|Ha	E2|Ha
.9	410	.03	.018	170	85	85	168	99	69

Also, we can plot values of the parameter of interest as a function of another parameter for given values of other study parameters. Suppose that we want to plot power as a function of the hazard ratio for, say, two values of sample size and other study parameters as given in the first example where we used `stpower logrank` (section 16.1.2):

```
. quietly stpower logrank, hratio(0.5(0.01)0.9) n(150 300) schoenfeld
> saving(mypower) columns(power n hr)
```

We specified the columns in `columns()` to be saved in the dataset `mypower.dta`, specified in option `saving()`. We also use `quietly` to avoid displaying a long table on the screen.

Finally, we use the saved dataset and `twoway line` to produce the power plot shown in figure 16.6:

```
. use mypower
. twoway line power hr if n==150 || line power hr if n==300,
> title("Power vs hazard ratio")
> subtitle("Log-rank test, Schoenfeld method")
> ytitle("Power") xtitle("Hazard ratio (experimental to control)")
> xlabel(0.5(0.05)0.9, grid) ylabel(#10, grid)
> note("Alpha = .05 (two sided); N2/N1 = 1")
> legend(label(1 "N = 150") label(2 "N = 300"))
```

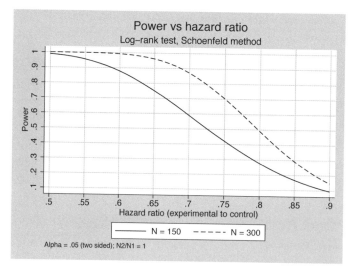

Figure 16.6. Power as a function of a hazard ratio for the log-rank test

References

Aalen, O. O. 1978. Nonparametric inference for a family of counting processes. *Annals of Statistics* 6: 701–726.

Akaike, H. 1974. A new look at the statistical model identification. *IEEE Transaction and Automatic Control* AC–19: 716–723.

Ambler, G., and P. Royston. 2001. Fractional polynomial model selection procedures: Investigation of type I error. *Journal of Statistical Computation and Simulation* 69: 89–108.

Andersen, P. K., and N. Keiding. 2006. *Survival and Event History Analysis*. Chichester: Wiley.

Barthel, F. M.-S., A. Babiker, P. Royston, and M. K. B. Parmar. 2006. Evaluation of sample size and power for multi-arm survival trials allowing for non-uniform accrual, non-proportional hazards, and loss to follow-up and cross-over. *Statistics in Medicine* 25: 2521–2542.

Barthel, F. M.-S., P. Royston, and A. Babiker. 2005. A menu-driven facility for complex sample size calculation in randomized controlled trials with a survival or a binary outcome: update. *Stata Journal* 5: 123–129.

Becketti, S. 1995. sg26.2: Calculating and graphing fractional polynomials. *Stata Technical Bulletin* 24: 14–16. Reprinted in *Stata Technical Bulletin Reprints*, vol. 4, pp. 129–132. College Station, TX: Stata Press.

Berger, U., J. Schäfer, and K. Ulm. 2003. Dynamic cox modelling based on fractional polynomials: Time—variations in gastric cancer prognosis. *Statistics in Medicine* 22: 1163–1180.

Breslow, N. E. 1970. A generalized Kruskal–Wallis test for comparing k samples subject to unequal patterns of censorship. *Biometrika* 57: 579–594.

———. 1974. Covariance analysis of censored survival data. *Biometrics* 30: 89–99.

Breslow, N. E., and N. E. Day. 1987. *Statistical Methods in Cancer Research: Volume 2 – The Design and Analysis of Cohort Studies*. Lyon: IARC.

Casella, G., and R. L. Berger. 2002. *Statistical Inference*. 2nd ed. Pacific Grove, CA: Duxbury.

Cleves, M. A. 1999. ssa13: Analysis of multiple failure-time data with Stata. *Stata Technical Bulletin* 49: 30–39. Reprinted in *Stata Technical Bulletin Reprints*, vol. 9, pp. 338–349. College Station, TX: Stata Press.

Cochran, W. G. 1977. *Sampling Techniques*. 3rd ed. New York: Wiley.

Collett, D. 2003. *Modelling Survival Data in Medical Research*. 2nd ed. London: Chapman & Hall/CRC.

Cox, C. S., M. E. Mussolino, S. T. Rothwell, M. A. Lane, C. D. Golden, J. H. Madans, and J. J. Feldman. 1997. Plan and operation of the NHANES I Epidemiologic Followup Study, 1992. In *Vital and Health Statistics*, vol. 1. Hyattsville, MD: National Center for Health Statistics.

Cox, D. R. 1972. Regression models and life-tables (with discussion). *Journal of the Royal Statistical Society, Series B* 34: 187–220.

Cox, D. R., and E. J. Snell. 1968. A general definition of residuals (with discussion). *Journal of the Royal Statistical Society, Series B* 30: 248–275.

Efron, B. 1977. The efficiency of Cox's likelihood function for censored data. *Journal of the American Statistical Association* 72: 557–565.

Engel, A., R. S. Murphy, K. Maurer, and E. Collins. 1978. Plan and operation of the HANES I augmentation survey of adults 25–74 years. In *Vital and Health Statistics*, vol. 1. Hyattsville, MD: National Center for Health Statistics.

Feiveson, A. H. 2002. Power by simulation. *Stata Journal* 2: 107–124.

Freedman, L. S. 1982. Tables of the number of patients required in clinical trials using the logrank test. *Statistics in Medicine* 1: 121–129.

Garrett, J. M. 1997. Graphical assessment of the Cox model proportional hazards assumption. *Stata Technical Bulletin* 35: 9–14. Reprinted in *Stata Technical Bulletin Reprints*, vol. 6, pp. 38–44. College Station, TX: Stata Press.

Gehan, E. A. 1965. A generalized Wilcoxon test for comparing arbitrarily singly censored data. *Biometrika* 52: 203–223.

Grambsch, P. M., and T. M. Therneau. 1994. Proportional hazards tests and diagnostics based on weighted residuals. *Biometrika* 81: 515–526.

Gray, R. J. 1990. Some diagnostic methods for Cox regression models through hazard smoothing. *Biometrics* 46: 93–102.

Greenwood, M. 1926. The natural duration of cancer. *Reports on Public Health and Medical Subjects* 33: 1–26.

Gutierrez, R. G. 2002. Parametric frailty and shared frailty survival models. *Stata Journal* 2: 22–44.

Harrell, F. E., R. M. Califf, D. B. Pryor, K. L. Lee, and R. A. Rosati. 1982. Evaluating the yield of medical tests. *Journal of the American Medical Association* 247: 2543–2546.

Harrell, F. E., K. L. Lee, and D. B. Mark. 1996. Multivariable prognostic models: Issues in developing models, evaluating assumptions and adequacy, and measuring and reducing errors. *Statistics in Medicine* 15: 361–387.

Harrington, D. P., and T. R. Fleming. 1982. A class of rank test procedures for censored survival data. *Biometrika* 69: 553–566.

Hess, K. R. 1995. Graphical methods for assessing violations of the proportional hazards assumption in Cox regression. *Statistics in Medicine* 14: 1707–1723.

Hess, K. R., D. M. Serachitopol, and B. W. Brown. 1999. Hazard function estimators: a simulation study. *Statistics in Medicine* 18: 3075–3088.

Hosmer, D. W., Jr., S. Lemeshow, and S. May. 2008. *Applied Survival Analysis: Regression Modeling of Time to Event Data*. 2nd ed. New York: Wiley.

Hougaard, P. 1984. Life table methods for hetergeneous populations: distributions describing the heterogeneity. *Biometrika* 71: 75–83.

———. 1986a. A class of multivariate failure time distributions. *Biometrika* 73: 671–678.

———. 1986b. Survival models for heterogeneous populations derived from stable distributions. *Biometrika* 73: 387–396.

Hsieh, F. Y. 1999. Comparing sample size formulae for trials with unbalanced allocation using the logrank test. *Statistics in Medicine* 11: 1091–1098.

Hsieh, F. Y., and P. W. Lavori. 2000. Sample size calculations for the Cox proportional hazards regression models with nonbinary covariates. *Controlled Clinical Trials* 21: 552–560.

Kalbfleisch, J. D., and R. L. Prentice. 2002. *The Statistical Analysis of Failure Time Data*. 2nd ed. New York: Wiley.

Kalish, L. A., and D. P. Harrington. 1988. Efficiency of balanced treatment allocation for survival analysis. *Biometrics* 44: 815–821.

Kaplan, E. L., and P. Meier. 1958. Nonparametric estimation from incomplete observations. *Journal of the American Statistical Association* 53: 457–481.

Klein, J. P., and M. L. Moeschberger. 2003. *Survival Analysis: Techniques for Censored and Truncated Data*. 2nd ed. New York: Springer.

Korn, E. L., and B. I. Graubard. 1999. *Analysis of Health Surveys*. New York: Wiley.

Korn, E. L., and R. Simon. 1990. Measures of explained variation for survival data. *Statistics in Medicine* 9: 487–503.

Krall, J. M., V. A. Uthoff, and J. B. Harley. 1975. A step-up procedure for selecting variables associated with survival. *Biometrics* 31: 49–57.

Lachin, J. M. 2000. *Biostatistical Methods: The Assessment of Relative Risks*. New York: Wiley.

Lachin, J. M., and M. A. Foulkes. 1986. Evaluation of sample size and power for analysis of survival with allowance for nonuniform patient entry, losses to follow-up, noncompliance, and stratification. *Biometrics* 42: 507–519.

Lakatos, E. 1988. Sample size based on the log-rank statistic in complex clinical trials. *Biometrics* 44: 229–241.

Lancaster, T. 1979. Econometric methods for the duration of unemployment. *Econometrica* 47: 939–956.

Lehr, S., and M. Schemper. 2007. Parsimonious analysis of time-dependent effects in the Cox model. *Statistics in Medicine* 26: 2686–2698.

Levy, P., and S. Lemeshow. 1999. *Sampling of Populations: Methods and Applications*. 3rd ed. New York: Wiley.

Lin, D. Y., and L. J. Wei. 1989. The robust inference for the Cox proportional hazards model. *Journal of the American Statistical Association* 84: 1074–1078.

Mantel, N., and W. Haenszel. 1959. Statistical aspects of the analysis of data from retrospective studies of disease. *Journal of the National Cancer Institute* 22: 719–748.

McGilchrist, C. A., and C. W. Aisbett. 1991. Regression with frailty in survival analysis. *Biometrics* 47: 461–466.

Miller, H. W. 1973. Plan and operation of the Health and Nutrition Examination Survey: United States 1971–1973. In *Vital and Health Statistics*, vol. 1. Hyattsville, MD: National Center for Health Statistics.

Muller, H. G., and J. L. Wang. 1994. Hazard rate estimation under random censoring with varying kernels and bandwidths. *Biometrics* 50: 61–76.

Nelson, W. 1972. Theory and applications of hazard plotting for censored failure data. *Technometrics* 14: 945–965.

Peto, R., and J. Peto. 1972. Asymptotically efficient rank invariant test procedures (with discussion). *Journal of the Royal Statistical Society* 135: 185–206.

Prentice, R. L. 1978. Linear rank tests with right-censored data. *Biometrika* 65: 167–179.

Royston, P., and D. G. Altman. 1994a. sg26: Using fractional polynomials to model curved regression relationships. *Stata Technical Bulletin* 21: 11–23. Reprinted in *Stata Technical Bulletin Reprints*, vol. 4, pp. 110–128. College Station, TX: Stata Press.

————. 1994b. Regression using fractional polynomials of continuous covariates: Parsimonious parametric modelling (with discussion). *Applied Statistics* 43: 429–467.

Royston, P., M. Reitz, and J. Atzpodien. 2006. An approach to estimating prognosis using fractional polynomials in metastatic renal carcinoma. *British Journal of Cancer* 94: 1785–1788.

Royston, P., and W. Sauerbrei. 2005. Building multivariable regression models with continuous covariates in clinical epidemiology—with an emphasis on fractional polynomials. *Methods of Information in Medicine* 44: 561–571.

————. 2007a. Improving the robustness of fractional polynomial models by preliminary covariate transformation: A pragmatic approach. *Computational Statistics and Data Analysis* 51: 4240–4253.

————. 2007b. Multivariable modeling with cubic regression splines: A principled approach. *Stata Journal* 7: 45–70.

Sauerbrei, W., and P. Royston. 1999. Building multivariable prognostic and diagnostic models: Transformation of the predictors by using fractional polynomials. *Journal of the Royal Statistical Society, Series A* 162: 71–94.

————. 2002. Corrigendum: Building multivariable prognostic and diagnostic models: Transformation of the predictors by using fractional polynomials. *Journal of the Royal Statistical Society, Series A* 166: 299–300.

Sauerbrei, W., P. Royston, and M. Look. 2007. A new proposal for multivariable modelling of time-varying effects in survival data based on fractional polynomial time-transformation. *Biometrical Journal* 49: 453–473.

Schoenfeld, D. 1981. The asymptotic properties of nonparametric tests for comparing survival distributions. *Biometrika* 68: 316–319.

————. 1982. Partial residuals for the proportional hazards regression model. *Biometrika* 69: 239–241.

————. 1983. Sample-size formula for the proportional-hazards regression model. *Biometrics* 39: 499–503.

Tarone, R. E., and J. H. Ware. 1977. On distribution-free tests for equality of survival distributions. *Biometrika* 64: 156–160.

Therneau, T. M., and P. M. Grambsch. 2000. *Modeling Survival Data: Extending the Cox Model*. New York: Springer.

Vaupel, J. W., K. Manton, and E. Stallard. 1979. The impact of heterogeneity in individual frailty on the dynamics of mortality. *Demography* 16: 439–454.

Author index

Subject index